Advanced Photocatalytic Materials for Environmental and Energy Applications

Advanced Photocatalytic Materials for Environmental and Energy Applications

Editors

Tongming Su
Xingwang Zhu

Basel • Beijing • Wuhan • Barcelona • Belgrade • Novi Sad • Cluj • Manchester

Editors
Tongming Su
School of Chemistry and
Chemical Engineering,
Guangxi University
Nanning, China

Xingwang Zhu
College of Environmental
Science and Engineering,
Institute of Technology for
Carbon Neutralization,
Yangzhou University
Yangzhou, China

Editorial Office
MDPI
St. Alban-Anlage 66
4052 Basel, Switzerland

This is a reprint of articles from the Special Issue published online in the open access journal *Materials* (ISSN 1996-1944) (available at: www.mdpi.com/journal/materials/special_issues/WRD97T9D5B).

For citation purposes, cite each article independently as indicated on the article page online and as indicated below:

Lastname, A.A.; Lastname, B.B. Article Title. *Journal Name* **Year**, *Volume Number*, Page Range.

ISBN 978-3-0365-9649-5 (Hbk)
ISBN 978-3-0365-9648-8 (PDF)
doi.org/10.3390/books978-3-0365-9648-8

© 2023 by the authors. Articles in this book are Open Access and distributed under the Creative Commons Attribution (CC BY) license. The book as a whole is distributed by MDPI under the terms and conditions of the Creative Commons Attribution-NonCommercial-NoDerivs (CC BY-NC-ND) license.

Contents

About the Editors vii

Tongming Su and Xingwang Zhu
Advanced Photocatalytic Materials for Environmental and Energy Applications
Reprinted from: *Materials* **2023**, *16*, 7197, doi:10.3390/ma16227197 1

Jianjian Yi, Guoxiang Zhang, Yunzhe Wang, Wanyue Qian and Xiaozhi Wang
Recent Advances in Phase-Engineered Photocatalysts: Classification and
Diversified Applications
Reprinted from: *Materials* **2023**, *16*, 3980, doi:10.3390/ma16113980 5

Achraf Amir Assadi, Oussama Baaloudj, Lotfi Khezami, Naoufel Ben Hamadi, Lotfi Mouni, Aymen Amine Assadi and et al.
An Overview of Recent Developments in Improving the Photocatalytic Activity of TiO_2-Based Materials for the Treatment of Indoor Air and Bacterial Inactivation
Reprinted from: *Materials* **2023**, *16*, 2246, doi:10.3390/ma16062246 23

Yelin Dai, Ziyi Feng, Kang Zhong, Jianfeng Tian, Guanyu Wu, Qing Liu and et al.
Highly Efficient and Exceptionally Durable Photooxidation Properties on Co_3O_4/g-C_3N_4 Surfaces
Reprinted from: *Materials* **2023**, *16*, 3879, doi:10.3390/ma16103879 45

Mizael Luque Morales, Priscy Alfredo Luque Morales, Manuel de Jesús Chinchillas Chinchillas, Víctor Manuel Orozco Carmona, Claudia Mariana Gómez Gutiérrez, Alfredo Rafael Vilchis Nestor and et al.
Theoretical and Experimental Study of the Photocatalytic Properties of ZnO Semiconductor Nanoparticles Synthesized by *Prosopis laevigata*
Reprinted from: *Materials* **2023**, *16*, 6169, doi:10.3390/ma16186169 57

Manying Sun, Chuanwei Zhu, Su Wei, Liuyun Chen, Hongbing Ji, Tongming Su and et al.
Phosphorus-Doped Hollow Tubular g-C_3N_4 for Enhanced Photocatalytic CO_2 Reduction
Reprinted from: *Materials* **2023**, *16*, 6665, doi:10.3390/ma16206665 71

Lotfi Khezami and Aymen Amin Assadi
Treatment of Mixture Pollutants with Combined Plasma Photocatalysis in Continuous Tubular Reactors with Atmospheric-Pressure Environment: Understanding Synergetic Effect Sources
Reprinted from: *Materials* **2023**, *16*, 6857, doi:10.3390/ma16216857 87

Guansheng Ma, Zhigang Pan, Yunfei Liu, Yinong Lu and Yaqiu Tao
Hydrothermal Synthesis of MoS_2/SnS_2 Photocatalysts with Heterogeneous Structures Enhances Photocatalytic Activity
Reprinted from: *Materials* **2023**, *16*, 4436, doi:10.3390/ma16124436 103

Beata Tryba, Piotr Miądlicki, Piotr Rychtowski, Maciej Trzeciak and Rafał Jan Wróbel
The Superiority of TiO_2 Supported on Nickel Foam over Ni-Doped TiO_2 in the Photothermal Decomposition of Acetaldehyde
Reprinted from: *Materials* **2023**, *16*, 5241, doi:10.3390/ma16155241 119

Ruyu Yan, Xinyi Liu, Haijie Zhang, Meng Ye, Zhenxing Wang, Jianjian Yi and et al.
Carbon Quantum Dots Accelerating Surface Charge Transfer of 3D $PbBiO_2I$ Microspheres with Enhanced Broad Spectrum Photocatalytic Activity—Development and Mechanism Insight
Reprinted from: *Materials* **2023**, *16*, 1111, doi:10.3390/ma16031111 137

Siti Nurul Falaein Moridon, Khuzaimah Arifin, Mohamad Azuwa Mohamed, Lorna Jeffery Minggu, Rozan Mohamad Yunus and Mohammad B. Kassim
TiO$_2$ Nanotubes Decorated with Mo$_2$C for Enhanced Photoelectrochemical Water-Splitting Properties
Reprinted from: *Materials* **2023**, *16*, 6261, doi:10.3390/ma16186261 . . . . **151**

Xiaodong Xu, Wangping Wu and Qinqin Wang
Efficiency Improvement of Industrial Silicon Solar Cells by the POCl$_3$ Diffusion Process
Reprinted from: *Materials* **2023**, *16*, 1824, doi:10.3390/ma16051824 . . . . **163**

About the Editors

Tongming Su

Dr. Tongming Su has been an Associate Professor at the School of Chemistry and Chemical Engineering, Guangxi University, China, since 2022. He obtained his Ph.D. in industrial catalysis from the School of Chemistry and Chemical Engineering, Guangxi University, China, in 2018. From Sept. 2016 to Mar. 2018, he conducted scientific research on nanomaterials at Oak Ridge National Lab (ORNL) and the University of Tennessee, Knoxville, as a Joint PhD student supported by the China Scholarship Council. His research interests focus on the synthesis of nanomaterials and their applications in catalysis, energy conversion, and environmental remediation.

Xingwang Zhu

Dr. Xingwang Zhu has been an Associate Professor (Contracted) at the College of Environmental Science and Engineering and Technology Research Institute for Carbon Neutrality in Yangzhou University since 2022. He received his BS (2016) in Applied Chemistry from Hainan Normal University. He then earned his PhD degree (2021) at Jiangsu University and studied nanocluster chemistry at the National University of Singapore (2021–2022). His research activities are focused on photocatalytic CO_2 reduction, low-dimensional nanomaterials' synthesis, and theoretical calculations.

Editorial

Advanced Photocatalytic Materials for Environmental and Energy Applications

Tongming Su [1,*] and Xingwang Zhu [2,*]

1. School of Chemistry and Chemical Engineering, Guangxi University, Nanning 530004, China
2. School of Environmental Science and Engineering, Institute of Technology for Carbon Neutralization, Yangzhou University, Yangzhou 225009, China
* Correspondence: sutm@gxu.edu.cn (T.S.); zxw@yzu.edu.cn (X.Z.)

With the development of modern society, environmental pollution and energy shortage have become the focus of worldwide attention. A majority of the global energy supplies are generated from fossil fuel, which gives rise to environmental pollution and climate change. As an inexhaustible clean energy source, solar energy has been widely researched and utilized for decades. Photocatalytic technology, which can directly convert solar energy into high value-added chemical energy and chemical materials or degrade a wide range of organic pollutants into easily degradable intermediates or less toxic small molecular substances, is regarded as one of the most crucial ways to solve the global energy shortage and environmental pollution problem.

This Special Issue includes two reviews and nine original research articles prepared by scientists from different countries, and these works focus on advanced photocatalytic materials for the treatment of indoor air, photocatalytic bacterial inactivation, photocatalytic hydrogen evolution, photocatalytic oxygen evolution, photocatalytic CO_2 reduction, photocatalytic hazardous pollutant removal, photothermal decomposition of pollutants, and photoelectrochemical water splitting. This Special Issue provided a platform for scientists to present their original research on "Advanced Photocatalytic Materials for Environmental and Energy Applications". The following is a brief summary of each of the papers that we have had the honor of editing to highlight recent advances in the utilization of photocatalytic materials for environmental and energy applications.

Indoor air quality has become a significant public health concern, and photocatalytic technology has been widely studied for environmental remediation, especially for air treatment. Assadi et al. [1] summarized the recent developments on the TiO_2-based materials for indoor air treatment and bacterial inactivation. In addition, several strategies, such as doping, heterojunction techniques, and combined catalysts, for enhancing the photocatalytic activity of TiO_2-based catalysts was discussed. Moreover, the catalysts used to remove volatile organic compounds and microorganisms was reviewed. Finally, the reaction mechanism of pollutant elimination and microorganism inactivation using photocatalytic technology was also summarized.

The greenhouse effect exerts a great influence on mankind, and the photocatalytic reduction of CO_2 into valuable products such as CO and CH_4 is considered a promising technology for alleviating the greenhouse effect. Qin et al. [2] synthesized the phosphorus-doped hollow tubular g-C_3N_4 (x-P-HCN) and used it for photocatalytic CO_2 reduction. They found that the phosphorus-doped g-C_3N_4 can effectively activate the CO_2 adsorbed on the surface of photocatalysts, which greatly enhanced the CO production rate of photocatalytic CO_2 reduction. Among the x-P-HCN samples, the 1.0-P-HCN sample exhibited the largest CO production rate of 9.00 $\mu mol \cdot g^{-1} \cdot h^{-1}$, which was 10.22 times higher than that of bulk g-C_3N_4. This work provided an in-depth research perspective for accelerating the photocatalytic CO_2 reduction rate of g-C_3N_4.

Citation: Su, T.; Zhu, X. Advanced Photocatalytic Materials for Environmental and Energy Applications. *Materials* **2023**, *16*, 7197. https://doi.org/10.3390/ma16227197

Received: 6 November 2023
Accepted: 13 November 2023
Published: 17 November 2023

Copyright: © 2023 by the authors. Licensee MDPI, Basel, Switzerland. This article is an open access article distributed under the terms and conditions of the Creative Commons Attribution (CC BY) license (https://creativecommons.org/licenses/by/4.0/).

Outdoor air pollutants originating from familiar anthropogenic sources, including industry, transportation, heating, and agriculture, exhibit adverse effects on health, the environment, the planet, and climate. In recent years, the dielectric barrier discharge (DBD) reactors have been widely used for hazardous pollutant removal. Khezami et al. [3] investigated the pilot-scale combination of non-thermal plasma and photocatalysis for removing toluene and dimethyl sulfur and revealed the effects of plasma energy and initial pollutant concentration on the performance and by-product formation in both pure compounds and mixtures. They found that coupling DBD plasma with a TiO_2 catalyst can achieve a synergetic effect, which resulted in improved toluene and dimethyl sulfur removal. In addition, the ozone can be reduced, and CO_2 selectivity was enhanced by combining plasma with photocatalysis.

Phase engineering is an effective strategy for tuning the electronic states and catalytic performance of photocatalysts, such as the light absorption range, charge separation efficiency, and surface redox reactivity. Yi et al. [4] provided a critical insight on the classification and diversified applications of phase-engineered photocatalysts. In this review, the classification of phase engineering for photocatalysis was summarized, and the applications of phase-engineered photocatalysts in hydrogen evolution, oxygen evolution, CO_2 reduction, and organic pollutant removal was reviewed. In addition, the synthesis and characterization methodologies for unique phase structures and the correlation between phase structure and photocatalytic performance was introduced. Finally, the current opportunities and challenges of phase engineering for photocatalysis was also discussed.

Photothermal catalysis, which combines photochemical and thermochemical contributions of sunlight, has attracted a great amount of attention in the field of pollutant decomposition. Tryba et al. [5] prepared the Ni-TiO_2 photocatalyst with nickel foam as the support and used it for the photothermal decomposition of acetaldehyde. They found that with the nickel foam as the support, the photocatalytic acetaldehyde decomposition performance over TiO_2 can be enhanced from 31% to 52% under room temperature, and from 40% to 85% at 100 °C. In addition, the mineralization of acetaldehyde into CO_2 doubled in the presence of nickel foam. This enhanced performance was attributed to the synergistic effect between nickel foam and TiO_2, which can enhance the separation of free carriers and provide more space for the interaction between the reactant and the photocatalyst.

In recent years, the photoelectrochemical process was considered as an effective way to generated clean, green hydrogen (H_2) directly from water using solar energy. Arifin et al. [6] successfully decorated Mo_2C on TiO_2 nanotube arrays (NTs) and used them for photoelectrochemical water splitting. They found that the photocurrent density was greatly increased from 0.21 mA cm^{-2} to 1.4 mA cm^{-2} when TiO_2 NTs were decorated with Mo_2C. This enhanced photocurrent density may have been due to the fact that the Mo_2C cocatalyst greatly improved the photocatalytic characteristics of the TiO_2 NTs. This work provides an effective method for developing the photoelectrochemical water splitting devices.

As already well known, process optimization is a feasible strategy to enhance the efficiency of polycrystalline silicon solar cells in the photovoltaic industry. To improve the efficiency of polycrystalline silicon solar cells, Wang et al. [7] developed a "low-high-low" temperature step of the $POCl_3$ diffusion process. Compared with the online low-temperature diffusion process, the open-circuit voltage and fill factor of solar cells increased up to 1 mV and 0.30%, respectively, in the optimization diffusion process. In addition, the efficiency of solar cells and the power of PV cells were also increased by 0.1% and 1 W, respectively. These enhancements can be ascribed to the low surface concentration of P doping and the strong impurity absorption effect of Si wafers obtained from the low-high-low temperature diffusion process.

Harnessing solar energy has been revealed to be a green strategy for solving the environmental pollution and energy crises. To make full use of solar energy, visible light-responsive photocatalysts have attracted a wide range of attention. Hu et al. [8] synthesized a novel carbon quantum dots (CQDs)-modified $PbBiO_2I$ photocatalyst and used it for the photocatalytic degradation of organic contaminants under visible/near-infrared light irradiation. After PbBiO$_2$I was modified with CQDs, the photocatalytic performance for the degradation of rhodamine B and ciprofloxacin was significantly enhanced under visible/near-infrared light irradiation. The enhanced photocatalytic performance can be ascribed to the following reasons: CQDs can absorb light in the near-infrared region, and the modification of CQDs enhanced the separation of photogenerated electrons and holes and increased the contact between the catalyst and the organic molecules. This work expanded the application of CQDs in the fields of photocatalysis and solar energy conversion.

Environmental issues, such as air pollution and water pollution, directly affect our lives. The photocatalytic degradation of organic pollutants in water is an effective method to deal with water pollution. Dai et al. [9] prepared a novel Co$_3$O$_4$/g-C$_3$N$_4$ composite photocatalyst and used it for the photocatalytic degradation of rhodamine B in water. They found that the type II heterojunction was formed between the Co$_3$O$_4$ and g-C$_3$N$_4$, and that the type II heterojunction significantly enhanced the transfer and separation of photogenerated electrons and holes in the Co$_3$O$_4$/g-C$_3$N$_4$ photocatalyst. Therefore, the photocatalytic performance for the degradation of rhodamine B over optimized 5% Co$_3$O$_4$/g-C$_3$N$_4$ photocatalyst was 5.8 times higher than that of g-C$_3$N$_4$. In addition, they found that the main active species for the degradation of rhodamine B are •O^{2-} and h$^+$.

In order to promote the photocatalytic degradation of methylene blue in water, Tao et al. [10] successfully synthesized the MoS$_2$/SnS$_2$ composite photocatalysts via a facile hydrothermal method. They found that the SnS$_2$ nanosheets were grown on MoS$_2$ nanoparticles under a smaller size, and that a heterojunction was formed between MoS$_2$ and SnS$_2$. The optimal MoS$_2$/SnS$_2$ composite exhibited a methylene blue degradation efficiency of 83.0%, which was 8.3 times and 16.6 times higher than that of MoS$_2$ and SnS$_2$, respectively. The enhanced photocatalytic performance of the MoS$_2$/SnS$_2$ composite can be attributed to the improved visible light absorption, more active sites at the exposed edges of MoS$_2$, and the enhanced separation of photogenerated electrons and holes.

Mizael et al. [11] synthesized ZnO nanoparticles with *Prosopis laevigata* extract as a stabilizing agent, the prepared ZnO nanoparticles were used for the photocatalytic degradation of methylene blue, and the degradation kinetics were investigated using the LHHW model. They found that the LHHW model sufficiently fits the experimental data, and that the size of the ZnO nanoparticles can be controlled using different concentrations of *Prosopis laevigata* as a reducing and stabilizing agent. In addition, the number of active sites increased with the decreasing nanoparticle size, which was demonstrated using the LHHW model. This work indicated that the extract of *Prosopis laevigata* can be used as a reducing and stabilizing agent to control the size and the photocatalytic performance of ZnO nanoparticles.

Funding: This Special Issue was founded by the National Natural Science Foundation of China (22208065, 22308300), Guangxi Natural Science Foundation (2022GXNSFBA035483), Natural Science Foundation of Jiangsu Province (BK20220598), Key Laboratory of Electrochemical Energy Storage and Energy Conversion of Hainan Province (KFKT2022001), Special Funding for 'Guangxi Bagui Scholars'.

Conflicts of Interest: The authors declare no conflict of interest.

References

1. Assadi, A.A.; Baaloudj, O.; Khezami, L.; Ben Hamadi, N.; Mouni, L.; Assadi, A.A.; Ghorbal, A. An Overview of Recent Developments in Improving the Photocatalytic Activity of TiO$_2$-Based Materials for the Treatment of Indoor Air and Bacterial Inactivation. *Materials* **2023**, *16*, 2246. [CrossRef] [PubMed]
2. Sun, M.; Zhu, C.; Wei, S.; Chen, L.; Ji, H.; Su, T.; Qin, Z. Phosphorus-Doped Hollow Tubular g-C$_3$N$_4$ for Enhanced Photocatalytic CO$_2$ Reduction. *Materials* **2023**, *16*, 6665. [CrossRef] [PubMed]
3. Khezami, L.; Assadi, A.A. Treatment of Mixture Pollutants with Combined Plasma Photocatalysis in Continuous Tubular Reactors with Atmospheric-Pressure Environment: Understanding Synergetic Effect Sources. *Materials* **2023**, *16*, 6857. [CrossRef] [PubMed]
4. Yi, J.; Zhang, G.; Wang, Y.; Qian, W.; Wang, X. Recent Advances in Phase-Engineered Photocatalysts: Classification and Diversified Applications. *Materials* **2023**, *16*, 3980. [CrossRef] [PubMed]

5. Tryba, B.; Miądlicki, P.; Rychtowski, P.; Trzeciak, M.; Wróbel, R.J. The Superiority of TiO_2 Supported on Nickel Foam over Ni-Doped TiO_2 in the Photothermal Decomposition of Acetaldehyde. *Materials* **2023**, *16*, 5241. [CrossRef] [PubMed]
6. Moridon, S.N.F.; Arifin, K.; Mohamed, M.A.; Minggu, L.J.; Mohamad Yunus, R.; Kassim, M.B. TiO_2 Nanotubes Decorated with Mo_2C for Enhanced Photoelectrochemical Water-Splitting Properties. *Materials* **2023**, *16*, 6261. [CrossRef] [PubMed]
7. Xu, X.; Wu, W.; Wang, Q. Efficiency Improvement of Industrial Silicon Solar Cells by the $POCl_3$ Diffusion Process. *Materials* **2023**, *16*, 1824. [CrossRef] [PubMed]
8. Yan, R.; Liu, X.; Zhang, H.; Ye, M.; Wang, Z.; Yi, J.; Gu, B.; Hu, Q. Carbon Quantum Dots Accelerating Surface Charge Transfer of 3D $PbBiO_2I$ Microspheres with Enhanced Broad Spectrum Photocatalytic Activity—Development and Mechanism Insight. *Materials* **2023**, *16*, 1111. [CrossRef] [PubMed]
9. Dai, Y.; Feng, Z.; Zhong, K.; Tian, J.; Wu, G.; Liu, Q.; Wang, Z.; Hua, Y.; Liu, J.; Xu, H.; et al. Highly Efficient and Exceptionally Durable Photooxidation Properties on Co_3O_4/g-C_3N_4 Surfaces. *Materials* **2023**, *16*, 3879. [CrossRef] [PubMed]
10. Ma, G.; Pan, Z.; Liu, Y.; Lu, Y.; Tao, Y. Hydrothermal Synthesis of MoS_2/SnS_2 Photocatalysts with Heterogeneous Structures Enhances Photocatalytic Activity. *Materials* **2023**, *16*, 4436. [CrossRef] [PubMed]
11. Luque Morales, M.; Luque Morales, P.A.; Chinchillas Chinchillas, M.d.J.; Orozco Carmona, V.M.; Gómez Gutiérrez, C.M.; Vilchis Nestor, A.R.; Villarreal Sánchez, R.C. Theoretical and Experimental Study of the Photocatalytic Properties of ZnO Semiconductor Nanoparticles Synthesized by *Prosopis laevigata*. *Materials* **2023**, *16*, 6169. [CrossRef] [PubMed]

Disclaimer/Publisher's Note: The statements, opinions and data contained in all publications are solely those of the individual author(s) and contributor(s) and not of MDPI and/or the editor(s). MDPI and/or the editor(s) disclaim responsibility for any injury to people or property resulting from any ideas, methods, instructions or products referred to in the content.

Review

Recent Advances in Phase-Engineered Photocatalysts: Classification and Diversified Applications

Jianjian Yi [1,2,*], Guoxiang Zhang [1], Yunzhe Wang [1], Wanyue Qian [1] and Xiaozhi Wang [1,*]

[1] College of Environmental Science and Engineering, Yangzhou University, Yangzhou 225127, China; 211604220@stu.yzu.edu.cn (W.Q.)
[2] College of Chemistry and Chemical Engineering, Yangzhou University, Yangzhou 225127, China
* Correspondence: jjyi@yzu.edu.cn (J.Y.); xzwang@yzu.edu.cn (X.W.)

Abstract: Phase engineering is an emerging strategy for tuning the electronic states and catalytic functions of nanomaterials. Great interest has recently been captured by phase-engineered photocatalysts, including the unconventional phase, amorphous phase, and heterophase. Phase engineering of photocatalytic materials (including semiconductors and cocatalysts) can effectively affect the light absorption range, charge separation efficiency, or surface redox reactivity, resulting in different catalytic behavior. The applications for phase-engineered photocatalysts are widely reported, for example, hydrogen evolution, oxygen evolution, CO_2 reduction, and organic pollutant removal. This review will firstly provide a critical insight into the classification of phase engineering for photocatalysis. Then, the state-of-the-art development of phase engineering toward photocatalytic reactions will be presented, focusing on the synthesis and characterization methodologies for unique phase structure and the correlation between phase structure and photocatalytic performance. Finally, personal understanding of the current opportunities and challenges of phase engineering for photocatalysis will also be provided.

Keywords: phase engineering; photocatalysis; water splitting; CO_2 reduction; pollutant degradation

Citation: Yi, J.; Zhang, G.; Wang, Y.; Qian, W.; Wang, X. Recent Advances in Phase-Engineered Photocatalysts: Classification and Diversified Applications. *Materials* **2023**, *16*, 3980. https://doi.org/10.3390/ma16113980

Academic Editor: Fotios Katsaros

Received: 20 April 2023
Revised: 17 May 2023
Accepted: 24 May 2023
Published: 26 May 2023

Copyright: © 2023 by the authors. Licensee MDPI, Basel, Switzerland. This article is an open access article distributed under the terms and conditions of the Creative Commons Attribution (CC BY) license (https://creativecommons.org/licenses/by/4.0/).

1. Introduction

The crystal phase is an emerging structural parameter of solid materials which holds the key to affecting the functionalities and properties of solid catalytic materials [1–3]. With phase transition, different atom arrangements over bulk and surface will lead to the change of physical and chemical properties. As a result, properties such as optical adsorption range, electrochemical conductivity, and molecular adsorption ability can be adjusted, which allows the modulation of function-oriented behaviors for various catalytic applications [4–6]. Phase engineering can therefore be defined as a strategy that constructs a specific phase for typical investigative purposes. Many catalytic materials exist in more than one crystal phase, thus providing the possibility of optimizing the performance or broadening the scope of applications. After decades of rapid progress in this research area, great achievements have been made in the phase engineering of nanomaterials, including synthetic methods to realize controllable synthesis of desired phase-engineered materials, fine characterization techniques to clearly analyze the phase structures, and development in various applications [7–10].

Photocatalysis is considered as an emerging concept of catalysis that can realize the conversion of solar energy to chemical energy [11–14]. As a green sustainable catalytic technology, the main working mechanism of photocatalysis is exciting semiconductors to generate electron-hole pairs, which can participate in different redox reactions, for example, water splitting [15], CO_2 reduction [16], and organic pollutant degradation [17]. Nevertheless, the development of photocatalysis up to practical applications still confronts a big challenge, since the performance of the traditional photocatalytic systems is far from satisfactory. Generally, the total photocatalytic process can be divided into six steps, (i) photon absorption, (ii) exciton separation, (iii) carrier diffusion, (iv) carrier transport, (v) catalytic conversion (redox), and (vi) mass transfer of reactants. Additionally, the total conversion efficiency in photocatalytic processes could be generally determined by the efficiency of three main steps, i.e., light absorption efficiency, charge separation efficiency, and surface redox reactivity [18]. The phase engineering of photocatalytic nanomaterials has recently captured great interest in photocatalysis since there are many unique superiorities of phase engineering for tuning photocatalytic behavior regarding the three main steps during photocatalysis. Firstly, phase engineering of semiconducting photocatalysts can provide possibility to adjust the electronic band structures for tunable light absorption ranges. Secondly, phase engineering can change the atomic arrangement in bulk and surface of materials, resulting in the optimization of built-in electric fields. As a result, the charge separation behavior can further be affected. Thirdly, the surface redox reactivity strongly depends on the adsorption/activation ability of reactant molecules, which can be finely tuned by phase-engineering-induced surface atomic reconfiguration [19]. In recent years, various investigations have resulted in improved photocatalysis by rational phase engineering, not limited to phase engineering for pristine semiconductors [20,21] but also for cocatalysts [22–24]. Moreover, phase engineering of nanomaterials is not just limited to phase transition from one phase to another. The formations of the amorphous phase [25] and heterophase (e.g., phase junction) [26,27] have also been widely studied.

In this review, we focus on presenting recent advances in phase engineering for photocatalytic reactions. The classification of phase engineering toward photocatalysis is firstly clarified. Then, an overview of state-of-the-art developments in photocatalytic applications based on phase engineering is presented, including but not limited to photocatalytic water splitting, CO_2 reduction, and pollutant removal. As a focus, the synthesis and characterization methods for desired phase structures and the correlation between phase structure and photocatalytic performance are emphasized. Finally, the current challenges and further opportunities of phase engineering for photocatalysis are envisioned. We try to summarize phase engineering for photocatalytic materials to tackle limited catalytic efficiency and highlight the importance of phase engineering for photocatalysis. The description of representative samples in each catalytic reaction would be sufficient for a general review for researchers who are not familiar with phase-engineered photocatalysts.

2. Classification of Phase-Engineered Nanostructures

Considering that the emerging phase engineering strategies are complicated and diversified, the classification of phase engineering is important for the discussion about phase engineering for photocatalytic applications. As illustrated in Figure 1, we classify the phase-engineered nanostructures into the following three types: the unconventional phase, amorphous phase, as well as the heterophase.

The definition of the unconventional phase is a relative concept to the conventional phase. In general, nanomaterials exist in the form of thermodynamically stable phases in bulky components, which can be defined as the conventional phase. However, nanomaterials with different phases can be obtained by adjusting reaction kinetics and surface energy under certain experimental conditions. The obtained phases with different atom arrangements compared to the conventional phases can be denoted as the unconventional phase. Taking metal as an example, Au is usually crystallized in the conventional face-centered cubic (fcc) phase, but Zhang et al. demonstrated that Au can also be crystallized into 2H and 4H phases by controlling the synthetic parameters [28,29]. In this case, the fcc phase is the conventional phase for Au, while the 2H and 4H phases are unconventional phases. Until now, unconventional phases have been found in a variety of nanomaterials, including metals, metal oxides, and transition metal chalcogenides. The obtained nanomaterials with unconventional phases show unique and enhanced performance effects in many applications.

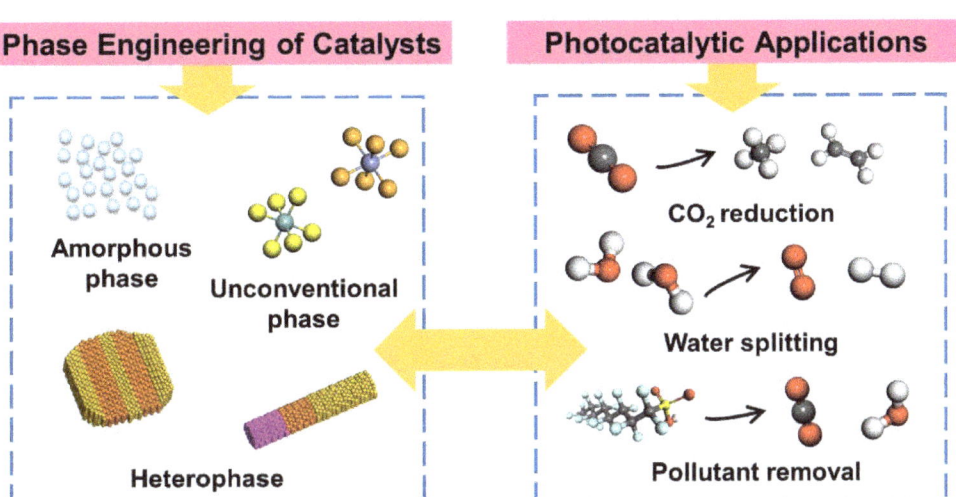

Figure 1. Schematic illustration of different types of phase engineering and the applications in photocatalysis.

The amorphous phase is a relative concept to the crystalline phase, with unique structural characteristics. The key feature of the amorphous phase is the short-range order but long-range disorder atomic structure. In contrast, the crystalline phase generally is in the form of short-range orders and long-range orders [30,31]. Due to the disordered arrangement of atoms and high entropy resulting from unsaturated bonds, amorphous materials are usually metastable and readily change to crystalline states under external heat or pressure. The synthesis of amorphous phase nanomaterials can be realized through direct preparation by finely controlling the crystallization process or indirect preparation by breaking the long-range order structure of the crystalline phase [32].

The heterophase structure is different from the mentioned unconventional phase and amorphous phase. It refers to the multi-phase structure composed of two or more crystalline phases of the same material.

The most classical example for the heterophase structure is the commercial TiO_2 materials, namely P25, which is composed of mixed anatase and rutile phases [33]. Compared to the pristine anatase phase and rutile phase, P25 can exhibit improved photocatalytic activity in many reactions due to the improved charge separation efficiency with the formation of a space-charge layer. In addition to the phase I/phase II heterophase structure, crystalline/amorphous structures can also be defined as heterophase structures [34,35].

With the rational design of material structural engineering based on constructing different phase structures, the electronic structure and catalytic functions can be finely tuned for improved photocatalysis. Different components in a typical photocatalytic material, including light-harvesting semiconductors and surface active cocatalysts, can be tuned based on phase engineering to meet different requirements in different reactions. With the development of phase engineering, phase-dependent properties and photocatalytic applications (e.g., water splitting, CO_2 reduction and pollutant removal) have been witnessed. Detailed discussion in terms of recent advances in phase engineering for photocatalysis will be provided in the following section.

3. Phase Engineering for Photocatalytic Applications

Phase-engineered nanomaterials have endowed them with unique electronic structures and catalytic properties for various applications such as hydrogen evolution, oxygen evolution, CO_2 reduction, and pollutant removal. Phase engineering can enhance the catalytic performance of typical reactions by broadening the light absorption or steering charge transfer kinetic or by maneuvering the surface redox reaction. In addition, photocatalysts or cocatalysts with different phase structures can also affect the reaction selectivity and stability. In this section, an overview of some recent advances in phase engineering for photocatalytic reactions will be provided.

3.1. Hydrogen Evolution

Developing clean and renewable fuels with high energy density is the common pursuit of the academic community. Hydrogen energy is one of the candidates to meet the requirements mentioned above. Photocatalytic water splitting to produce hydrogen is the "holy grail" in solar energy conversion, which is still restricted by the activity, stability, and economic cost [13,36,37]. It has been widely found that phase engineering can significantly boost the photocatalytic hydrogen evolution performance.

Recently, phase-engineered semiconducting photocatalysts have been widely designed to modulate their photocatalytic hydrogen evolution performance. Some important electronic structures such as light absorption edges and strength of built-in electric fields can be tuned by phase engineering. Taking the most classic photocatalyst, TiO_2, as an example, it was reported that the photocatalytic hydrogen evolution performance of brookite phase TiO_2 was significantly higher than that of anatase phase TiO_2 (Figure 2a) [38]. The author explained that the conduction band (CB) edge of brookite phase TiO_2 was more negative than that of anatase TiO_2 supported by experimental characterizations. Electrons excited at the CB of brookite phase TiO_2 with higher reduction ability can more effectively reduce H^+ to H_2, leading to higher hydrogen evolution activity. Similar cases can also be found in sulfide photocatalysts such as ZnS. Feng et al. reported that the phase structure of ZnS can be regulated by ambient S annealing [39]. The photocatalytic hydrogen evolution measurement showed that wurtzite phase ZnS showed better hydrogen evolution activity than sphalerite phase ZnS. The reason for the phase dependence was owing to the strengthened inter-polar electric field of wurtzite phase ZnS, which could promote the electron-hole separation.

Apart from the formation of a typical phase, the construction of hybrid photocatalysts with heterophase (e.g., phase junction) structure is considered as another effective phase engineering strategy for improved hydrogen evolution catalysis. For instance, CdS, a photocatalyst with desirable bandgap and availability limited by photo-corrosion and inefficiency, was reported to have photocatalytic hydrogen evolution performance that can be optimized by constructing phase junctions [20,40]. Experimental results revealed that the phase junction composed of cubic and hexagonal phase CdS (denoted as c-CdS/h-CdS) showed a high hydrogen evolution rate (4.9 mmol h^{-1} g^{-1}) and external quantum efficiency (EQE) of 41.5% at 420 nm (Figure 2b) [40]. The hydrogen evolution rate was 60 times higher than those of c-CdS and h-CdS. Notably, photo-corrosion can also be inhibited over the phase junction. The origin of improved activity and stability was owing to the greatly enhanced charge separation by the regulation of bonding region between cubic and hexagonal phases. Recently, Yu et al. demonstrated an efficient heterophase red P photocatalyst with a hydrogen evolution rate over 1280 μmol h^{-1} g^{-1} (Figure 2c) [41]. The formation of red P heterophases consisting of fibrous and Hittorf's phases can be realized by Bi-mediated catalytic synthetic method. From the fact that each phase red P possessed different band alignments, the intimate heterophase junction afforded an effective built-in driving force for efficient charge transport, thus achieving high catalytic performance. Similar case studies were also demonstrated over TiO_2 [34], phosphorus [42], $Cd_{1-x}Zn_xS$ [43], In_2O_3 [44], and $ZnIn_2S_4$ [45], highlighting the advance of heterophase structures.

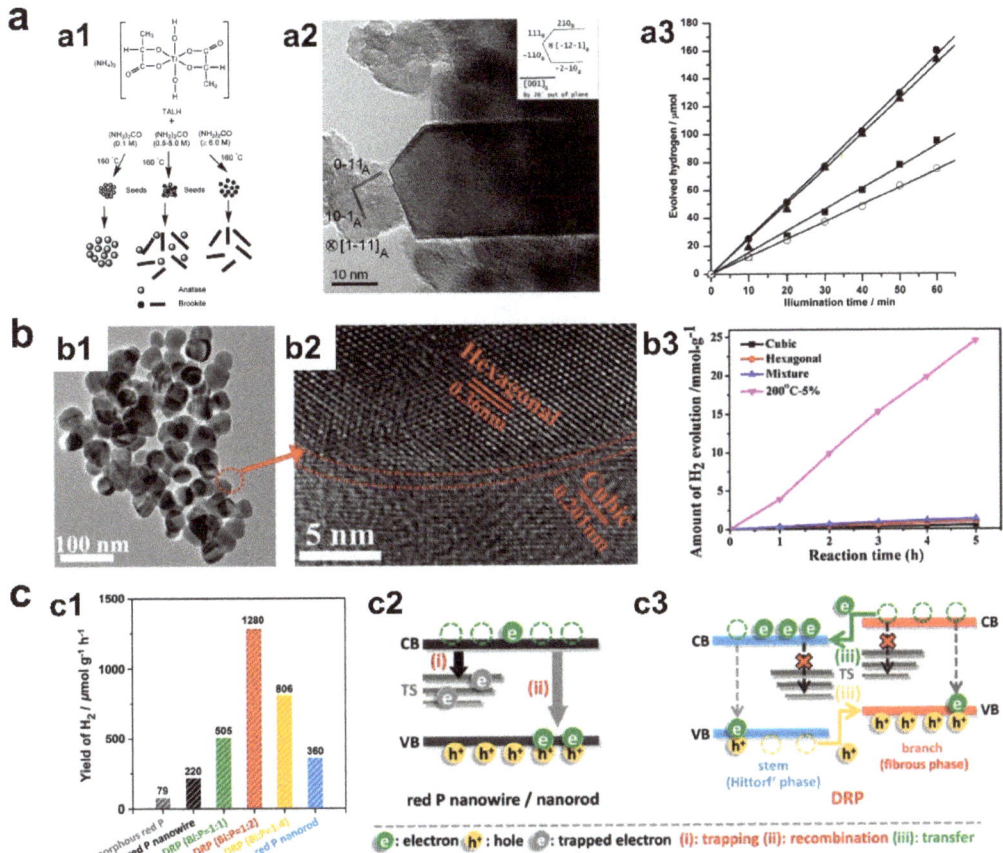

Figure 2. (**a1**) Proposed routes for the formation of anatase and brookite phase TiO$_2$. (**a2**) HRTEM image of TiO$_2$ with mixed anatase and brookite phase. (**a3**) Photocatalytic hydrogen evolution activity of TiO$_2$ with different phase structures using Pt as cocatalyst; ■, ●, ▲, and ○ indicate anatase phase, anatase/brookite mixed phase, brookite phase, and P25, respectively (copyright from Ref. [38], 2010, American Chemical Society). (**b1,b2**) TEM and HRTEM images of CdS with mixed hexagonal and cubic phases. (**b3**) Photocatalytic hydrogen evolution activity of the catalysts (copyright from Ref. [40], 2018, Elesvier). (**c1**) Hydrogen product yield of red P samples with different phase structures under visible light irradiation. (**c2,c3**) Schematic representation of proposed charge dynamics in (**c2**) red P nanowire/nanorod and (**c3**) heterophase red P with fibrous and Hittorf's phases (copyright from Ref. [41], 2022, Elesvier).

Transitional metal dichalcogenides (TMDs) represented by MoS$_2$ show potential to improve photocatalytic hydrogen evolution performance as cocatalysts on semiconductor surfaces. TMD nanomaterials show various phases owing to the different electronic structures of transition metal atoms with different d orbital filling states [22,46,47]. For example, 2H-MoS$_2$ and 2H-WS$_2$ are semiconducting with band gaps, whereas 1T-MoS$_2$ and 1T-WS$_2$ are metallic with good conductivity for hydrogen evolution reactions. Our group found that 1T-MoS$_2$/O-g-C$_3$N$_4$ demonstrated a greatly higher photocatalytic hydrogen evolution performance compared to 2H-MoS$_2$/O-g-C$_3$N$_4$ (Figure 3a) [48]. The optimal 1T-MoS$_2$/O-g-C$_3$N$_4$ sample showed hydrogen evolution rate over 1800 μmol/g/h with external quantum efficiency of 7.11% at 420 nm. The origin for the high catalytic activity of 1T-MoS$_2$ can be ascribed to the metal-like conductivity and the active edge and basal sites for hydrogen evolution. By contrast, low conductivity and limited active sites at edge sites leads to poor performance. Recent studies also reported that 1T-MoS_2, 1T-WS_2, and 1T-$MoSe_2$ can improve the catalytic activity of hosting semiconductors not limited to g-C_3N_4 but also other semiconductors such as TiO_2 and CdS [23,24,49–51].

In addition to MoS_2 with layered structure, TMD materials with non-layered structures such as $CoSe_2$ and $NiSe_2$ also show phase-dependent catalytic hydrogen evolution performance. Our group demonstrated the phase-dependent photocatalytic hydrogen evolution catalysis of $CoSe_2$. In a practical photocatalytic process, it was observed that $CoSe_2$ with orthorhombic phase (o-$CoSe_2$) can better improve the hydrogen evolution rate of g-C_3N_4 semiconductors than $CoSe_2$ with cubic phase (c-$CoSe_2$) (Figure 3b) [52]. It was revealed by density functional theory (DFT) calculations that the Co site on o-$CoSe_2$ surface showed more appropriate hydrogen adsorption Gibbs free energy (ΔG_{H^*} = 0.27 eV) than the c-$CoSe_2$ surface, resulting in improved catalysis. Despite c-$CoSe_2$ possessing better conductivity than o-$CoSe_2$, the charge separation efficiency may not be the rate-determining step in this case. Interestingly, unlike the higher catalytic performance of o-$CoSe_2$/g-C_3N_4 compared to c-$CoSe_2$/g-C_3N_4, we recently found that the hydrogen evolution of o-$CoSe_2$/TiO_2 (2.601 μmol/h) was lower than that of c-$CoSe_2$/TiO_2 (12.001 μmol/h) [53]. We propose the reason for this phenomenon would be that the interfacial charge transfer between TiO_2 and c-$CoSe_2$ is the dominating factor but not the surface redox reactivity. Impressively, phase engineering of TMD-based cocatalysts show not only phase-dependent activity but also stability. Very recently, we reported that the phase structure of $NiSe_2$ played important role in determining the photocatalytic hydrogen evolution stability instead of stability (Figure 3c) [54]. Upon light irradiation on m-$NiSe_2$/CN and p-$NiSe_2$/CN in TEOA/H_2O, comparable photocatalytic hydrogen evolution rates of 3.26 μmol h^{-1} and 3.75 μmol h^{-1} can be observed. Importantly, we found that $NiSe_2$ exhibited phase-dependent stability, i.e., m-$NiSe_2$ can evolve H_2 steadily, but p-$NiSe_2$ showed a ~57.1% rate decrease after 25 h of reaction. After fine characterization, we proposed the origin of phase-dependent stability. The chemical structure of m-$NiSe_2$ can be well preserved in a catalytic process, but partial p-$NiSe_2$ tends to be converted to NiOOH, resulting in different catalytic stability.

3.2. Oxygen Evolution

Photocatalytic water splitting to produce hydrogen or CO_2 reduction to obtain high value-added chemical products are ideal pathways to achieve solar energy storage and conversion. In the overall reactions, water oxidation is the most important half reaction of these two energy photocatalytic reactions [55–57]. In water splitting and CO_2 reduction reactions, water oxidation provides protons and electrons, which is the premise of the reduction reaction. It should be noticed that the water oxidation reaction is a four-electron reaction, and the overpotential of this reaction is very high. It is thus considered as the rate-determining step of the overall reactions, which holds the key to the proceeding of hydrogen evolution from water or CO_2 reduction [58,59]. Phase engineering of nanomaterials could improve photocatalytic oxygen evolution performance by enhancing charge separation, or decreasing the reaction energy barrier.

Phase-dependent photocatalytic performance was reported over a classical oxygen evolution semiconducting photocatalyst, i.e., $BiVO_4$. Kudo et al. synthesized tetragonal phase and monoclinic phase $BiVO_4$ successfully, characterized the phase and optical structures, and measured the photocatalytic O_2 evolution performance in $AgNO_3$ solution (Figure 4a) [60]. Given the similar light harvesting capacity of tetragonal $BiVO_4$ and monoclinic $BiVO_4$, it was interesting that negligible O_2 gas product can be detected over tetragonal $BiVO_4$, but monoclinic $BiVO_4$ exhibited high O_2 evolution activity (over 120 μmol/h under visible light and over 70 μmol/h under UV light). Mechanism analysis revealed that distortion of a Bi-O polyhedron by a $6s^2$ lone pair of Bi^{3+} in monoclinic $BiVO_4$ was beneficial for the surface conversion from H_2O to O_2, leading to high photocatalytic activity. In another research work conducted by Amal et al., the authors investigated the amorphous and crystalline evolution of $BiVO_4$ during the synthesis by flame spray pyrolysis. In terms of the photocatalytic test, the first finding was that amorphous BiVO$_4$ cannot produce O$_2$ by photocatalysis. For crystalline BiVO$_4$, the photocatalytic oxygen evolution rate increased with the increased content of monoclinic phase in BiVO$_4$, highlighting the important role of monoclinic phase [61].

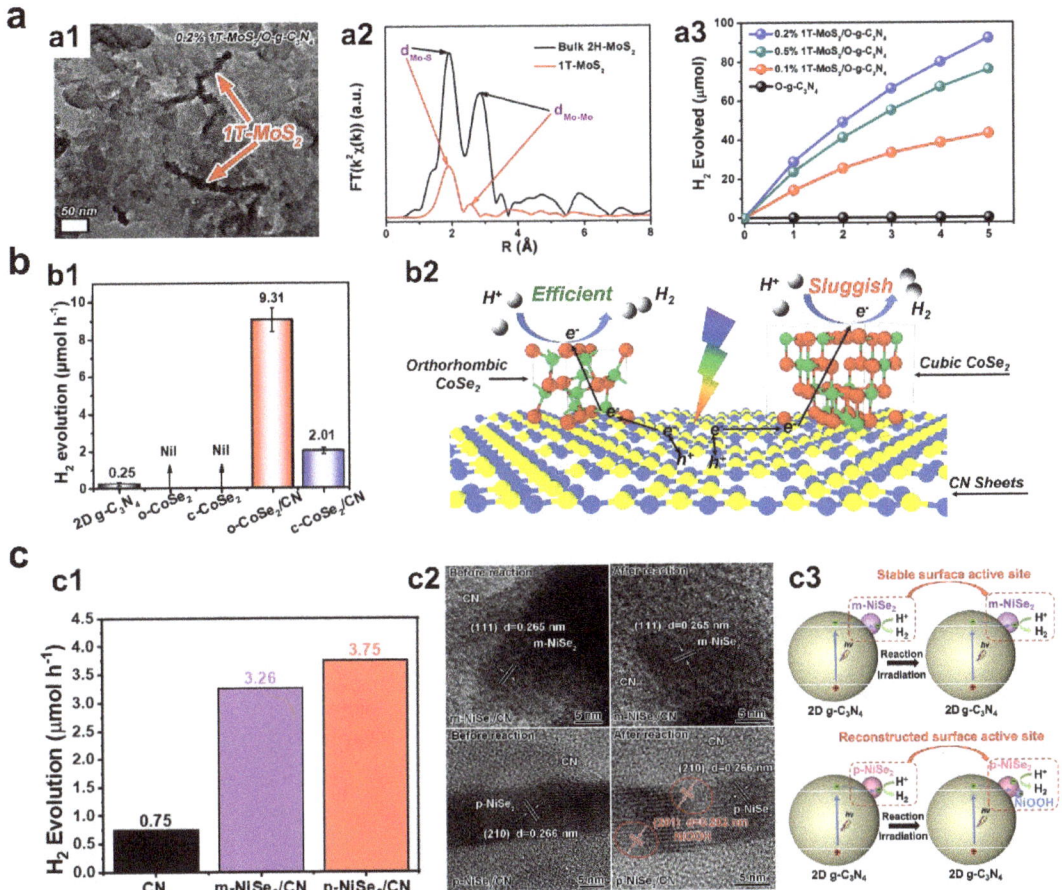

Figure 3. (**a1**) TEM image of 1T-MoS$_2$/O-g-C$_3$N$_4$. (**a2**) Fitted EXAFS results of prepared 1T-MoS$_2$ sample and commercial 2H-MoS$_2$. (**a3**) Time-dependent hydrogen evolution rate of the samples (copyright from Ref. [48], 2018, Elesvier). (**b1**) A comparison of light-driven hydrogen evolution rate of CoSe$_2$-based catalysts. (**b2**) Charge behavior of o-CoSe$_2$/CN and c-CoSe$_2$/CN (copyright from Ref. [52], 2020, Elesvier). (**c1**) A comparison of light-driven hydrogen evolution rate of NiSe$_2$-based catalysts. (**c2**) HRTEM images of NiSe$_2$/CN samples before and after photocatalysis. (**c3**) Mechanism illustration of phase-dependent stability of NiSe$_2$ (copyright from Ref. [54], 2023, Elesvier).

Except for phase engineering of semiconductors, phase-dependent oxygen evolution photocatalysis was also found in surface active cocatalysts. For example, our research group demonstrated that the phase structure of CoSe$_2$ cocatalysts plays an important role in determining the oxygen evolution performance of Fe$_2$O$_3$ semiconductors (Figure 4b) [62]. Experimental results found that orthorhombic phase CoSe$_2$ (o-CoSe$_2$) showed better potential than cubic phase CoSe$_2$ (c-CoSe$_2$) in enhancing photocatalytic oxygen evolution performance of Fe$_2$O$_3$. o-CoSe$_2$/Fe$_2$O$_3$ can realize the qualitative changes of oxygen evolution rate from "0" to "1" under visible light irradiation, using AgNO$_3$ as sacrificial agent and La$_2$O$_3$ as pH balance agent. However, c-CoSe$_2$/Fe$_2$O$_3$ cannot work in photocatalytic oxygen evolution processes under the same conditions. Combined with photoelectrochemical characterization and theoretical simulations, we proposed that the key factor for the superior activity of o-CoSe$_2$ was the decreased activation barrier of H$_2$O on its surface.

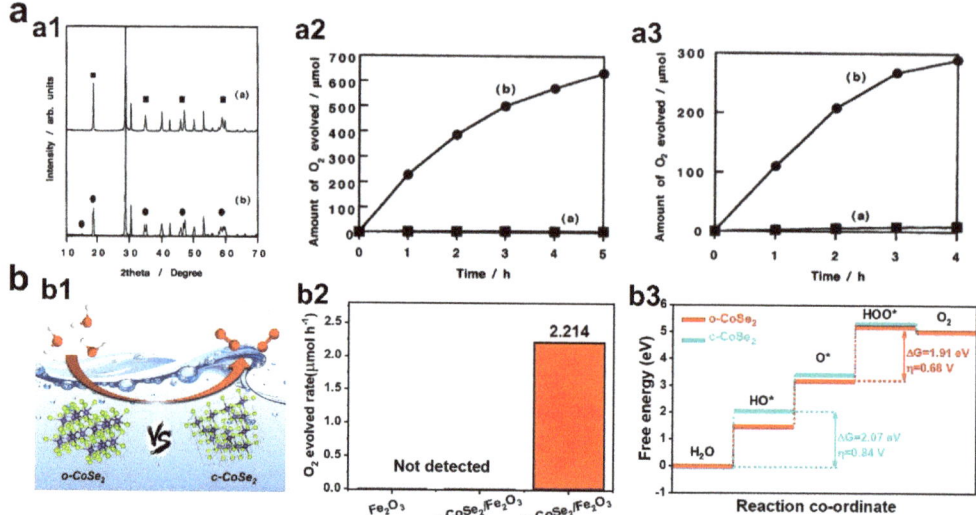

Figure 4. (**a1**) XRD patterns of BiVO$_4$ with different phases, ■ and ● indicate tetragonal and monoclinic phase respectively. (**a2**,**a3**) Photocatalytic oxygen evolution performance of BiVO$_4$ samples under visible light irradiation (λ > 420 nm) and under ultraviolet light (300 nm < λ < 380 nm). Inset: (a)—tetragonal phase, (b)—monoclinic phase (copyright from Ref. [60], 2001, American Chemical Society). (**b1**) Schematic of oxygen evolution process over CoSe$_2$ with different phase structures. (**b2**) Photocatalytic oxygen evolution performance of the catalysts. (**b3**) Diagram of the calculated free energy change in each reaction step over o-CoSe$_2$ and c-CoSe$_2$ (copyright from Ref. [62], 2022, Elsevier).

3.3. CO$_2$ Reduction

Converting CO$_2$ into valuable chemical feedstocks or liquid fuels under mild conditions is of great significance for reducing the greenhouse effect and achieving carbon neutrality [63–65]. Among them, photocatalysis is considered a green technology that converts CO$_2$ from solar energy [66,67]. The key steps of photocatalytic CO$_2$ reduction include the generation of electron-hole (e-h) pairs, electron transfer to catalytic active sites, and catalytic CO$_2$ reduction. In order to meet these requirements, the rational structural design of catalysts is required [68]. Due to the ability to optimize the multiple electronic structures of catalysts, especially the regulation of CO$_2$ molecular adsorption activation behavior, phase engineering in the field of CO$_2$ photoreduction has also attracted extensive attention [69].

Phase engineering of pristine semiconductor has shown potential in tuning the photocatalytic CO$_2$ reduction performance. Li et al. reported the phase-controlled synthesis of three types of TiO$_2$ (anatase, rutile, and brookite), and the evaluation of photocatalytic CO$_2$ reduction performance with water vapor (Figure 5a) [70]. Experimental results found that the production of CO and CH$_4$ using anatase and brookite phase TiO$_2$ were enhanced by 10-fold in contrast to rutile phase TiO$_2$. The new finding in this study is that brookite phase TiO$_2$, with less attention among the three TiO$_2$ polymorphs, exhibited the highest CO and CH$_4$ production rates. Detailed mechanism studies revealed that the superior activity of brookite phase TiO$_2$ may be owing to the existence of surface oxygen vacancies, faster reaction rate of CO_2^- with surface adsorbed H_2O or OH groups, and a new reaction pathway involving an HCOOH intermediate.

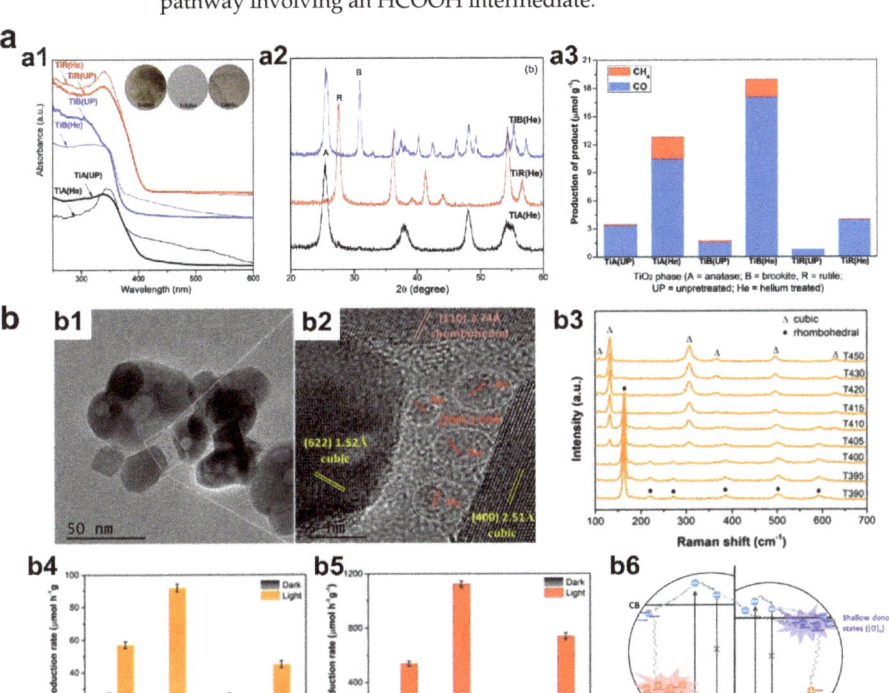

Figure 5. (**a1**) UV-visible diffuse reflectance spectra of unpretreated and He-pretreated TiO_2 (anatase, rutile, and brookite). (**a2**) XRD patterns of He-pretreated TiO_2 (anatase, rutile, and brookite). (**a3**) Photocatalytic CO_2 reduction performance of TiO_2 with different phases (copyright from ref. [70], 2012, American Chemical Society). (**b1,b2**) TEM and HRTEM images of $In_2O_{3-x}(OH)_y$ composed of rhombohedral and cubic phase. (**b3**) XRD patterns illustrating the phase transition from rhombohedral to cubic phase with increased temperature. (**b4,b5**) CH_3OH and CO evolution rates over the catalysts with different phase structures. (**b6**) Schematic illustration of charge carrier separation and recombination pathways in polymorphic heterostructures of rh/c-$In_2O_{3-x}(OH)_y$ (copyright from ref. [71], 2020, Royal Society of Chemistry).

In addition to the complete phase transition in semiconductor photocatalyst, the construction of heterophase photocatalyst recently attracts intensive attention in CO_2 reduction. Yan et al. reported a heterophase photocatalyst based on $In_2O_{3-x}(OH)_y$ to optimize the CO_2 reduction performance (Figure 5b) [71]. The continuous transition from cubic to rhombohedral can be realized by temperature variety so that the heterophase structure can be obtained by partial phase transition at an appropriate temperature. The charge separation efficiency can be greatly improved due to the formation of the cubic/rhombohedral interface. As a result, the optimized cubic/rhombohedral $In_2O_{3-x}(OH)_y$ exhibited improved CO_2 reduction activity compared to that of the pure cubic or rhombohedral phase, with a CH_3OH evolution rate of 92 µmol/g/h and CO evolution rate of 1120 µmol/g/h. This strategy can also be extended to other material systems with similar working mechanisms for CO_2 reduction, for example, $CuInS_2$ [72] and CdS [73].

In photocatalytic hybrid structures, cocatalysts are widely used to improving the activity and selectivity of CO_2 reduction by promoting the electron-hole separation and providing highly active catalytic sites for CO_2 catalytic conversion. Phase structure engineered metal cocatalysts are recently known as efficient cocatalysts for CO_2 reduction. For instance, Bai et al. reported that the Ru cocatalyst with hexagonal close-packed (hcp) phase (hcp Ru) can more effectively boost CO_2 reduction efficiency of C_3N_4 semiconductor in contrast to face-centered cubic (fcc) phase Ru (fcc Ru) (Figure 6a) [74]. Thanks to phase engineering of Ru, not only CO and CH_4 evolution rates but also selectivity for carbon-based products can be improved over C_3N_4-hcp Ru. To uncover the mechanism for phase-dependent activity and selectivity, the authors explained this phenomenon supported by electrochemical characterizations and theoretical calculations. The improved performance was not owing to the altered interfacial electron transfer from C_3N_4 to Ru, but the higher CO_2 adsorption energy on $(10\bar{1}1)$ face of hcp Ru was higher than that on the (111) face of fcc Ru. Except for the design of an unconventional phase, amorphous phase cocatalysts can also work in CO_2 reduction reaction with optimized performance. In another case, also reported by Bai et al., the effect of crystallinity on photocatalytic CO_2 reduction performance was systematically studied by using Pd nanosheets as model cocatalysts (Figure 6b) [75]. When Pd nanosheets were assembled with CdS quantum dots (QDs), it was found that Pd with high crystallity and good lattice periodicity was more conducive to electron transfer from CdS to Pd, which more effectively inhibited H_2 production on the surface of CdS. In contrast, low-crystallinity Pd provided a large number of surface unsaturated atoms and defects as highly active centers for efficient CO_2-CO/CH_4 conversion. The formation rates of CO (23.93 µmol/g/h) and CH_4 (0.35 µmol/g/h) in CdS-Pd-48 with low-crystallinity Pd were 10.3 and 5.9 times of those of the pristine CdS QDs, respectively. This result highlighted that amorphous-phase Pd contributed to efficient CO_2 adsorption and activation, and crystalline Pd favored charge migration. Taken together, amorphous Pd was a better cocatalyst for CO_2 reduction in terms of the yield and selectivity of carbon products.

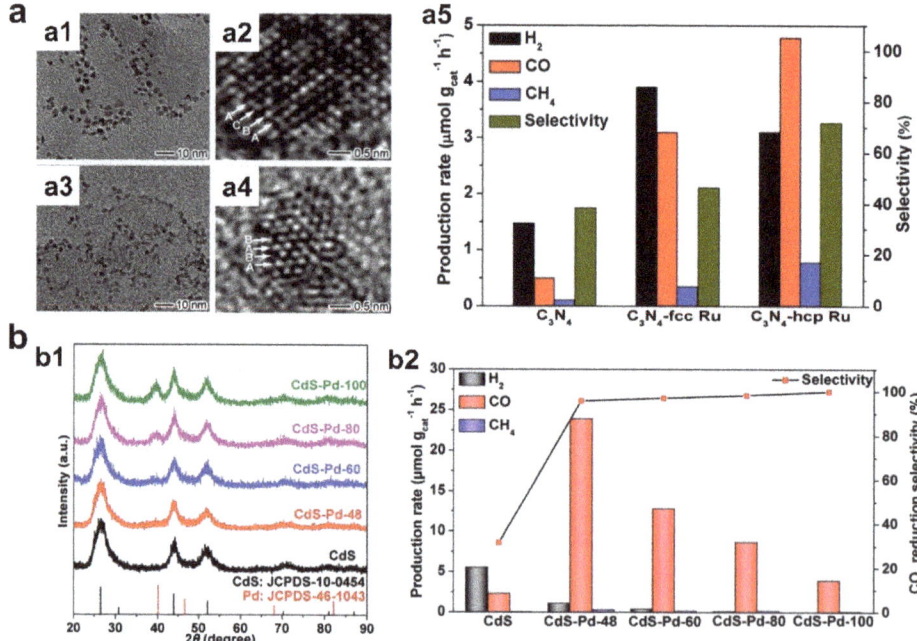

Figure 6. (**a1,a2**) TEM and HRTEM images of C_3N_4-fcc Ru. (**a3,a4**) TEM and HRTEM images of C_3N_4-hcp Ru. (**a5**) Photocatalytic CO_2 reduction activity and selectivity (copyright from ref. [74], 2018, Elesvier). (**b1**) XRD patterns of CdS-Pd samples. (**b2**) Photocatalytic CO_2 reduction activity and selectivity (copyright from ref. [75], 2020, Royal Society of Chemistry).

3.4. Pollutant Removal

Semiconductor photocatalysis has been widely applied in environmental remediation. Upon light irradiation, photocatalysts can harvest solar energy to generate electron-hole (e-h) pairs for pollutant removal. Various reactive oxygen species, including O_2^-, OH, and $^1O_2$, can be generated by different redox reactions. In addition, the photogenerated holes can also directly oxidize organic pollutants. All the mentioned reactive species generated by photocatalysis can contribute to efficient pollutant removal [18,76,77]. Phase engineering of photocatalysts can improve the pollutant removal performance by several working mechanisms, for example, decreasing the formation energies of oxygen species, improving the light absorption range, promoting charge separation, improving reduction and oxidation potentials of electrons or holes, and so on.

Phase-controlled synthesis of TiO_2 nanorods can be realized by a hydrothermal method using peroxide titanic acid solution with different pH values, forming rutile, anatase, and brookite phase TiO_2, respectively (Figure 7a) [78]. The phase-dependent photocatalytic activities of the samples were evaluated by reduction of Cr (VI) and degradation of methylene blue (MB). Experimental results indicated that rutile TiO_2 possessed best photo-degradation performance of MB, while brookite TiO_2 showed best activity for Cr (VI) photo-reduction. The improved photo-oxidation performance can be explained that rutile TiO_2 can expose more {111} facets with high surface energy, which can boost the oxidation reaction more easily. For photo-reduction reaction, the more negative CB potential of brookite TiO_2 contributed to more effective Cr (IV) reduction. This case study highlights the importance of phase design for different environmental photocatalytic reactions.

In many recent studies, it has been proven that amorphization of catalysts is an effective method to tune the physical/chemical properties and thus modulate their photocatalytic degradation performance [79,80]. For example, Mao et al. reported a classical study about increasing light absorption of TiO_2 by amorphization, forming the so-called black TiO_2 (Figure 7b) [81]. The researchers demonstrated that the color of white TiO_2 changed to black for surface amorphization by hydrogenation treatment, leading to light absorption broadening from ultraviolet region to near-infrared region. In a practical photocatalytic degradation of methylene blue (MB) solution, black TiO_2 with crystalline core and amorphous surface showed significantly improved degradation rate with high stability compared to that of white TiO_2. The origin for the phase-dependent performance was mainly owing to the change of electronic and optical properties of black TiO_2, especially the greatly reduced band gap promoted by the formation of midgap states. Similarly, highly efficient photocatalytic degradation can also be demonstrated over other semiconductors though phase transition to amorphous phase by virtue of the high surface energy and outstanding adsorption/desorption properties. Hu et al. reported the synthesis of a series of phase-engineered Sb_2S_3 photocatalysts with different degrees of amorphization by adjusting the concentration of hydrochloric acid in hydrothermal processes [82]. The amorphous Sb_2S_3 exhibited the best photocatalytic methyl orange degradation activity, which was 13 times higher than that of crystalline Sb_2S_3.

Except for the direct photocatalytic degradation of pollutants, phase engineering in photo-assisted environmental catalysis has recently attracted wide attention [83–85]. For instance, the Qu research group reported the construction of heterophase MoS_2, i.e., 2H/1T MoS_2, and its application in photo-assisted permonosulfate (PMS) activation for water pollutant degradation (Figure 8) [86]. Although MoS_2 has been proven as an efficient activator of PMS, it is still restricted by the loss of low-valence Mo during the catalytic process. The authors found that the integration of semiconducting 2H phase MoS_2 and 1T phase MoS_2 forming 2H/1T MoS_2 can favor the catalytic reaction, by introducing photogenerated electrons of 2H MoS_2 under light irradiation to trigger the formation of low-valence Mo. In this case, the key of this reaction is the regeneration of low valence Mo in 1T MoS_2. 2H MoS_2 with semiconductor characteristic in heterophase MoS_2 can transfer electrons to 1T MoS_2 under light irradiation, leading to the reduction of high-valence Mo to low-valence Mo (active center in PMS activation). As a result, 2H/1T phase MoS_2 showed efficient and continuous degradation of organic pollutants in the existence of PMS and light. The discovery in this work highlighted the merits of constructing heterophase structure in heterogeneous photocatalytic degradation reactions.

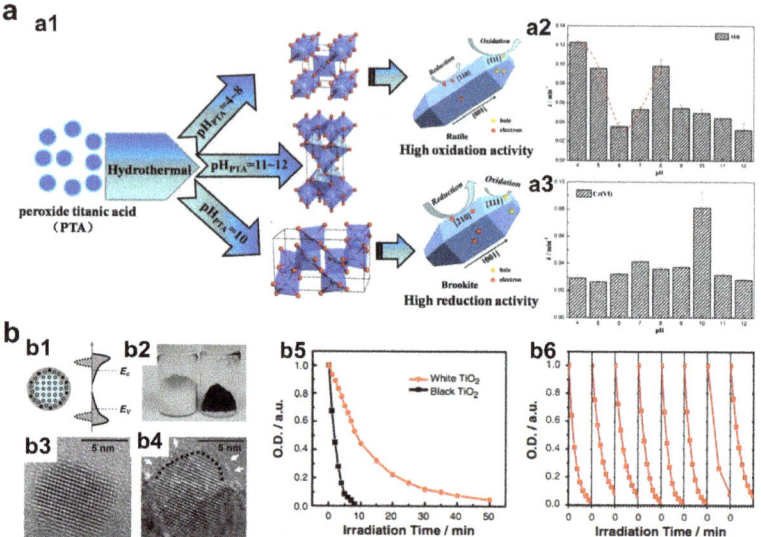

Figure 7. (**a1**) Schematic illustration of synthetic process of TiO_2 with different phases. (**a2**) Photocatalytic performance of MB degradation. (**a3**) Photocatalytic performance of Cr (VI) reduction (copyright from ref. [78], 2015, American Chemical Society). (**b1**) Schematic illustration of the structure and electronic DOS of a semiconductor in the form of a disorder engineered nanocrystal with dopant incorporation. (**b2**) Digital graphs of unmodified white TiO_2 and black TiO_2 with amorphous surface layer. (**b3,b4**) HRTEM images of unmodified white TiO_2 and black TiO_2 with amorphous surface layer. (**b5**) Photocatalytic MB degradation performance. (**b6**) Stability tests (copyright from ref. [81], 2011, American Association for the Advancement of Science).

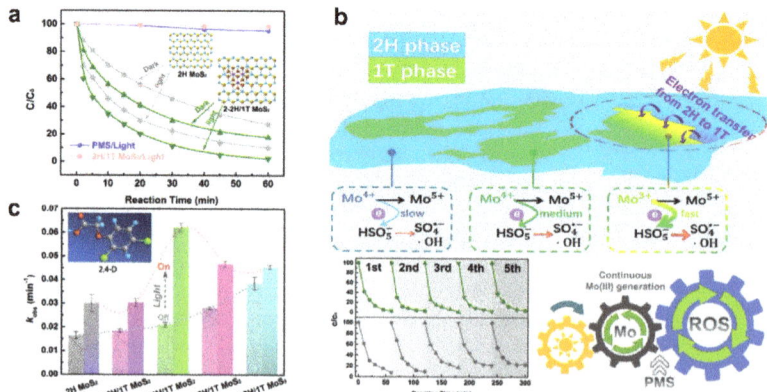

Figure 8. (**a**) Removal efficiency of 2,4-D in different reaction systems. (**b**) Rate constants of 2,4-D degradation using different MoS_2 samples with (red dashed line) and without (gray dashed line) light irradiation. (**c**) Proposed mechanism for the photoinduced PMS activation on the surface of multiphase MoS_2 (copyright from ref. [86], 2019, American Chemical Society).

4. Conclusions and Perspectives

In summary, this review summarizes the advances in phase engineering for photocatalytic applications, aiming to outline the route for designing efficient photocatalysts based on phase engineering. Table 1 summarizes some representative nanomaterials with phase engineering and their applications in photocatalysis. Phase-engineered materials can exist in different forms, including the unconventional phase, amorphous phase, and heterophase. Phase engineering can also be performed on both light-harvesting semiconductor and surface active cocatalyst. With optimized electronic structures and physico-chemical properties, the light absorption, charge separation, or surface redox reaction behavior can be tuned, providing the possibility for improved photocatalysis. Phase engineering has gained huge success in the photocatalysis research community, owing to the positive role in various energy and environmental applications.

Table 1. Summary of some representative phase-engineered photocatalysts and their applications.

Material	Phase	Application	Performance	Ref.
TiO_2	Brookite	HER	Brookite > Anatase	[38]
CdS	Hexagonal	HER	Hexagonal > Cubic	[40]
Red P	Hittorf/fibrous	HER	Heterophase > Single phase	[41]
MoS_2	1T	HER	1T > 2H	[48]
$CoSe_2$	Orthorhombic	HER	Orthorhombic > Cubic	[52]
$NiSe_2$	Marcasite	HER	Marcasite > Pyrite (stability)	[54]
$BiVO_4$	Monoclinic	OER	Monoclinic > Tetragonal	[60]
$CoSe_2$	Orthorhombic	OER	Orthorhombic > Cubic	[62]
TiO_2	Brookite	CO_2 RR	Brookite > Anatase > Rutile	[70]
$In_2O_{3-x}(OH)_y$	Rhombohedra/cubic	CO_2 RR	Heterophase > Single phase	[71]
Ru	hcp	CO_2 RR	hcp > fcc	[74]
Pd	Amorphous	CO_2 RR	Amorphous > Crystalline	[75]
TiO_2	Amorphous/Crystalline	MB Degradation	Heterophase > Single phase	[81]
MoS_2	1T/2H	PMS actication	Heterophase > Single phase	[86]

Although considerable success has been witnessed, there are still some challenges in the research field. Firstly, phase engineering has made surprising progress in enhancing catalyst efficiency, including activity, selectivity, and stability, but record-breaking high-performance photocatalytic efficiency is rarely reported. For example, 1T-MoS_2 has shown great potential in hydrogen evolution, but the performance is still far from noble metal catalysts such as Pt. Secondly, the phase purity is a general concern in this research field. The incomplete phase transition or unsatisfied synthetic methods are still limiting the phase purity of phase-engineered nanostructures. For instance, 1T-MoS_2 with pure 1T phase is hard to obtain. Most of the reported studies are 1T/2H mixed phase. Additionally, the phase stability should also be focused. Thirdly, the delicate control over other structural parameters in addition to phase is still challenging. Different morphology, size, or other structural parameters of phase-engineered materials make it difficult to reveal the mechanism. For example, TiO_2 with different phase structures but also different sizes would make it difficult to determine whether the increase in catalytic activity is due to the crystal phase structure or size.

Despite the presented challenges, there are plenty of opportunities in phase engineering for photocatalytic applications. Firstly, based on the fast development of material science, synthesizing novel nanomaterials such as MOFs and COFs with different phase structures is promising. Secondly, given that phase-engineered photocatalysts are now mainly used in limited applications, exploring new energy and environmental applications such as organic synthesis, H_2O_2 production, and N_2/NO_3 reduction may broaden the functions. Thirdly, the integration of phase engineering to other structural engineering strategies may make it possible to construct record-breaking high performance photocatalytic systems, and fourth, the development of facile and scale-up synthetic methods is of great significance in future research, with aims to realizing large scale synthesis of phase-engineered materials with high pure purity and stability. Last but not least, from the mechanism point of view, using advanced characterization technologies such as in situ XRD, in-situ Raman spectroscopy, and in situ TEM to uncover the phase evolution mechanism is still under development.

Author Contributions: J.Y.: investigation, data curation, writing—original draft preparation, funding acquisition; G.Z.: writing—review and editing; W.Q.: writing—review and editing; Y.W.: writing—review and editing; X.W.: project administration, supervision. All authors have read and agreed to the published version of the manuscript.

Funding: The authors appreciate for the financial support by Natural Science Foundation of Jiangsu Province (BK20210827) and China Postdoctoral Science Foundation (2021M700117).

Institutional Review Board Statement: Not applicable.

Informed Consent Statement: Not applicable.

Data Availability Statement: Raw data are available upon request.

Conflicts of Interest: The authors declare no conflict of interest.

References

1. Lu, S.; Liang, J.; Long, H.; Li, H.; Zhou, X.; He, Z.; Chen, Y.; Sun, H.; Fan, Z.; Zhang, H. Crystal Phase Control of Gold Nanomaterials by Wet-Chemical Synthesis. *Acc. Chem. Res.* **2020**, *53*, 2106–2118. [CrossRef] [PubMed]
2. Ge, Y.; Shi, Z.; Tan, C.; Chen, Y.; Cheng, H.; He, Q.; Zhang, H. Two-Dimensional Nanomaterials with Unconventional Phases. *Chem* **2020**, *6*, 1237–1253. [CrossRef]
3. Chen, Y.; Lai, Z.; Zhang, X.; Fan, Z.; He, Q.; Tan, C.; Zhang, H. Phase engineering of nanomaterials. *Nat. Rev. Chem.* **2020**, *4*, 243–256. [CrossRef]
4. Sokolikova, M.S.; Mattevi, C. Direct synthesis of metastable phases of 2D transition metal dichalcogenides. *Chem. Soc. Rev.* **2020**, *49*, 3952–3980. [CrossRef] [PubMed]
5. Li, H.; Zhou, X.; Zhai, W.; Lu, S.; Liang, J.; He, Z.; Long, H.; Xiong, T.; Sun, H.; He, Q.; et al. Phase Engineering of Nanomaterials for Clean Energy and Catalytic Applications. *Adv. Energy Mater.* **2020**, *10*, 2002019. [CrossRef]
6. Wang, J.; Wei, Y.; Li, H.; Huang, X.; Zhang, H. Crystal phase control in two-dimensional materials. *Sci. China Chem.* **2018**, *61*, 1227–1242. [CrossRef]
7. Lai, Z.; Yao, Y.; Li, S.; Ma, L.; Zhang, Q.; Ge, Y.; Zhai, W.; Chi, B.; Chen, B.; Li, L.; et al. Salt-Assisted 2H-to-1T' Phase Transformation of Transition Metal Dichalcogenides. *Adv. Mater.* **2022**, *34*, e2201194. [CrossRef]
8. Han, X.; Wu, G.; Ge, Y.; Yang, S.; Rao, D.; Guo, Z.; Zhang, Y.; Yan, M.; Zhang, H.; Gu, L.; et al. In situ Observation of Structural Evolution and Phase Engineering of Amorphous Materials during Crystal Nucleation. *Adv. Mater.* **2022**, *34*, e2206994. [CrossRef]
9. Zhou, J.; Zhu, C.; Zhou, Y.; Dong, J.; Li, P.; Zhang, Z.; Wang, Z.; Lin, Y.C.; Shi, J.; Zhang, R.; et al. Composition and phase engineering of metal chalcogenides and phosphorous chalcogenides. *Nat. Mater.* **2023**, *22*, 450–458. [CrossRef]
10. Kim, J.; Kim, H.J.; Ruqia, B.; Kim, M.J.; Jang, Y.J.; Jo, T.H.; Baik, H.; Oh, H.S.; Chung, H.S.; Baek, K.; et al. Crystal Phase Transition Creates a Highly Active and Stable RuC$_X$ Nanosurface for Hydrogen Evolution Reaction in Alkaline Media. *Adv. Mater.* **2021**, *33*, e2105248. [CrossRef]
11. Shi, Y.; Li, J.; Mao, C.; Liu, S.; Wang, X.; Liu, X.; Zhao, S.; Liu, X.; Huang, Y.; Zhang, L. Van Der Waals gap-rich BiOCl atomic layers realizing efficient, pure-water CO_2-to-CO photocatalysis. *Nat. Commun.* **2021**, *12*, 5923. [CrossRef]
12. Liu, D.; Xue, C. Plasmonic Coupling Architectures for Enhanced Photocatalysis. *Adv. Mater.* **2021**, *33*, e2005738. [CrossRef]
13. Wang, Z.; Li, C.; Domen, K. Recent developments in heterogeneous photocatalysts for solar-driven overall water splitting. *Chem. Soc. Rev.* **2019**, *48*, 2109–2125. [CrossRef] [PubMed]
14. Wang, Q.; Domen, K. Particulate Photocatalysts for Light-Driven Water Splitting: Mechanisms, Challenges, and Design Strategies. *Chem. Rev.* **2019**, *120*, 919–985. [CrossRef] [PubMed]
15. Zhou, P.; Navid, I.A.; Ma, Y.; Xiao, Y.; Wang, P.; Ye, Z.; Zhou, B.; Sun, K.; Mi, Z. Solar-to-hydrogen efficiency of more than 9% in photocatalytic water splitting. *Nature* **2023**, *613*, 66–70. [CrossRef]
16. Wang, B.; Zhang, W.; Liu, G.; Chen, H.; Weng, Y.X.; Li, H.; Chu, P.K.; Xia, J. Excited Electron-Rich Bi$^{(3-x)+}$ Sites: A Quantum Well-Like Structure for Highly Promoted Selective Photocatalytic CO_2. *Adv. Funct. Mater.* **2022**, *32*, 2202885. [CrossRef]
17. Kumar, A.; Khan, M.; He, J.; Lo, I.M.C. Recent developments and challenges in practical application of visible-light-driven TiO_2-based heterojunctions for PPCP degradation: A critical review. *Water Res.* **2020**, *170*, 115356. [CrossRef]
18. Xiong, J.; Li, H.M.; Zhou, J.D.; Di, J. Recent progress of indium-based photocatalysts: Classification, regulation and diversified applications. *Coord. Chem. Rev.* **2022**, *473*, 214819. [CrossRef]
19. Bai, S.; Gao, C.; Low, J.; Xiong, Y. Crystal phase engineering on photocatalytic materials for energy and environmental applications. *Nano Res.* **2018**, *12*, 2031–2054. [CrossRef]

20. Zhao, Y.; Shao, C.; Lin, Z.; Jiang, S.; Song, S. Low-Energy Facets on CdS Allomorph Junctions with Optimal Phase Ratio to Boost Charge Directional Transfer for Photocatalytic H_2 Fuel Evolution. *Small* **2020**, *16*, e2000944. [CrossRef]
21. Liu, F.; Shi, R.; Wang, Z.; Weng, Y.; Che, C.M.; Chen, Y. Direct Z-Scheme Hetero-phase Junction of Black/Red Phosphorus for Photocatalytic Water Splitting. *Angew. Chem. Int. Ed.* **2019**, *58*, 11791–11795. [CrossRef] [PubMed]
22. Lei, Z.; Zhan, J.; Tang, L.; Zhang, Y.; Wang, Y. Recent Development of Metallic (1T) Phase of Molybdenum Disulfide for Energy Conversion and Storage. *Adv. Energy Mater.* **2018**, *8*, 1703482. [CrossRef]
23. Gebruers, M.; Wang, C.; Saha, R.A.; Xie, Y.; Aslam, I.; Sun, L.; Liao, Y.; Yang, X.; Chen, T.; Yang, M.Q.; et al. Crystal phase engineering of Ru for simultaneous selective photocatalytic oxidations and H_2 production. *Nanoscale* **2023**, *15*, 2417–2424. [CrossRef] [PubMed]
24. Hai, X.; Chang, K.; Pang, H.; Li, M.; Li, P.; Liu, H.; Shi, L.; Ye, J. Engineering the Edges of MoS_2 (WS_2) Crystals for Direct Exfoliation into Monolayers in Polar Micromolecular Solvents. *J. Am. Chem. Soc.* **2016**, *138*, 14962–14969. [CrossRef] [PubMed]
25. Lin, Z.; Du, C.; Yan, B.; Wang, C.; Yang, G. Two-dimensional amorphous NiO as a plasmonic photocatalyst for solar H2 evolution. *Nat. Commun.* **2018**, *9*, 4036. [CrossRef] [PubMed]
26. Ge, J.; Yin, P.; Chen, Y.; Cheng, H.; Liu, J.; Chen, B.; Tan, C.; Yin, P.F.; Zheng, H.X.; Li, Q.Q.; et al. Ultrathin Amorphous/Crystalline Heterophase Rh and Rh Alloy Nanosheets as Tandem Catalysts for Direct Indole Synthesis. *Adv. Mater.* **2021**, *33*, e2006711. [CrossRef]
27. Ge, Y.; Huang, Z.; Ling, C.; Chen, B.; Liu, G.; Zhou, M.; Liu, J.; Zhang, X.; Cheng, H.; Liu, G.; et al. Phase-Selective Epitaxial Growth of Heterophase Nanostructures on Unconventional 2H-Pd Nanoparticles. *J. Am. Chem. Soc.* **2020**, *142*, 18971–18980. [CrossRef]
28. Fan, Z.; Luo, Z.; Huang, X.; Li, B.; Chen, Y.; Wang, J.; Hu, Y.; Zhang, H. Synthesis of 4H/fcc Noble Multimetallic Nanoribbons for Electrocatalytic Hydrogen Evolution Reaction. *J. Am. Chem. Soc.* **2016**, *138*, 1414–1419. [CrossRef]
29. Fan, Z.; Bosman, M.; Huang, Z.; Chen, Y.; Ling, C.; Wu, L.; Akimov, Y.A.; Laskowski, R.; Chen, B.; Ercius, P.; et al. Heterophase fcc-2H-fcc gold nanorods. *Nat. Commun.* **2020**, *11*, 3293. [CrossRef]
30. Li, Z.; Li, B.; Yu, M.; Yu, C.; Shen, P. Amorphous metallic ultrathin nanostructures: A latent ultra-high-density atomic-level catalyst for electrochemical energy conversion. *Int. J. Hydrogen Energy* **2022**, *47*, 26956–26977. [CrossRef]
31. He, Y.; Liu, L.; Zhu, C.; Guo, S.; Golani, P.; Koo, B.; Tang, P.; Zhao, Z.; Xu, M.; Zhu, C.; et al. Amorphizing noble metal chalcogenide catalysts at the single-layer limit towards hydrogen production. *Nat. Catal.* **2022**, *5*, 212–221. [CrossRef]
32. Guo, C.; Shi, Y.; Lu, S.; Yu, Y.; Zhang, B. Amorphous nanomaterials in electrocatalytic water splitting. *Chin. J. Catal.* **2021**, *42*, 1287–1296. [CrossRef]
33. Zhao, Z.; Tian, J.; Sang, Y.; Cabot, A.; Liu, H. Structure, synthesis, and applications of TiO_2 nanobelts. *Adv. Mater.* **2015**, *27*, 2557–2582. [CrossRef] [PubMed]
34. Zhang, K.; Wang, L.; Kim, J.K.; Ma, M.; Veerappan, G.; Lee, C.-L.; Kong, K.-j.; Lee, H.; Park, J.H. An order/disorder/water junction system for highly efficient co-catalyst-free photocatalytic hydrogen generation. *Energy Environ. Sci.* **2016**, *9*, 499–503. [CrossRef]
35. Guo, Z.Y.; Sun, R.; Huang, Z.; Han, X.; Wang, H.; Chen, C.; Liu, Y.Q.; Zheng, X.; Zhang, W.; Hong, X.; et al. Crystallinity engineering for overcoming the activity-stability tradeoff of spinel oxide in Fenton-like catalysis. *Proc. Natl. Acad. Sci. USA* **2023**, *120*, e2220608120. [CrossRef]
36. Takata, T.; Jiang, J.; Sakata, Y.; Nakabayashi, M.; Shibata, N.; Nandal, V.; Seki, K.; Hisatomi, T.; Domen, K. Photocatalytic water splitting with a quantum efficiency of almost unity. *Nature* **2020**, *581*, 411–414. [CrossRef]
37. Yi, J.; El-Alami, W.; Song, Y.; Li, H.; Ajayan, P.M.; Xu, H. Emerging surface strategies on graphitic carbon nitride for solar driven water splitting. *Chem. Eng. J.* **2020**, *382*, 122812. [CrossRef]
38. Kandiel, T.A.; Feldhoff, A.; Robben, L.; Dillert, R.; Bahnemann, D.W. Tailored Titanium Dioxide Nanomaterials: Anatase Nanoparticles and Brookite Nanorods as Highly Active Photocatalysts. *Chem. Mater.* **2010**, *22*, 2050–2060. [CrossRef]
39. Hong, Y.; Zhang, J.; Wang, X.; Wang, Y.; Lin, Z.; Yu, J.; Huang, F. Influence of lattice integrity and phase composition on the photocatalytic hydrogen production efficiency of ZnS nanomaterials. *Nanoscale* **2012**, *4*, 2859–2862. [CrossRef]
40. Ai, Z.; Zhao, G.; Zhong, Y.; Shao, Y.; Huang, B.; Wu, L.; Hao, X. Phase junction CdS: High efficient and stable photocatalyst for hydrogen generation. *Appl. Catal. B Environ.* **2018**, *221*, 179–186. [CrossRef]
41. Wu, C.; Zhu, R.; Teoh, W.Y.; Liu, Y.; Deng, J.; Dai, H.; Jing, L.; Ng, Y.H.; Yu, J.C. Hetero-phase dendritic elemental phosphorus for visible light photocatalytic hydrogen generation. *Appl. Catal. B Environ.* **2022**, *312*, 121428. [CrossRef]
42. Wang, X.; Ma, M.; Zhao, X.; Jiang, P.; Wang, Y.; Wang, J.; Zhang, J.; Zhang, F. Phase Engineering of 2D Violet/Black Phosphorus Heterostructure for Enhanced Photocatalytic Hydrogen Evolution. *Small Struct.* **2023**, 2300123. [CrossRef]
43. Lv, G.; Long, L.; Wu, X.; Qian, Y.; Zhou, G.; Pan, F.; Li, Z.; Wang, D. Realizing highly efficient photoelectrochemical performance for vertically aligned 2D $ZnIn_2S_4$ array photoanode via controlled facet and phase modulation. *Appl. Surf. Sci.* **2023**, *609*, 155335. [CrossRef]
44. Han, L.; Jing, F.; Zhang, J.; Luo, X.-Z.; Zhong, Y.-L.; Wang, K.; Zang, S.-H.; Teng, D.-H.; Liu, Y.; Chen, J.; et al. Environment friendly and remarkably efficient photocatalytic hydrogen evolution based on metal organic framework derived hexagonal/cubic In_2O_3 phase-junction. *Appl. Catal. B Environ.* **2021**, *282*, 119602. [CrossRef]
45. Wang, J.; Chen, Y.; Zhou, W.; Tian, G.; Xiao, Y.; Fu, H.; Fu, H. Cubic quantum dot/hexagonal microsphere $ZnIn_2S_4$ heterophase junctions for exceptional visible-light-driven photocatalytic H_2 evolution. *J. Mater. Chem. A* **2017**, *5*, 8451–8460. [CrossRef]

46. Li, Z.; Meng, X.; Zhang, Z. Recent development on MoS$_2$-based photocatalysis: A review. *J. Photochem. Photobiol. C* **2018**, *35*, 39–55. [CrossRef]
47. Huang, H.H.; Fan, X.; Singh, D.J.; Zheng, W.T. Recent progress of TMD nanomaterials: Phase transitions and applications. *Nanoscale* **2020**, *12*, 1247–1268. [CrossRef]
48. Xu, H.; Yi, J.; She, X.; Liu, Q.; Song, L.; Chen, S.; Yang, Y.; Song, Y.; Vajtai, R.; Lou, J.; et al. 2D heterostructure comprised of metallic 1T-MoS$_2$/Monolayer O-g-C$_3$N$_4$ towards efficient photocatalytic hydrogen evolution. *Appl. Catal. B Environ.* **2018**, *220*, 379–385. [CrossRef]
49. Yi, J.; Li, H.; Gong, Y.; She, X.; Song, Y.; Xu, Y.; Deng, J.; Yuan, S.; Xu, H.; Li, H. Phase and interlayer effect of transition metal dichalcogenide cocatalyst toward photocatalytic hydrogen evolution: The case of MoSe$_2$. *Appl. Catal. B Environ.* **2019**, *243*, 330–336. [CrossRef]
50. Yi, J.; She, X.; Song, Y.; Mao, M.; Xia, K.; Xu, Y.; Mo, Z.; Wu, J.; Xu, H.; Li, H. Solvothermal synthesis of metallic 1T-WS$_2$: A supporting co-catalyst on carbon nitride nanosheets toward photocatalytic hydrogen evolution. *Chem. Eng. J.* **2018**, *335*, 282–289. [CrossRef]
51. Chang, K.; Hai, X.; Pang, H.; Zhang, H.; Shi, L.; Liu, G.; Liu, H.; Zhao, G.; Li, M.; Ye, J. Targeted Synthesis of 2H- and 1T-Phase MoS$_2$ Monolayers for Catalytic Hydrogen Evolution. *Adv. Mater.* **2016**, *28*, 10033–10041. [CrossRef] [PubMed]
52. Yi, J.; Zhu, X.; Zhou, M.; Zhang, S.; Li, L.; Song, Y.; Chen, H.; Chen, Z.; Li, H.; Xu, H. Crystal phase dependent solar driven hydrogen evolution catalysis over cobalt diselenide. *Chem. Eng. J.* **2020**, *396*, 125244. [CrossRef]
53. Zhou, Z.; Yang, Y.; Hu, L.; Zhou, G.; Xia, Y.; Hu, Q.; Yin, W.; Zhu, X.; Yi, J.; Wang, X. Phase Control of Cobalt Selenide: Unraveling the Relationship Between Phase Property and Hydrogen Evolution Catalysis. *Adv. Mater. Interfaces* **2022**, *9*, 2201473. [CrossRef]
54. Yi, J.; Zhou, Z.; Xia, Y.; Zhou, G.; Zhang, G.; Li, L.; Wang, X.; Zhu, X.; Wang, X.; Pang, H. Unraveling the role of phase engineering in tuning photocatalytic hydrogen evolution activity and stability. *Chin. Chem. Lett.* **2023**, 108328. [CrossRef]
55. Dou, Y.; Yuan, D.; Yu, L.; Zhang, W.; Zhang, L.; Fan, K.; Al-Mamun, M.; Liu, P.; He, C.T.; Zhao, H. Interpolation between W Dopant and Co Vacancy in CoOOH for Enhanced Oxygen Evolution Catalysis. *Adv. Mater.* **2022**, *34*, e2104667. [CrossRef]
56. Liu, J.; Liu, Y.; Liu, N.; Han, Y.; Zhang, X.; Huang, H.; Lifshitz, Y.; Lee, S.-T.; Zhong, J.; Kang, Z. Metal-free efficient photocatalyst for stable visible water splitting via a two-electron pathway. *Science* **2015**, *347*, 970–974. [CrossRef]
57. Li, Y.-F.; Selloni, A. Pathway of Photocatalytic Oxygen Evolution on Aqueous TiO$_2$ Anatase and Insights into the Different Activities of Anatase and Rutile. *ACS Catal.* **2016**, *6*, 4769–4774. [CrossRef]
58. Zhao, C.X.; Liu, J.N.; Wang, C.D.; Wang, J.; Song, L.; Li, B.Q.; Zhang, Q. An anionic regulation mechanism for the structural reconstruction of sulfide electrocatalysts under oxygen evolution conditions. *Energy Environ. Sci.* **2022**, *15*, 3257–3264. [CrossRef]
59. Xing, W.; Yin, S.; Tu, W.; Liu, G.; Wu, S.; Wang, H.; Kraft, M.; Wu, G.; Xu, R. Rational Synthesis of Amorphous Iron-Nickel Phosphonates for Highly Efficient Photocatalytic Water Oxidation with Almost 100 % Yield. *Angew. Chem. Int. Ed.* **2020**, *59*, 1171–1175. [CrossRef]
60. Tokunaga, S.; Kato, H.; Kudo, A. Selective Preparation of Monoclinic and Tetragonal BiVO$_4$ with Scheelite Structure and Their Photocatalytic Properties. *Chem. Mater.* **2001**, *13*, 4624–4628. [CrossRef]
61. Kho, Y.K.; Teoh, W.Y.; Iwase, A.; Madler, L.; Kudo, A.; Amal, R. Flame preparation of visible-light-responsive BiVO$_4$ oxygen evolution photocatalysts with subsequent activation via aqueous route. *ACS Appl. Mater. Interfaces* **2011**, *3*, 1997–2004. [CrossRef] [PubMed]
62. Li, L.; Zhu, X.; Zhou, Z.; Wang, Z.; Song, Y.; Mo, Z.; Yuan, J.; Yang, J.; Yi, J.; Xu, H. Crystal phase engineering boosted photo-electrochemical kinetics of CoSe$_2$ for oxygen evolution catalysis. *J. Colloid. Interface Sci.* **2022**, *611*, 22–28. [CrossRef] [PubMed]
63. Di, J.; Lin, B.; Tang, B.J.; Guo, S.S.; Zhou, J.D.; Liu, Z. Engineering Cocatalysts onto Low-Dimensional Photocatalysts for CO$_2$ Reduction. *Small Struct.* **2021**, *2*, 2100046. [CrossRef]
64. Deng, B.; Huang, M.; Li, K.; Zhao, X.; Geng, Q.; Chen, S.; Xie, H.; Dong, X.; Wang, H.; Dong, F. The Crystal Plane is not the Key Factor for CO$_2$-to-Methane Electrosynthesis on Reconstructed Cu$_2$O Microparticles. *Angew. Chem. Int. Ed.* **2021**, *61*, e202114080. [CrossRef]
65. Zhang, H.X.; Hong, Q.L.; Li, J.; Wang, F.; Huang, X.; Chen, S.; Tu, W.; Yu, D.; Xu, R.; Zhou, T.; et al. Isolated Square-Planar Copper Center in Boron Imidazolate Nanocages for Photocatalytic Reduction of CO$_2$ to CO. *Angew. Chem. Int. Ed.* **2019**, *58*, 11752–11756. [CrossRef]
66. Zhu, X.; Zhou, G.; Yi, J.; Ding, P.; Yang, J.; Zhong, K.; Song, Y.; Hua, Y.; Zhu, X.; Yuan, J.; et al. Accelerated Photoreduction of CO$_2$ to CO over a Stable Heterostructure with a Seamless Interface. *ACS Appl. Mater. Interfaces* **2021**, *13*, 39523–39532. [CrossRef]
67. Zhu, X.; Cao, Y.; Song, Y.; Yang, Y.; She, X.; Mo, Z.; She, Y.; Yu, Q.; Zhu, X.; Yuan, J.; et al. Unique Dual-Sites Boosting Overall CO$_2$ Photoconversion by Hierarchical Electron Harvesters. *Small* **2021**, *17*, e2103796. [CrossRef]
68. Kong, T.; Low, J.; Xiong, Y. Catalyst: How Material Chemistry Enables Solar-Driven CO$_2$ Conversion. *Chem* **2020**, *6*, 1035–1038. [CrossRef]
69. Chen, Y.; Fan, Z.; Wang, L.; Ling, C.; Niu, W.; Huang, Z.; Liu, G.; Chen, B.; Lai, Z.; Liu, X.; et al. Ethylene Selectivity in Electrocatalytic CO$_2$ Reduction on Cu Nanomaterials: A Crystal Phase-Dependent Study. *J. Am. Chem. Soc.* **2020**, *142*, 12760–12766. [CrossRef]
70. Liu, L.; Zhao, H.; Andino, J.M.; Li, Y. Photocatalytic CO$_2$ Reduction with H$_2$O on TiO$_2$ Nanocrystals: Comparison of Anatase, Rutile, and Brookite Polymorphs and Exploration of Surface Chemistry. *ACS Catal.* **2012**, *2*, 1817–1828. [CrossRef]
71. Yan, T.; Li, N.; Wang, L.; Liu, Q.; Jelle, A.; Wang, L.; Xu, Y.; Liang, Y.; Dai, Y.; Huang, B.; et al. How to make an efficient gas-phase heterogeneous CO$_2$ hydrogenation photocatalyst. *Energy Environ. Sci.* **2020**, *13*, 3054–3063. [CrossRef]

72. Li, H.; Li, W.; Li, W.; Chen, M.; Snyders, R.; Bittencourt, C.; Yuan, Z. Engineering crystal phase of polytypic $CuInS_2$ nanosheets for enhanced photocatalytic and photoelectrochemical performance. *Nano Res.* **2020**, *13*, 583–590. [CrossRef]
73. Chai, Y.; Lu, J.; Li, L.; Li, D.; Li, M.; Liang, J. TEOA-induced in situ formation of wurtzite and zinc-blende CdS heterostructures as a highly active and long-lasting photocatalyst for converting CO_2 into solar fuel. *Catal. Sci. Technol.* **2018**, *8*, 2697–2706. [CrossRef]
74. Ye, F.; Wang, F.; Meng, C.; Bai, L.; Li, J.; Xing, P.; Teng, B.; Zhao, L.; Bai, S. Crystalline phase engineering on cocatalysts: A promising approach to enhancement on photocatalytic conversion of carbon dioxide to fuels. *Appl. Catal. B Environ.* **2018**, *230*, 145–153. [CrossRef]
75. Chen, Q.; Wu, S.; Zhong, S.; Gao, B.; Wang, W.; Mo, W.; Lin, H.; Wei, X.; Bai, S.; Chen, J. What is the better choice for Pd cocatalysts for photocatalytic reduction of CO_2 to renewable fuels: High-crystallinity or amorphous? *J. Mater. Chem. A* **2020**, *8*, 21208–21218. [CrossRef]
76. Wang, L.; Bahnemann, D.W.; Bian, L.; Dong, G.; Zhao, J.; Wang, C. Two-Dimensional Layered Zinc Silicate Nanosheets with Excellent Photocatalytic Performance for Organic Pollutant Degradation and CO_2 Conversion. *Angew. Chem. Int. Ed.* **2019**, *58*, 8103–8108. [CrossRef]
77. Han, Y.; Wang, S.; Li, M.; Gao, H.; Han, M.; Yang, H.; Fang, L.; Angadi, V.J.; El-Rehim, A.F.A.; Ali, A.M.; et al. Strontium-induced phase, energy band and microstructure regulation in $Ba_{1-x}Sr_xTiO_3$ photocatalysts for boosting visible-light photocatalytic activity. *Catal. Sci. Technol.* **2023**, *13*, 2841–2854. [CrossRef]
78. Yang, Z.; Wang, B.; Cui, H.; An, H.; Pan, Y.; Zhai, J. Synthesis of Crystal-Controlled TiO_2 Nanorods by a Hydrothermal Method: Rutile and Brookite as Highly Active Photocatalysts. *J. Phys. Chem. C* **2015**, *119*, 16905–16912. [CrossRef]
79. Liu, W.; Liu, H.; Dang, L.; Zhang, H.; Wu, X.; Yang, B.; Li, Z.; Zhang, X.; Lei, L.; Jin, S. Amorphous Cobalt-Iron Hydroxide Nanosheet Electrocatalyst for Efficient Electrochemical and Photo-Electrochemical Oxygen Evolution. *Adv. Funct. Mater.* **2017**, *27*, 1603904. [CrossRef]
80. Kang, Y.; Yang, Y.; Yin, L.C.; Kang, X.; Liu, G.; Cheng, H.M. An Amorphous Carbon Nitride Photocatalyst with Greatly Extended Visible-Light-Responsive Range for Photocatalytic Hydrogen Generation. *Adv. Mater.* **2015**, *27*, 4572–4577. [CrossRef]
81. Chen, X.; Liu, L.; Yu, P.Y.; Mao, S.S. Increasing solar absorption for photocatalysis with black hydrogenated titanium dioxide nanocrystals. *Science* **2011**, *331*, 746–750. [CrossRef] [PubMed]
82. Li, F.; Zhang, L.; Hu, C.; Xing, X.; Yan, B.; Gao, Y.; Zhou, L. Enhanced azo dye decolorization through charge transmission by σ-Sb^{3+}-azo complexes on amorphous Sb_2S_3 under visible light irradiation. *Appl. Catal. B Environ.* **2019**, *240*, 132–140. [CrossRef]
83. Zuo, S.; Jin, X.; Wang, X.; Lu, Y.; Zhu, Q.; Wang, J.; Liu, W.; Du, Y.; Wang, J. Sandwich structure stabilized atomic Fe catalyst for highly efficient Fenton-like reaction at all pH values. *Appl. Catal. B Environ.* **2021**, *282*, 119551. [CrossRef]
84. Yang, J.; Zhu, M.; Dionysiou, D.D. What is the role of light in persulfate-based advanced oxidation for water treatment? *Water Res.* **2021**, *189*, 116627. [CrossRef]
85. Yi, X.H.; Ji, H.D.; Wang, C.C.; Li, Y.; Li, Y.H.; Zhao, C.; Wang, A.O.; Fu, H.F.; Wang, P.; Zhao, X.; et al. Photocatalysis-activated SR-AOP over PDINH/MIL-88A(Fe) composites for boosted chloroquine phosphate degradation: Performance, mechanism, pathway and DFT calculations. *Appl. Catal. B Environ.* **2021**, *293*, 120229. [CrossRef]
86. Chen, Y.; Zhang, G.; Ji, Q.; Liu, H.; Qu, J. Triggering of Low-Valence Molybdenum in Multiphasic MoS_2 for Effective Reactive Oxygen Species Output in Catalytic Fenton-like Reactions. *ACS Appl. Mater. Interfaces* **2019**, *11*, 26781–26788. [CrossRef]

Disclaimer/Publisher's Note: The statements, opinions and data contained in all publications are solely those of the individual author(s) and contributor(s) and not of MDPI and/or the editor(s). MDPI and/or the editor(s) disclaim responsibility for any injury to people or property resulting from any ideas, methods, instructions or products referred to in the content.

Review

An Overview of Recent Developments in Improving the Photocatalytic Activity of TiO$_2$-Based Materials for the Treatment of Indoor Air and Bacterial Inactivation

Achraf Amir Assadi [1,2], Oussama Baaloudj [3,4], Lotfi Khezami [5,*], Naoufel Ben Hamadi [5], Lotfi Mouni [6], Aymen Amine Assadi [7,*] and Achraf Ghorbal [2]

1. Center for Research on Microelectronics and Nanotechnology, CRMN Sousse Techno Park, Sahloul BP 334, Sousse 4054, Tunisia
2. Research Unit Advanced Materials, Applied Mechanics, Innovative Processes and Environment, Higher Institute of Applied Sciences and Technology of Gabes (ISSAT), University of Gabes, Gabes 6029, Tunisia
3. Laboratory of Reaction Engineering, Faculty of Mechanical Engineering and Process Engineering, Université des Sciences et de la Technologie Houari Boumediene, BP 32, Algiers 16111, Algeria
4. Laboratory of Advanced Materials for Energy and Environment, Université du Québec à Trois-Rivières (UQTR), 3351, Boul. des Forges, C.P. 500, Trois-Rivières, QC G9A 5H7, Canada
5. Chemistry Department, College of Science, Imam Mohammad Ibn Saud Islamic University (IMSIU), Riyadh 11432, Saudi Arabia
6. Laboratoire de Gestion et Valorisation des Ressources Naturelles et Assurance Qualité, Faculté SNVST, Université Bouira, Bouira 10000, Algeria
7. École Nationale Supérieure de Chimie de Rennes (ENSCR), Université de Rennes, UMR CNRS 6226, 11 Allée de Beaulieu, 35700 Rennes, France
* Correspondence: lhmkhezami@imamu.edu.sa (L.K.); aymen.assadi@ensc-rennes.fr (A.A.A.); Tel.: +966-11-2594-659 (L.K.); +33-(0)-223-238-152 (A.A.A.)

Abstract: Indoor air quality has become a significant public health concern. The low cost and high efficiency of photocatalytic technology make it a natural choice for achieving deep air purification. Photocatalysis procedures have been widely investigated for environmental remediation, particularly for air treatment. Several semiconductors, such as TiO$_2$, have been used for photocatalytic purposes as catalysts, and they have earned a lot of interest in the last few years owing to their outstanding features. In this context, this review has collected and discussed recent studies on advances in improving the photocatalytic activity of TiO$_2$-based materials for indoor air treatment and bacterial inactivation. In addition, it has elucidated the properties of some widely used TiO$_2$-based catalysts and their advantages in the photocatalytic process as well as improved photocatalytic activity using doping and heterojunction techniques. Current publications about various combined catalysts have been summarized and reviewed to emphasize the significance of combining catalysts to increase air treatment efficiency. Besides, this paper summarized works that used these catalysts to remove volatile organic compounds (VOCs) and microorganisms. Moreover, the reaction mechanism has been described and summarized based on literature to comprehend further pollutant elimination and microorganism inactivation using photocatalysis. This review concludes with a general opinion and an outlook on potential future research topics, including viral disinfection and other hazardous gases.

Keywords: semiconductor; photocatalysis; indoor air treatment; volatile organic compounds; microorganism

1. Introduction

Air pollution and the degradation of air quality are becoming severe issues to deal with, but these notions often remain abstract and complex to surround or identify [1]. The contamination of food and drink raises a lot of interest since they are linked to vital daily elements for everyone [2]. Understanding how airborne particles can affect food and beverage quality is the first step to understanding how air filtration systems can address this issue [3]. Therefore, it is natural that the agri-food sector has a significant challenge in protecting its employees and processes against harmful atmospheric pollutants [4].

The primary pollutants confronted in indoor air include carbon monoxide (CO), microorganisms (fungi, bacteria, and viruses), nitrogen oxides (NOx), and a multitude of varieties of volatile organic compounds (VOCs) [5]. Given that in France, agri-food companies represent 15.3% of manufacturing industries with more than 17,647 companies [6], it is therefore essential and urgent to employ the purification system technology more effectively [7,8].

Numerous developing and encouraging technologies currently supply a solution to this issue [9,10]. Among them, heterogeneous photocatalysis in visible light proves its interest in compounds' degradation and/or mineralization [11,12]. However, these technologies do not make it possible to effectively guarantee constant purification over time of the microorganisms without the need for frequent maintenance operations due to their excessive bulk [13,14]. Advanced Oxidation Processes (AOPs) are processes that produce highly oxidizing species such as hydroxyl radicals ($^\bullet OH$) and other reactive oxygen species (ROS), including the anion superoxide radical ($^\bullet O_2^-$) and hydrogen peroxide (H_2O_2) capable of degrading target pollutants present in effluents [15,16]. The semiconductor material TiO_2 is considered a reference photocatalyst and an antibacterial agent due to its physicochemical properties [11,17,18]. However, TiO_2 has a wide bandgap, which limits its practical application in environmental remediation under visible light irradiation, including a wide range of the solar spectrum [16,19]. Many strategies have been implemented to overcome this concern, such as doping TiO_2 with metallic or non-metallic elements [20] and coupling with other semiconductors [21–23] to increase their absorption in the visible and improve the lifetime of electron-hole pairs [24]. It is possible to improve the redox process of pollutant degradation by doping TiO_2 with a metal oxide, which produces photoexcited charge carriers [25].

Indoor air quality has emerged as a significant public health problem. Photocatalytic technology is a natural solution for deep air filtration due to its low cost and excellent efficiency. Photocatalysis methods have been extensively researched for environmental remediation, notably for air treatment. Several semiconductors, such as TiO_2, have been used as photocatalytic catalysts, and they have gained a lot of attention in recent years due to their remarkable properties. For indoor air purification and bacterial inactivation, this review has compiled and evaluated current findings on improvements in the photocatalytic activity of TiO_2-based photocatalytic materials. The characteristics of various popular TiO_2-based catalysts and their benefits in the photocatalytic process have also been clarified, as well as how doping and heterojunction approaches might increase photocatalytic activity. Recent articles regarding diverse combined catalysts have been summarized and examined to underline the relevance of combining catalysts to boost efficiency. The studies that employ these catalysts to remove microorganisms and volatile organic compounds (VOCs) were also covered in this publication. Based on the literature, the reaction mechanism has also been defined and summarized to understand better pollutant removal and microorganism inactivation utilizing photocatalysis. Finally, this review's conclusion includes a summary and prognosis on prospective future study areas, such as viral disinfection and other dangerous gases. To our knowledge, there are few studies on the catalytic activity of alternative materials for indoor air treatment by eliminating both pollutants types, microorganisms, and VOCs.

2. Photocatalysis and Mass Transfer

Heterogeneous photocatalytic oxidation (HPO) is one of the active investigations in environmental treatment and purification [26–28]. It is widely applied in air pollution treatment, especially volatile organic compounds [29,30]. The resourceful technology is reserved for decomposing gaseous contaminants by employing photocatalysts under UV or solar light free of additional energy expenses [13,31].

2.1. Principle of Photocatalysis

Photocatalysis is generally described as the process of employing light (UV or visible light) to activate a substrate (such as a semiconductor photocatalyst) so that photo-reaction can be accelerated or facilitated with the catalyst remaining unconsumed [5]. The process can be divided into five steps (Figure 1):

(1) Transfer the reactants to the air phase.
(2) Adsorption of the reactants on the surface of the catalyst.
(3) Reaction in the adsorbed phase.
(3.1) Absorption of a photon by the catalyst.
(3.2) Generation of the electron-hole pairs.
(3.3) Separation of the pair.
(4) The oxidation and reduction with the adsorbed substrate.
(5) Desorption of the intermediate product.

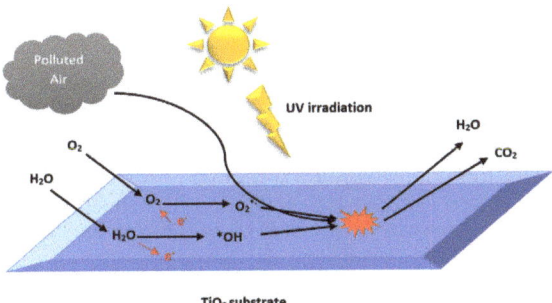

Figure 1. Polluted air photocatalysis treatment by TiO_2.

Among these five steps, the photocatalytic reaction is of crucial significance. It is initiated by the electron's excitation from the filled valence band (V_B) to the empty conduction band (C_B) of the photocatalyst when the energy carried by the absorbed photon equals or exceeds the band gap of the photocatalyst (Figure 2). In addition, the reaction results in the creation of a negative electron in the C_B and a positive hole in the V_B is called an electron-hole pair [32–34]. The positive hole oxidizes the hydroxide ion to yield hydroxyl radical ($^\bullet OH$), a potent oxidant of organic pollutants. The photo-excited electron is reduced to form the superoxide radical anion ($O_2^{\bullet -}$). These radicals are keys to the degradation of organic compounds [35].

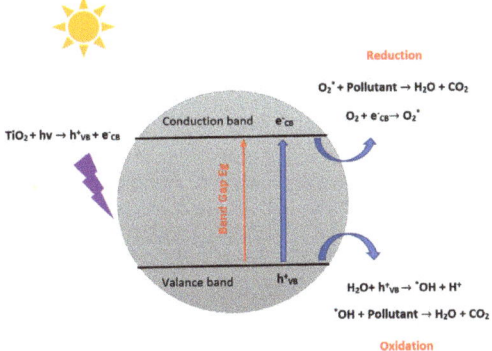

Figure 2. Schematic illustration of the photocatalytic reaction mechanism.

2.2. Development of Heterogeneous Photocatalytic Oxidation

Among all semiconductors, Titanium dioxide (TiO_2)-based materials have received particular attention in the photocatalysis field for their light absorption ability and high-efficiency treatment for both water and air; it was discovered by Fujishima and Honda in 1972 [36]. According to previous investigations, TiO_2-based photocatalysts also provide the advantages of high stability, availability, nontoxicity, excellent photoactivity, and low cost [35]. The photocatalysis of TiO_2 depends on variables such as specific surface area, crystallinity and surface hydroxyl groups of the TiO_2 [37]. This particular material has relatively polar surfaces that allow easy adsorption of hydrophilic pollutants. Nevertheless, titanium dioxide (TiO_2)-based materials have rapid recombination of electron-hole pairs, which, to some degree, suppresses the reaction efficiency [38,39].

Moreover, the band gap of materials is wide (3.0–3.2 eV), so the reaction is only activated with the irradiation of ultraviolet; hence the utilization of visible light irradiation is limited [14,40]. In order to improve the activity of photocatalysts under solar or artificial light at a lower energy cost and under more economic conditions, several strategies and investigations have been carried out to enhance the performance of TiO_2 [41]. These strategies include chemical modification, dye sensitization, and coupling with other semiconductor materials by introducing impurity atoms into pure TiO_2 to change electron-hole pairs concentrations in TiO_2 [35].

Metal doping is a method in which traces of foreign elements are introduced within the crystal lattice, and researchers widely use this strategy to reduce the band gap of titanium dioxide-based materials [42]. Noble metallic metals such as Ag, Au, Pt, and Pd have been researched extensively for years because of their properties and contribution to visible light absorption [5].

Ag is of particular interest as it acts as an electron trap and leads to retard the recombination of the electron-hole pair through the improvement of the transfer of interfacial charge [43]. Yi et al. studied a composite of Ag–AgI–TiO_2/CNFs; the Ag and I (Iodine) oxidation generated the reactive oxygen species (ROS) in the visible light range [44], doping TiO_2 with Ag and I increase its light range, increasing the photocatalytic activity. Yangfeng Chen et al. proposed a composite of heterostructured g-C_3N_4/Ag/TiO_2 microspheres by using the properties of Ag to delay the recombination of electron-hole pairs [45]. Apart from Ag, other metallic oxides can be composited with titanium dioxide to make the photo-reaction work under visible light. For example, halogens (X: Cl, Br, or I) bound to Bismuth oxide to form BiO_X, as a new class of promising catalyst has also drawn significant attention due to their attractive physicochemical characteristics, such as unique micro/nanostructures, bandgaps, optical and electrical properties and many other physicochemical characteristics [46–49]. Wendong Zhang et al. [49] found that the nanoplate BiOBr was highly efficient under visible light for NO photoreduction. Those promising catalysts (halogens) can be used as heterostructured photocatalysts with TiO_2 in order to enhance their photocatalytic activity. Actually, There is a lot of work done in heterogeneous photocatalysts with TiO_2, such as TiO_2/Ag [50], TiO_2/SiO_2 [51], TiO_2/Fe_2O_3 [52,53], TiO_2/Graphene [54] which have been shown to improve the photocatalytic performance of TiO_2, especially in the degradation of organic pollutants. These are only a few examples of TiO_2-based heterostructured photocatalysts. TiO_2 may be mixed with a variety of different substances to improve its photocatalytic activity. Aguilera-Ruiz also stated that cuprous oxide (Cu_2O), a visible-light-driven photocatalyst, has a band gap of about 2.07 eV [55]. Meanwhile, the conduction and valence band boundaries of $BiVO_4$ are located at 0.11 V and 2.65 V NHE. Thus, the composite $Cu_2O/BiVO_4$ has a promising photocatalytic performance under visible light [56]. Those two interesting materials, CuO and $BiVO_4$, can be used as the heterojunction or heterostructure to enhance the photocatalytic activity of the TiO_2-based catalysts. Moreover, it has been shown that Ag- V- and Fe-doped TiO2 achieved by various routes are very efficient in the oxidation of VOCs (butyl acetate, hexane or gaseous toluene) [57].

Non-metal doping is another strategy established to increase titanium dioxide's activity under solar or visible light. This technique takes advantage of the possible electronic transition from the induced new electronic states above TiO_2 V_B (2p or 3p orbitals of the dopant) to TiO_2 C_B (3d orbitals of Ti). Several researchers reported that the doped photocatalyst activity increases after non-metal doping, as the electronic structure has been modified to extend the absorption of the photocatalyst into the visible-light region [57]. Various studies have shown non-metal doping of TiO_2, such as Nitrogen-doped TiO_2 [58], carbon-doped TiO_2 [59–61], sulfur-doped TiO_2 [62,63], boron-doped TiO_2 [64] and phosphorus-doped TiO_2 [65]. Vaiano et al. studied recyclable visible-light active N-doped TiO_2 photocatalysts coated on glass spheres using a simple sol-gel method. They obtained excellent photocatalytic activity with visible light irradiation [66].

2.3. Reactors and Configurations

The configuration of reactors for air treatment is a critical element in the efficiency of the process. It should promote effective contact between the catalyst and the photons on the one hand and between the catalyst and the pollutants on the other. Care must also be taken to limit pressure drops. This part will present different continuous-flow photoreactors used in the laboratory or on an industrial scale.

Usually, this type of reactor is made up of a perforated plate placed at the inlet to ensure the homogeneity of the airflow. The central box contains two fixing devices for the photocatalytic support on one hand and a UV lamp on the other. In this configuration, the polluted air, driven by a fan, passes through the photocatalytic support. Another reactor configuration is based on using porous monolithic supports with varying thicknesses. The structure is based on the successive use of several UV lamps and monolithic "honeycomb" type photocatalytic media. The lamps irradiate the front and back sides of the monolithic supports.

The photocatalytic medium is placed against the reactor's internal wall and irradiated by a lamp set in a central tube. The particularity of this type of pilot is that the distance between the two plates or the diameters carrying the media is variable, which makes it possible to test the effect of the gap on the performance of the process.

Flat and cylindrical configurations:

A schematic representation of this rectangular configuration is used in the works of Assadi and his co-workers [3,5,17]. This reactor is formed by a chamber containing two glass plates at a variable distance. Each plate carries the photocatalytic media. Lamps are positioned along the length of the reactor at an equal distance in the inter-plate space [3,5,17].

The reactor is formed by two cylindrical tubes. The catalyst is installed on the inner wall of the outer cylinder. A UV lamp is installed in the inner tube in order to have uniform radiation from the catalytic surface. The gaseous effluent circulates between the outer tube's inner wall and the inner tube's outer wall. Note, To demonstrate the effect of material transfer, the diameter of the inner tube is studied to vary the thickness of the gas film [3,5,17].

Another planar configuration is based on surface-degraded fiber optic sheets on which titanium dioxide has been deposited. The latter replaced the bulky UV lamps. Optical fibers are used to activate the catalyst and further optimize the supply of UV radiation compared to lamps. This configuration will make it possible to inactivate the pollutants while offering compactness of the solution, i.e., lower pressure drops and easy handling in use. Figure 3 shows images of a fiber optic photocatalytic reactor and a fiber optic shee [3,5,17].

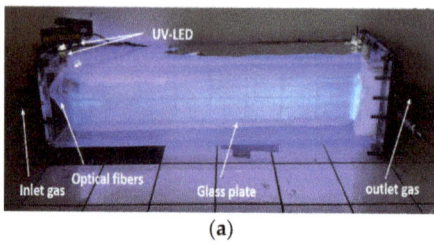 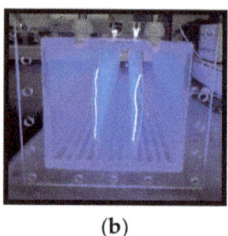

Figure 3. (a) Images of a photocatalytic reactor based on optical fibers, (b) of a side view (Range of processed flow: from 5 at 20 m^3/h with concentrations varying from 5 to 50 mg/m^3) [5].

3. Volatile Organic Compounds (VOCs)

Volatile organic compounds are substances containing organic carbon which vaporize at significant rates [67,68]. They are the second-most widespread and various emissions classes after particulates [7]. Besides, we can state that as an approximate rule, VOCs are the organic liquids or solids whose vapor pressures at room temperature exceed 0.01 psi (=0.0007 atm) with atmospheric boiling points equal to or less than 480 °F estimated at 101.3 kPa, i.e., standard atmospheric pressure. Among hundreds of VOCs that have been qualitatively identified in the indoor environment, the main compounds are alkanes, alkenes, carboxylic acids and alcohols, esters, and aromatics [69]. Organic compounds are primarily found in home items such as wax, varnishes, and paints. All these chemicals can emit organic byproducts when utilized and stored in a non-controlled method. Studies have revealed that levels of numerous organic chemicals indoors are 2 to 5 times greater than outside. Therefore, with a specific level of time exposure, these organic compounds may have short or long term adverse health effects such as headaches, eye and respiratory tract irritation and even cancers [70].

VOCs in the atmosphere or the environment are relatively at low concentrations; hence, they are detectable based on interactions between the sensor component and the organic compounds. In addition, ventilation is also a conventional dilution method. Still, it is not firmly recommended in current practice because of its limitation on outdoor air quality (OAQ) and energy consumption [11,17,18]. Accordingly, researchers are still developing technologies and efficient approaches to meet IAQ standards and reduce energy costs to avail a secure, healthful, livable environment. During their research, Abidi and his collaborators studied the elimination of chloroform $CHCl_3$ by using different catalysts and analyzing the removal efficiency under several initial concentrations of each catalyst type supported on polyester under certain conditions [5]. Many works have demonstrated the ability of some TiO_2-based photocatalysts to remove VOCs from the air due to their high photocatalytic activity and stability [71]. Tobaldi et al., 2021 have reported that TiO_2-graphene oxide composites exhibit enhanced photocatalytic activity for the removal of various VOCs, such as benzene, toluene, and formaldehyde [72]. Another work has shown that TiO_2-carbon nanotube composite photocatalysts have efficient and improved photocatalytic activity for the removal of various VOCs, such as xylene and toluene [73]. The efficiency of the TiO_2 can be enhanced in its photocatalytic activity for the elimination of VOCs in the air by doping it with metal oxides such as ZnO, Fe_2O_3, and WO_3 [74], as this doping can increase the surface area and prevent electron-hole recombination. Table 1 summarizes several recent studies on the removal of VOCs using TiO_2-metal.

Table 1. List of some studies on using TiO$_2$-metal for VOCs removal.

Target Pollutants	Reactors	Catalyst	Radical Species	Operating Conditions	Degradation Performance	Formed Products (Intermediate and Final)	Ref.
Propionic acid (PPA) and benzene (BENZ)	annular reactor + dielectric barrier discharge (DBD)	SiO$_2$-TiO$_2$ + UV	°OH, CH$_3$ CH$_2$°	SiO$_2$ = 6.5 g m^{-2} et TiO$_2$ = 6.5 g m^{-2} performance lamp UV-A (80 W /10) output intensity (25 W/m^2) Odor inlet concentrations 0.068 to 0.405 mmol m^{-3}, Q = 2 at 6 m^3 h^{-1}, relative Humidity: 5 to 90%, T = 20 °C	RE tested alone: 55% (APP) et 40% (BENZ) RE of mixture: 50% for APP and 30% for BENZ RE combined process: 60% for a voltage equal to 9 kV RE of mixture gaseous effluent (5% HR): 50% APP et 50% BENZ	BENZ: CO$_2$ dominating CO weak, O$_3$, CH$_3$CH$_2$OOH instable → Alcool + Aldéhyde → CO$_2$ PPA: CO$_2$, ethanoic acid (CH$_3$CH$_2$OOH), ethanol (CH$_3$CH$_2$OH), aldehyde (CH$_3$CHO), H$_2$O, O$_2$	[75]
Butane-2,3-dione and Heptane-2-one	Continuous Planar Reactor	TiO$_2$, TiO$_2$-Cu et TiO$_2$-Ag	•OH, O$_2$°$^-$	Q = 1–12 m^3 h^{-1} concentration of COV= 5–20 mg.m^{-3} Humidity level = 5–70%, under UV-A light oxidation.	RE of TiO$_2$ alone: 63% RE of TiO$_2$-Ag: 46% RE of TiO$_2$-Cu: 52%	acetone (C$_3$H$_5$O) propionic acid (C$_3$H$_6$O$_2$) butanoic acid (C$_4$H$_8$O$_2$) pentanoic acid (C$_5$H$_{10}$O$_2$) acetic acid (C$_2$H$_4$O$_2$) acetaldehyde (C$_2$H$_4$O) formic acid (HCOH) carbon dioxide (CO$_2$) and H$_2$O	[76]
Acetone and toluene	Surface DBD discharge	Pt/TiO$_2$ and MnO$_2$/CuO$_2$/Al$_2$O$_3$	NS	Concentration: 0.2 ppm flow rate: 38.42 m^3/h	100% toluene destruction of toluene at 0.2 ppm and 100% acetone destruction at 0.46 ppm	NS	[77]
Butane-2,3-dione (BUT) + E. coli	spherical batch reactor	Cu$_2$O/TiO$_2$ and TiO$_2$-Ag	•OH, HO$_2$° and O$_2$°$^-$	Concentration: 4.4 g/m$_3$ T = 50 at 100 °C λ = 380–420 nm, under UV-vis light irradiation.	99.7% E. coli inactivation and 100% VOC degradation within 60 min and 25 min with TiO$_2$-Ag for simultaneous treatment	CO$_2$, H$_2$O	[78]

Table 1. Cont.

Target Pollutants	Reactors	Catalyst	Radical Species	Operating Conditions	Degradation Performance	Formed Products (Intermediate and Final)	Ref.
methyl ethyl ketone (MEK) or 2-butanone	annular reactor	TiO_2 (fiberglass + Ahlström support)	•OH, $O_2^{-\circ}$, °H_2C-CH_3, °CH_3, H_3C-C°=O, °H_2C-CO-CH_2-CH_3	MEK concentration on glass fibers: 1.51 mg/L MEK concentration on Ahlström: 1.75 mg/L HR glass fibers: 0.11–3.94 mW/cm^2 HR Ahlström: 0.12–2.53 mW/cm^2 T = 30 °C and 20 vol.% O_2, under UV light source.	Deposition of TiO_2 on glass fibers leads to 10% degradation of MEK for 1.5 mg/L. TiO_2 Ahlström leads to the elimination of 40% of MEK for 1.5 mg/L.	acetaldehyde (C_2H_4O) ethane (C_2H_6) methane (CH_4) methanol (CH_3OH) acetone (C_3H_6O) methyl formate ($C_2H_4O_2$) carbon dioxide (CO_2) and H_2O	[79]
Acetone	annular reactor	TiO_2 (fiberglass + Ahlström support)	°CH_3, •OH, H_2C°-COOH, H_3C-°C=O	Concentration: 14.9 ng/L and 66.0 ng/L light power: 0.21 to 3.94 mW/cm^2 T = 30 °C, 20 vol.% O_2 Volume flow: 150 to 300 mL/min, under UV light.	90% of Acetone conversion has been obtained for low initial concentrations with TiO_2 photocatalyst deposited on fiberglass for simultaneous treatment	acetaldehyde (C_2H_4O) methyl alcohol (CH_3OH) isopropyl alcohol (C_3H_8O) methyl ethyl ketone (C_4H_8O) acetic acid (CH_3COOH) mesityl oxide ($C_6H_{10}O$) diacetone-alcohol ($C_6H_{12}O_2$)	[80]
Benzene	the outer surface of the rectangular SiC ceramic membrane	Pt/SiC@Al_2O_3	NS	0.176% by mass of Pt	90% reduction at 215 °C with a space velocity of 6000 mg^{-1} h^{-1}	CO_2, H_2O	[81]
n-butanol and acetic acid	fixed-bed tubular reactor	Pt/CeO_2-AlO_3	NS	1000 ppm of COV T = 50–350 °C 0, 7, 15, 23 et 51% by weight of CeO_2	100% reduction for n-butanol at T < 250 °C 50 or 90% reduction for a reduction of 80 or 20 °C.	Butanal (C_4H_8O) methanol (CH_4OH) propanol (C_3H_8O) isopropanol (C_3H_8O) formaldehyde (HCOH) propanal (C_3H_6O) carbon dioxide (CO_2)	[82]

Table 1. Cont.

Target Pollutants	Reactors	Catalyst	Radical Species	Operating Conditions	Degradation Performance	Formed Products (Intermediate and Final)	Ref.
Formaldehyde	organic glass reactor	Pt/AlOOH/, Pt/AlOOH-c, Pt/c-Al$_2$O$_3$ and Pt/TiO$_2$	NS	HCHO concentration: 127 ppm for adsorption and 139 ppm for catalytic oxidation, fan: 5 W T: 35 °C HR: 25% oxidation time: 51 min.	Pt/AlOOH > Pt/AlOOH-c > Pt/c-Al$_2$O$_3$ > Pt/TiO$_2$	surface formate carbon dioxide (CO$_2$) water (H$_2$O)	[83]
Formaldehyde	fixed-bed quartz flow reactor	Ag/TiO$_2$, Ag/Al$_2$O$_3$ et Ag/CeO$_2$	NS	Concentration: 110 ppm T = 35 to 125 °C Debit: 100 mL min^{-1}, under light containing ultraviolet.	Ag/TiO$_2$ > Ag/Al$_2$O$_3$ > Ag/CeO$_2$ 100% HCHO conversion with Ag/TiO$_2$ at T = 95 °C	carbon dioxide (CO$_2$) another carbon-containing compound	[84]
Formaldehyde	NS	Pt/TiO$_2$, Rh/TiO$_2$, Pd/TiO$_2$, Au/TiO$_2$ (noble metals/TiO$_2$)	NS	Concentration: 100 ppm 1% noble metals/TiO$_2$ O$_2$ 20 vol.% Debit: 50 cm^3 min^{-1} T: 20 °C GHSV: 5000 h^{-1}	Pt/TiO$_2$ ≫ Rh/TiO$_2$ > Pd/TiO$_2$ > Au/TiO$_2$	carbon dioxide (CO$_2$)carbon monoxyde (CO); water (H$_2$O)	[85]
Dimethyl disulfide (DMDS)	Continuous Flow Quartz Tubular Reactor	(Au + Pd)/TiO$_2$, Au/MCM-41, (AU + Rh)/MCM and Au/TiO$_2$, Pd/TiO$_2$	NS	3%Pd/TiO$_2$ and 1%Au/TiO$_2$ (1%Au + 3%Pd)/TiO$_2$ gas flow: 42,000 h^{-1} Temperature: 20–320 °C	Au/TiO$_2$ and Au-Pd/TiO$_2$ effectively remove DMDS for T < 155 °C Au/MCM-41 less effective in DMDS eliminating	methanol (CH$_3$OH) ethanol (C$_2$H$_6$O) methyl mercaptan (CH$_3$SH) ethyl mercaptan (CH$_3$SCH$_3$) hydrogen sulfur (H$_2$S) carbon dioxide (CO$_2$) carbon monoxide (CO) sulfur dioxide (SO$_2$) water (H$_2$O)	[86]

Table 1. *Cont.*

Target Pollutants	Reactors	Catalyst	Radical Species	Operating Conditions	Degradation Performance	Formed Products (Intermediate and Final)	Ref.
toluene + m-xylene + ethyl acetate or acetone	fixed-bed Quartz Continuous Flow Microreactor (ICP-AES)	0.91 wt.% $Au_{0.48}$ Pd/α-MnO_2 et α-MnO_2	α-, β- et γ-oxygène	1% (Au-Pd) Mixing flow: 17 mL/min concentration: 1000 ppm + O_2 + N_2 (solid) molar ratio COV/O_2 = 1/400 SV (space velocity) = 40,000 mL (g h) T = 320 °C	0.91 wt.% Au 0.48 Pd/α-MnO_2 > α-MnO_2	carbon dioxide (CO_2) water (H_2O)	[69]
Isovaleraldehyde	continuous annular plasma reactor DBD combined photocatalysis	TiO_2	•OH, $O_2^{\bullet-}$	concentration: 75 to 200 mg m^{-3} Debit: 2 m^3 h^{-1} HR: 5% T: 20 °C I: 20 W m^{-2} SE: 17 J L^{-1}, under UV light.	NS	propanoic acid (CH_3CH_2COOH) propanone (CH_3COCH_3) ethanoic acid (CH_3COOH) carbon dioxide (CO_2) carbon monoxide (CO) ozone (O_3)	[87]
Benzene	New UV-LED frontal flow photocatalytic reactor	TiO_2 deposited on luminous textiles	OH°, $O_2^{\circ-}$	concentration: 100 to 200 mg m^{-3} Debit: 1 m^3 h^{-1} HR: 5 to 80% T: 20 °C		CO_2 and H_2O	[72]

4. Microorganism Inactivation and Reactional Mechanisms

Understanding the mechanism of the bactericidal effect action of semiconductors is fundamental to improving its activity and, in particular, involves the analysis of the targets of TiO_2 at the bacterial level [88]. TiO_2 is a multifunctional photocatalyst that may be utilized to render microorganisms inactive [61]. The following steps are involved in the overall process for the inactivation of microorganisms using TiO_2. Step 1: TiO_2 is exposed to irradiation and undergoes a photocatalytic reaction that produces ROS like hydroxyl radicals (*OH) and superoxide radicals (O_2^{*-}). Step 2: ROS is formed in the photocatalytic reaction and interacts with bacterial cells and membranes and damaging DNA, proteins, and lipids. In addition, ROS can combine with water molecules to form more ROS, such as H_2O_2. Damages and harm caused by ROS interactions lead then to step 3 inactivation of the microorganism, in which the cell of the microorganism dies. The general microorganisms' photocatalytic inactivation mechanisms of TiO_2 can be summarized by the following equations and Figure 4:

$$TiO_2 + h\nu \rightarrow TiO_2\,(C_B\,e^-) + TiO_2\,(V_B\,h^+) \tag{1}$$

$$O_2 + e^- \rightarrow O_2^{-*} \tag{2}$$

$$H_2O + h^+ \rightarrow H^+ + {}^*OH \tag{3}$$

$$^*OH + O_2^{-*} + \text{Microorganism} \rightarrow \text{Inactivated Microorganism} \tag{4}$$

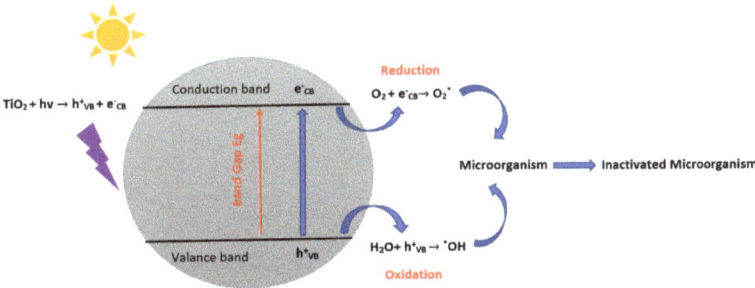

Figure 4. Schematic illustration of the antibacterial photocatalytic mechanisms of TiO_2 (inspired from ref. [89]).

Overall, because TiO_2 is ecologically neutral and doesn't produce toxic byproducts, using it as a photocatalyst to inactivate bacteria presents a viable substitute for conventional disinfection techniques that involve chemicals or heat [90]. The effectiveness of TiO_2-based photocatalysis, however, is dependent on several variables, including the characteristics of the TiO_2, the strength and wavelength of the light source, and the kind and quantity of bacteria present [91]. The inorganic semiconductors doping or adding a co-catalyst, such as TiO_2, with metals such as Cu, mainly accelerates bacterial inactivation kinetics [76,92]. Different reactions will likely be generated when the copper oxides are in contact with the catalyst's surface [76,92]. Indeed, CuO and Cu_2O are spawned when there is an interaction between copper and O_2 (air) under light irradiation. Cu_xO is found in two forms (CuO and Cu_2O) and exhibits the Cu(+I) and Cu(+II) oxidation states, of which the main form that interacts with bacteria and VOCs is Cu_2O, thus generating electrons at the level of the conduction band; $Cu_2O\,(C_B\,e^-)$ and holes in the valence band; $Cu_2O\,(V_B\,h^+)$ [78,92,93]. Under simulated sunlight, $Cu_2O\,(C_B\,e^-)$ enters a reduction reaction with TiO_2 to reduce Ti^{4+} to Ti^{3+} and yields Cu(+I) at the $V_B\,h^+$ level, which may lead to bacterial inactivation and/or VOCs to form CO_2, H_2O, N, S and inactivated bacteria [17,76,92]. The main antibacterial photocatalytic mechanisms of TiO_2 with Cu_2O suggested by previous research papers cited above can be summarized by the following equations and Figure 5 [78,92,93]:

$$Cu_2O + hv \rightarrow Cu_2O\ (C_B\ e^-) + Cu_2O\ (V_B\ h^+) \quad (5)$$

$$Cu_2O\ (C_B\ e^-) + TiO_2 \rightarrow TiO_2^- \text{ ou } (Ti^{3+}) + Cu_2O \quad (6)$$

$$TiO_2^- + O_2 \rightarrow TiO_2 + O_2^{-*} \quad (7)$$

$$°O_2^- + h^+ \rightarrow H_2O^* \quad (8)$$

$$H_2O^* + h^+ + e^-_{cb} \rightarrow H_2O_2 \quad (9)$$

$$H_2O_2 + e^-_{cb} \rightarrow OH + {}^*OH \quad (10)$$

$$Cu_2O\ (V_B\ h^+) + \text{bacteria} \rightarrow CO_2 \text{ and } H_2O \quad (11)$$

$$h^+ + \text{Bacteria} \rightarrow \text{Inactivated Bacteria} \quad (12)$$

$$H_2O_2 + \text{Bacteria} \rightarrow \text{Inactivated Bacteria} \quad (13)$$

$$^*OH + \text{Bacteria} \rightarrow \text{Inactivated Bacteria} \quad (14)$$

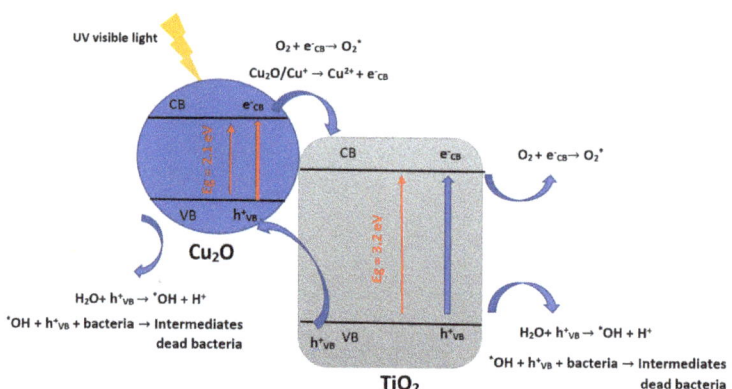

Figure 5. The main antibacterial photocatalytic mechanisms of TiO_2 with Cu_2O (inspired from refs. [78,92,93]).

The inactivation of bacteria cells can occur by many processes during the photocatalytic reactions, either the rupture of the cell membrane (Membrane disruption), the cell wall (Exposed cellular components), or the attack of the cells by ROS [19]. Where high levels of oxidative stress may be produced by ROS, which interacts with bacterial cells effectively and kills them by destroying the cell wall and a variety of bacterial cell components such as protein, lipids, carbohydrates, DNA, and amino acids [94]. Furthermore, when photocatalyst particles are deposited at the surface of bacterial cells, they can interact with them via diffusion and endocytosis mechanisms, which induce the destruction of membrane proteins or cell membranes owing to the phenomena of member permeability [19]. Moreover, both catalysts and generated ROS can interfere with the movement of electrons within the cell microorganisms, loss of protein motive force, depletion of intracellular ATP production with DNA replication disintegration, and intracellular outflow resulting in bacteria cell inactivation [95].

The hydroxyl radicals (*OH) produced on the surface of copper (Cu^+) in contact with H_2O with the holes generated at the level of $V_B\ h^+$ is the primary ROS involved in bacterial inactivation [29,96]. Cu_2O exhibits high bacterial inactivation capacity when light irradiation stimulates electron transfer between copper and bacterial cells and produces reactive oxygen species (ROS), resulting in bacterial cell inactivation [78,97,98]. Abidi et al.

investigated the effects of Cu_xO amounts at different sputtering times on the TiO_2-Polyester (PES) photocatalyst in the inactivation of microorganisms [17]. For sputtering intensities ranging from 20 to 80 A, it was regarded that the Cu_xO/TiO_2–PES catalyst sputtered at 80 A; the total inactivation of the bacteria was obtained after an hour of exposure to indoor light. Copper oxide showed high antibacterial activity, and the intrinsic activity of Cu(+I) can be enhanced by UV-vis illumination [17].

Additionally, Ag-NP is an excellent material used to improve the photocatalytic inactivation of microorganisms using TiO_2, which has recently been proven in previous works [76]. Ag particles could inactivate bacteria as Ag-NP is an essential factor that controls and regulates antimicrobial activity [92].On contact of Ag with TiO_2 under light irradiation, either Ag(0), Ag(+I), or Ag(+II) are yielded. The release of these different forms of Ag in contact with Escherichia coli induces bacterial inactivation [99]. The Ag used for TiO_2-NT decoration showed +1 and +2 oxidation states (Ag^+ and Ag^{2+}) [78].

In its metallic state, silver is oxidized in the air (O_2), breeding Ag_2O; this substance yields Ag^+ ions. This 4-electron process can be outlined by the following two equations [99]:

$$4\ Ag^0 + O_2 \rightarrow 2\ Ag_2O \tag{15}$$

$$2\ Ag_2O + 4H^+ \rightarrow 4\ Ag^+ + 2\ H_2O \tag{16}$$

Ag_2O is at the origin of the inactivation of bacteria when it generates the production of reactive oxygen species in contact with TiO_2-NTs. While Ag_2O is in contact with TiO_2 as a semiconductor, electrons (e^-) are photo-generated by the semiconductor under the action of bandgap radiation as indicated by the chemical reaction (Equation (20)) and photo-generated holes (h^+) react with H_2O (Equation (18)) to yield hydroxyl radicals ($OH°$) [76,92]:

$$Ag_2O + e^- \rightarrow 2\ Ag^+ + \frac{1}{2} O_2^- \tag{17}$$

$$h^+ + H_2O \rightarrow {}^*OH + H^+ \tag{18}$$

$$2H_2O + O_2 + 2e^- \rightarrow 2\ {}^*OH + 2\ OH^- \tag{19}$$

$$e^- + O_2 \rightarrow O_2^{*-} \tag{20}$$

The suggested bacterial inactivation mechanism with Ag/TiO_2 under light can be recapitulated in the following equations [76] and Figure 6:

$$Ag_2O + h\nu \rightarrow Ag_2O\ (C_B\ e^-) + Ag_2O\ (V_B\ h^+) \tag{21}$$

$$Ag_2O\ (e^- + h^+) + TiO_2 \rightarrow Ag_2O\ (V_B\ h^+) + TiO_2\ (C_B\ e^-) \tag{22}$$

$$TiO_2\ (C_B\ e^-) + O_2 \rightarrow TiO_2 + O_2 \tag{23}$$

$$2\ e^- + O_2 + 2H^+ \rightarrow H_2O_2 \tag{24}$$

$$H_2O_2 + O_2^- \rightarrow {}^*OH + OH^- + O_2 \tag{25}$$

$$Ag_2O\ (V_B\ h^+) + Bacteria \rightarrow Inactivated\ Bacteria \tag{26}$$

$$H_2O_2 + Bacteria \rightarrow Inactivated\ Bacteria \tag{27}$$

$$^*OH + Bacteria \rightarrow Inactivated\ Bacteria \tag{28}$$

$$^*OH + VOCs \rightarrow CO_2 + H_2O \tag{29}$$

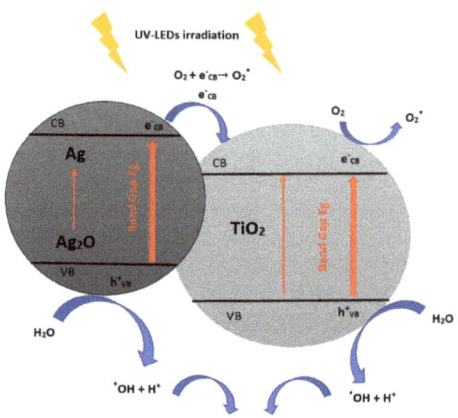

Figure 6. The main antibacterial photocatalytic mechanisms of TiO_2 with Ag_2O (inspired from ref [76]).

It is well known that the reaction of oxygen radicals in the cell causes its death [96]. Furthermore, different bacteria have different membrane structures [100]. For example, Gram − bacteria have peptidoglycan of the wall less thick than Gram + bacteria, which have an additional outer membrane composed of a double layer of lipids. This finding is in chains of different catalytic reactions and further disinfection efficiencies [101]. Accordingly, other bacteria's survival rates will differ under identical disinfection conditions [102].

The most critical mechanism in antibacterial activity is cell membrane damage. Oxidative stress generated by ROS is a second mechanism involved in antibacterial activity [103–105]. This stress inhibits DNA replication, protein synthesis, and cellular metabolism, causing cell death [106]. In order to demonstrate the effect of ROS on cell death, a study was conducted in the absence and presence of L-cysteine, a natural antioxidant, with *E. coli* bacteria. Indeed, growth was inhibited by Cu-TiO_2/GF with an efficiency of 79.4% in the absence of L-Cysteine compared to 65.1% in its presence. Similarly, Ag-TiO_2/GF, where the efficiency was 100% without the antioxidant and diminished to 84.7% in its presence [92]. The experiment is conducted on the following bacteria: *E. coli* and *Staphylococcus aureus* (*S. aureus*) on copper-doped TiO_2/GF and silver-doped TiO_2/GF synthesized by sol-gel method, and at different relative humidities (Table 2).

Overall, silver-doped TiO_2/GF performed best on both bacteria, followed closely by copper TiO_2/GF and [107] then TiO_2/GF alone. The yield was better at a relative humidity of 60% than 80%. They were significantly lower at 40% humidity. *E. coli* is eliminated reasonably than *S. aureus* since the latter is a Gram + bacterium with a more complex wall [76,92].

Table 2. List of some studies on using TiO$_2$-metal for bacterial inactivation.

Bio Contaminants	Reactor	Catalyst	Operations Parameters	Performance	Ref.
E. coli	Petri dishes	TiO$_2$-NT and Ag-TiO$_2$-NTs	Concentration: 4 × 10^6 UFC/mL volume: 100 mL diameter TiO$_2$: 100 nm at 70V diameter Ag: 8 nm	TiO$_2$: reduction of 1.6 log with 180 min Ag/TiO$_2$: reduction of 99.99% after 90 min	[107]
P. aeruginosa	Glass fiber tissue (GFT)	Poroux TiO$_2$ TiO$_2$ pur (TiO$_2$-PEG) and TiO$_2$-Ag	Concentration: 10^3 UFC/mL TiO$_2$ pur: 14.7 nm TiO$_2$-Ag-PEG:16.6 nm TiO$_2$-Ag: 25.3 nm, under UV light.	TiO$_2$-1Ag: 100% of inactivation after 10 min TiO$_2$ poroux: 57% TiO$_2$-PEG: 93%	[108]
E. coli K12	Agar matrix surface + blueberry skin + calyx	UV-TiO$_2$ & UV alone	Initial bacterial populations: 7 log CFU/g UV-Photocatalysis (4.5 mW/cm^2). UV alone (6.0 mW/cm^2). TiO$_2$-coated quartz tubes (38 cm length, 24.5 mm outer diameter, thickness 0.7–0.9 mm.	4.5 log CFU/g for UV alone and 5.3 log CFU/g for UV-TiO$_2$ in 30 s. 3.4 log and 4.6 log CFU/g, respectively, UV alone and UV-TiO$_2$ for the first 30 s. 4.0 log and 5.2 log CFU/g, respectively, UV alone and photocatalysis.	[109]
S. aureus. P. aeruginosa and E. coli	LB agar plates	TiO$_2$-Ag (TiO$_2$ (calcined at 300 °C) (CB300) at (500 °C), (CB500) et TiO$_2$ (not calcinated) (CB))	Concentration: 10 µL with 10^9 UFC/mL 5%w of TiO$_2$	TiO$_2$ (calcined 300 °C)-Ag: reduces bacterial growth by 95%, i.e., 1.05 × 10^8 CFU/mL with UV. TiO$_2$ (calcined 500 °C) without Ag: reduces bacterial growth by 30% with UV. TiO$_2$ (calcined at 300 °C) without Ag: reduces growth by 75%.	[110]
E. coli	Planar reactor	TiO$_2$, TiO$_2$-Ag and TiO$_2$-Cu deposited on optical fibers	Initial bacterial populations: 2.4 × 10^7 UFC/mL. The core of optical fibers is constructed of polymethyl methacrylate resin with a mean diameter of 480 m and coated with 10 m of a thick fluorinated polymer, under UVA-LEDs (365 nm, UVA-LED intensity = 1.5 W m^{-2}).	3 log of removal with TiO$_2$/Ag and TiO$_2$/Cu	[76]

Table 2. Cont.

Bio Contaminants	Reactor	Catalyst	Operations Parameters	Performance	Ref.
S. aureus CCM 3955 & S. aureus CCM 3953 (Gram+) E. coli & P. aeruginosa (Gram−)	Disposable plates	Ag NPs	Initial bacterial populations: from 10^5 to 10^6 UFC/mL, Particle size from 40 to 60 nm, Temperature 35 °C.	Higher activity at 7 ppm against P. aeruginosa. NP Ag synthesized based on AgNO$_3$: considerable antibacterial activity at 14 and 29 ppm (82.49% inactivation). NP Ag synthesized based on AgNO$_3$ and citrate: 88.56 inactivations.	[111]
E. coli	Batch reactor	Cu$_2$O-NPs/TiO$_2$-NTs catalyst	Initial bacterial populations: from 10^6 to 10^7 UFC/mL. Under visible light irradiation (380–720) nm. Temperature 37 °C.	Bacterial inactivation rate of 98% and a concomitant 99.7% VOC removal within 60 min and 25 min	[78]

5. Conclusions and Outlook

Recent studies on photocatalysis for indoor air purification and bacterial inactivation have shown promising results. Even though a variety of photocatalysts are available, TiO_2-based materials are the most effective or, at the very least, effective option for practical and financial reasons. This review has collected and covered recent research that has improved the photocatalytic activity of materials based on TiO_2 for VOC degradation in indoor air and bacterial inactivation. Coupling TiO_2 materials with other methods has been increasingly explored. This paper also reviewed the literature on the material aspects of photocatalysis based on $AgxO/TiO_2$ and $CuxO/TiO_2$ to treat air-containing chemical and biological pollution. A bibliographical synthesis of the type of catalyst and the operating conditions was detailed concerning the decontamination of VOCs. Moreover, the different types of microorganisms treated by TiO_2-based photocatalysts have been listed. In-depth explanations of the reaction mechanisms for photocatalytic degradation and inactivation have been provided. As a look ahead to future research, we believe more study and testing are needed to clarify and comprehend the benefits of TiO_2-based materials on photocatalytic applications. There are only a few works on the combined treatment of chemical and biological pollution using photocatalysis at the same time. The investigations do not consider evaluating removal or if mineralization is complete, which is an essential criterion because it can generate more harmful intermediates than the pollutant. Finally, experiments in real cases of air pollution, such as hospital air pollution, are required to apply this process.

Author Contributions: Conceptualization, A.A.A. (Achraf Amir Assadi) and O.B.: methodology and writing—review, A.A.A. (Achraf Amir Assadi), N.B.H., L.K. and A.A.A. (Aymen Amine Assadi); writing—review and editing, A.A.A. (Achraf Amir Assadi), A.A.A. (Aymen Amine Assadi) and L.M.; conceptualization, funding acquisition, L.M.; methodology, L.K.; resources, project administration, supervision, A.A.A. (Aymen Amine Assadi): writing-review and editing, A.G. and L.M. All authors have read and agreed to the published version of the manuscript.

Funding: This research received no external funding.

Institutional Review Board Statement: Not applicable.

Informed Consent Statement: Not applicable.

Acknowledgments: The authors extend their appreciation to the Deanship of Scientific Research at Imam Mohammad Ibn Saud Islamic University (IMSIU) for funding and supporting this work through Research Partnership Program no RP-21-09-66.

Conflicts of Interest: The authors declare no conflict of interest.

References

1. Cincinelli, A.; Martellini, T. Indoor air quality and health. *Int. J. Environ. Res. Public Health* **2017**, *14*, 1286. [CrossRef]
2. Capolongo, S.; Settimo, G.; Gola, M. (Eds.) *Indoor Air Quality (IAQ) in Healthcare Facilities*; Springer: Berlin/Heidelberg, Germany, 2017; Volume 4386.
3. Assadi, I.; Guesmi, A.; Baaloudj, O.; Zeghioud, H.; Elfalleh, W.; Benhammadi, N. Review on inactivation of airborne viruses using non-thermal plasma technologies: From MS_2 to coronavirus. *Environ. Sci. Pollut. Res.* **2021**, *29*, 4880–4892. [CrossRef]
4. Heinsohn, R.J.; John, M.C. (Eds.) *Indoor Air Quality Engineering: Enviromental Health and Control Indoor*; CRC Press: Boca Raton, FL, USA, 2003; ISBN 0-8247-4061-0.
5. Abidi, M.; Hajjaji, A.; Bouzaza, A.; Lamaa, L.; Peruchon, L.; Brochier, C.; Rtimi, S.; Wolbert, D.; Bessais, B.; Assadi, A.A. Modeling of indoor air treatment using an innovative photocatalytic luminous textile: Reactor compactness and mass transfer enhancement. *Chem. Eng. J.* **2022**, *430*, 132636. [CrossRef]
6. Salmon, D.G. Annual Exporter Guide France. Available online: https://www.google.com.hk/url?sa=i&rct=j&q=&esrc=s&source=web&cd=&ved=0CAQQw7AJahcKEwjw49TzpND9AhUAAAAAHQAAAAAQAg&url=https%3A%2F%2Fapps.fas.usda.gov%2Fnewgainapi%2Fapi%2Freport%2Fdownloadreportbyfilename%3Ffilename%3DExporter%2520Guide_Paris_France_12-8-2014.pdf&psig=AOvVaw0bBETSpKjn6_G28dOcdT4j&ust=1678500724709337 (accessed on 25 December 2022).
7. Erisman, J.W. Air Pollution Science for the 21st Century. *Environ. Sci. Policy* **2003**, *6*, 396. [CrossRef]
8. Quyen, N.T.; Traikool, T.; Nitisoravut, R.; Onjun, T. Improvement of water quality using dielectric barrier discharge plasma. *J. Phys. Conf. Ser.* **2017**, *860*, 12031. [CrossRef]

9. Pichat, P. Some views about indoor air photocatalytic treatment using TiO_2: Conceptualization of humidity effects, active oxygen species, problem of C1-C3 carbonyl pollutants. *Appl. Catal. B Environ.* **2010**, *99*, 428–434. [CrossRef]
10. Ghezzi, S.; Pagani, I.; Poli, G.; Perboni, S.; Vicenzi, E. Rapid Inactivation of Severe Acute Respiratory Syndrome Coronavirus 2 (SARS-CoV-2) by Tungsten Trioxide-Based (WO_3) Photocatalysis. *bioRxiv* **2020**. [CrossRef]
11. Assadi, A.A.; Karoui, S.; Trabelsi, K.; Hajjaji, A.; Elfalleh, W.; Ghorbal, A.; Maghzaoui, M.; Assadi, A.A. Synthesis and Characterization of TiO2 Nanotubes (TiO2-NTs) with Ag Silver Nanoparticles (Ag-NPs): Photocatalytic Performance for Wastewater Treatment under Visible Light. *Materials* **2022**, *15*, 1463. [CrossRef]
12. Malayeri, M.; Haghighat, F.; Lee, C.S. Kinetic modeling of the photocatalytic degradation of methyl ethyl ketone in air for a continuous-flow reactor. *Chem. Eng. J.* **2021**, *404*, 126602. [CrossRef]
13. Zhang, Z.; Gamage, J. Applications of photocatalytic disinfection. *Int. J. Photoenergy* **2010**, *2010*, 764870. [CrossRef]
14. Bono, N.; Ponti, F.; Punta, C.; Candiani, G. Effect of UV irradiation and TiO_2-photocatalysis on airborne bacteria and viruses: An overview. *Materials* **2021**, *14*, 1075. [CrossRef] [PubMed]
15. Ali, I.; Al-hammadi, S.A.; Saleh, T.A. Simultaneous sorption of dyes and toxic metals from waters using synthesized titania-incorporated polyamide. *J. Mol. Liq.* **2018**, *269*, 564–571. [CrossRef]
16. Ali, T.; Ahmed, A.; Alam, U.; Uddin, I.; Tripathi, P.; Muneer, M. Enhanced photocatalytic and antibacterial activities of Ag-doped TiO_2 nanoparticles under visible light. *Mater. Chem. Phys.* **2018**, *212*, 325–335. [CrossRef]
17. Abidi, M.; Assadi, A.A.; Bouzaza, A.; Hajjaji, A.; Bessais, B.; Rtimi, S. Photocatalytic indoor/outdoor air treatment and bacterial inactivation on CuxO/TiO_2 prepared by HiPIMS on polyester cloth under low intensity visible light. *Appl. Catal. B Environ.* **2019**, *259*, 118074. [CrossRef]
18. Zhang, Y.; Zhao, X.; Fu, S.; Lv, X.; He, Q.; Li, Y.; Ji, F.; Xu, X. Preparation and antibacterial activity of Ag/TiO_2-functionalized ceramic tiles. *Ceram. Int.* **2022**, *48*, 4897–4903. [CrossRef]
19. Baaloudj, O.; Assadi, I.; Nasrallah, N.; El, A.; Khezami, L. Simultaneous removal of antibiotics and inactivation of antibiotic-resistant bacteria by photocatalysis: A review. *J. Water Process Eng.* **2021**, *42*, 102089. [CrossRef]
20. Wu, Y.; Chen, X.; Cao, J.; Zhu, Y.; Yuan, W.; Hu, Z.; Ao, Z.; Brudvig, G.W.; Tian, F.; Yu, J.C.; et al. Photocatalytically recovering hydrogen energy from wastewater treatment using MoS_2@TiO_2 with sulfur/oxygen dual-defect. *Appl. Catal. B Environ.* **2022**, *303*, 120878. [CrossRef]
21. Assadi, A.A.; Bouzaza, A.; Wolbert, D. Study of synergetic effect by surface discharge plasma/TiO_2 combination for indoor air treatment: Sequential and continuous configurations at pilot scale. *J. Photochem. Photobiol. A Chem.* **2015**, *310*, 148–154. [CrossRef]
22. Karoui, S.; Ben Arfi, R.; Mougin, K.; Ghorbal, A.; Assadi, A.A.; Amrane, A. Synthesis of novel biocomposite powder for simultaneous removal of hazardous ciprofloxacin and methylene blue: Central composite design, kinetic and isotherm studies using Brouers-Sotolongo family models. *J. Hazard. Mater.* **2020**, *387*, 121675. [CrossRef]
23. Zeghioud, H.; Khellaf, N.; Amrane, A.; Djelal, H.; Elfalleh, W.; Assadi, A.A.; Rtimi, S. Photocatalytic performance of TiO_2 impregnated polyester for the degradation of Reactive Green 12: Implications of the surface pretreatment and the microstructure. *J. Photochem. Photobiol. A Chem.* **2017**, *346*, 493–501. [CrossRef]
24. Baaloudj, O.; Kenfoud, H.; Badawi, A.K.; Assadi, A.A.; El Jery, A.; Assadi, A.A.; Amrane, A. Bismuth Sillenite Crystals as Recent Photocatalysts for Water Treatment and Energy Generation: A Critical Review. *Catalysts* **2022**, *12*, 500. [CrossRef]
25. Kappadan, S.; Gebreab, T.W.; Thomas, S.; Kalarikkal, N. Tetragonal $BaTiO_3$ nanoparticles: An efficient photocatalyst for the degradation of organic pollutants. *Mater. Sci. Semicond. Process.* **2016**, *51*, 42–47. [CrossRef]
26. Malato, S.; Fernández-Ibáñez, P.; Maldonado, M.I.; Blanco, J.; Gernjak, W. Decontamination and disinfection of water by solar photocatalysis: Recent overview and trends. *Catal. Today* **2009**, *147*, 1–59. [CrossRef]
27. Koe, W.S.; Lee, J.W.; Chong, W.C.; Pang, Y.L.; Sim, L.C. An overview of photocatalytic degradation: Photocatalysts, mechanisms, and development of photocatalytic membrane. *Environ. Sci. Pollut. Res.* **2020**, *27*, 2522–2565. [CrossRef]
28. Akerdi, A.G.; Bahrami, S.H. Application of heterogeneous nano-semiconductors for photocatalytic advanced oxidation of organic compounds: A review. *J. Environ. Chem. Eng.* **2019**, *7*, 103283. [CrossRef]
29. Almomani, F.; Rene, E.R.; Veiga, M.C.; Bhosale, R.R.; Kennes, C. Treatment of waste gas contaminated with dichloromethane using photocatalytic oxidation, biodegradation and their combinations. *J. Hazard. Mater.* **2021**, *405*, 123735. [CrossRef]
30. Mohseni, M.; Prieto, L. Biofiltration of hydrophobic VOCs pretreated with UV photolysis and photocatalysis. *Int. J. Environ. Technol. Manag.* **2008**, *9*, 47–58. [CrossRef]
31. Khezami, L.; Nguyen-Tri, P.; Saoud, W.A.; Bouzaza, A.; El Jery, A.; Duc Nguyen, D.; Gupta, V.K.; Assadi, A.A. Recent progress in air treatment with combined photocatalytic/plasma processes: A review. *J. Environ. Manag.* **2021**, *299*, 113588. [CrossRef]
32. Kuwahara, Y.; Yamashita, H. Efficient photocatalytic degradation of organics diluted in water and air using TiO_2 designed with zeolites and mesoporous silica materials. *J. Mater. Chem.* **2011**, *21*, 2407–2416. [CrossRef]
33. Baaloudj, O.; Nasrallah, N.; Bouallouche, R.; Kenfoud, H.; Khezami, L.; Assadi, A.A. High efficient Cefixime removal from water by the sillenite $Bi_{12}TiO_{20}$: Photocatalytic mechanism and degradation pathway. *J. Clean. Prod.* **2022**, *330*, 129934. [CrossRef]
34. Bolton, J.R.; Bircher, K.G.; Tumas, W.; Tolman, C.A. Figures-of-merit for the technical development and application of advanced oxidation technologies for both electric- and solar-driven systems. *Pure Appl. Chem.* **2001**, *73*, 627–637. [CrossRef]
35. Serhane, Y.; Belkessa, N.; Bouzaza, A.; Wolbert, D.; Assadi, A.A. Continuous air purification by front flow photocatalytic reactor: Modelling of the influence of mass transfer step under simulated real conditions. *Chemosphere* **2022**, *295*, 133809. [CrossRef] [PubMed]

36. Pelaez, M.; Nolan, N.T.; Pillai, S.C.; Seery, M.K.; Falaras, P.; Kontos, A.G.; Dunlop, P.S.M.; Hamilton, J.W.J.; Byrne, J.A.; O'Shea, K.; et al. A review on the visible light active titanium dioxide photocatalysts for environmental applications. *Appl. Catal. B Environ.* **2012**, *125*, 331–349. [CrossRef]
37. Kang, X.; Liu, S.; Dai, Z.; He, Y.; Song, X.; Tan, Z. Titanium dioxide: From engineering to applications. *Catalysts* **2019**, *9*, 191. [CrossRef]
38. Hodgson, A.T.; Destaillats, H.; Sullivan, D.P.; Fisk, W.J. Performance of ultraviolet photocatalytic oxidation for indoor air cleaning applications. *Indoor Air* **2007**, *17*, 305–316. [CrossRef]
39. Muscetta, M.; Russo, D. Photocatalytic applications in wastewater and air treatment: A patent review (2010–2020). *Catalysts* **2021**, *11*, 834. [CrossRef]
40. Riaz, N.; Fen, D.A.C.S.; Khan, M.S.; Naz, S.; Sarwar, R.; Farooq, U.; Bustam, M.A.; Batiha, G.E.S.; El Azab, I.H.; Uddin, J.; et al. Iron-zinc co-doped titania nanocomposite: Photocatalytic and photobiocidal potential in combination with molecular docking studies. *Catalysts* **2021**, *11*, 1112. [CrossRef]
41. Malliga, P.; Pandiarajan, J.; Prithivikumaran, N.; Neyvasagam, K. Effect of film thickness on structural and optical properties of TiO2 thin films. In Proceedings of the International Conference on Advanced Nanomaterials & Emerging Engineering Technologies, Chennai, India, 24–26 July 2013; Volume 2, pp. 488–491. [CrossRef]
42. Fagan, R.; McCormack, D.E.; Dionysiou, D.D.; Pillai, S.C. A review of solar and visible light active TiO2 photocatalysis for treating bacteria, cyanotoxins and contaminants of emerging concern. *Mater. Sci. Semicond. Process.* **2016**, *42*, 2–14. [CrossRef]
43. Rabhi, S.; Belkacemi, H.; Bououdina, M.; Kerrami, A.; Ait Brahem, L.; Sakher, E. Effect of Ag doping of TiO2 nanoparticles on anatase-rutile phase transformation and excellent photodegradation of amlodipine besylate. *Mater. Lett.* **2019**, *236*, 640–643. [CrossRef]
44. Yi, J.; Huang, L.; Wang, H.; Yu, H.; Peng, F. AgI/TiO2 nanobelts monolithic catalyst with enhanced visible light photocatalytic activity. *J. Hazard. Mater.* **2015**, *284*, 207–214. [CrossRef]
45. Chen, Y.; Huang, W.; He, D.; Situ, Y.; Huang, H. Construction of heterostructured g-C3N4/Ag/TiO2 microspheres with enhanced photocatalysis performance under visible-light irradiation. *ACS Appl. Mater. Interfaces* **2014**, *6*, 14405–14414. [CrossRef]
46. Monga, D.; Basu, S. Single-crystalline 2D BiOCl nanorods decorated with 2D MoS2 nanosheets for visible light-driven photocatalytic detoxification of organic and inorganic pollutants. *FlatChem* **2021**, *28*, 100267. [CrossRef]
47. Guan, Z.; Li, Q.; Shen, B.; Bao, S.; Zhang, J.; Tian, B. Fabrication of Co3O4 and Au co-modified BiOBr flower-like microspheres with high photocatalytic efficiency for sulfadiazine degradation. *Sep. Purif. Technol.* **2020**, *234*, 116100. [CrossRef]
48. Raizada, P.; Thakur, P.; Sudhaik, A.; Singh, P.; Thakur, V.K.; Hosseini-Bandegharaei, A. Fabrication of dual Z-scheme photocatalyst via coupling of BiOBr/Ag/AgCl heterojunction with P and S co-doped g-C3N4 for efficient phenol degradation. *Arab. J. Chem.* **2020**, *13*, 4538–4552. [CrossRef]
49. Zhang, W.; Zhang, Q.; Dong, F. Visible-light photocatalytic removal of NO in air over BiOX (X = Cl, Br, I) single-crystal nanoplates prepared at room temperature. *Ind. Eng. Chem. Res.* **2013**, *52*, 6740–6746. [CrossRef]
50. Gao, F.; Yang, Y.; Wang, T. Preparation of porous TiO2/Ag heterostructure films with enhanced photocatalytic activity. *Chem. Eng. J.* **2015**, *270*, 418–427. [CrossRef]
51. Xie, Z.; Yang, J.; Wang, K.; Meng, Q.; Tang, Y.; Zhao, K. Facile fabrication of TiO2-SiO2-C composite with anatase/rutile heterostructure via sol-gel process and its enhanced photocatalytic activity in the presence of H2O2. *Ceram. Int.* **2022**, *48*, 9114–9123. [CrossRef]
52. Eskandari, P.; Farhadian, M.; Solaimany Nazar, A.R.; Jeon, B.H. Adsorption and Photodegradation Efficiency of TiO2/Fe2O3/PAC and TiO2/Fe2O3/Zeolite Nanophotocatalysts for the Removal of Cyanide. *Ind. Eng. Chem. Res.* **2019**, *58*, 2099–2112. [CrossRef]
53. Pal, B.; Sharon, M.; Nogami, G. Preparation and characterization of TiO2/Fe2O3 binary mixed oxides and its photocatalytic properties. *Mater. Chem. Phys.* **1999**, *59*, 254–261. [CrossRef]
54. Hou, F.; Lu, K.; Liu, F.; Xue, F.; Liu, M. Manipulating a TiO2-graphene-Ta3N5 heterojunction for efficient Z-scheme photocatalytic pure water splitting. *Mater. Res. Bull.* **2022**, *150*, 111782. [CrossRef]
55. Li, H.; Hong, W.; Cui, Y.; Hu, X.; Fan, S.; Zhu, L. Enhancement of the visible light photocatalytic activity of Cu2O/BiVO4 catalysts synthesized by ultrasonic dispersion method at room temperature. *Mater. Sci. Eng. B* **2014**, *181*, 1–8. [CrossRef]
56. Aguilera-Ruiz, E.; García-Pérez, U.M.; De La Garza-Galván, M.; Zambrano-Robledo, P.; Bermúdez-Reyes, B.; Peral, J. Efficiency of Cu2O/BiVO4 particles prepared with a new soft procedure on the degradation of dyes under visible-light irradiation. *Appl. Surf. Sci.* **2015**, *328*, 361–367. [CrossRef]
57. Sun, S.; Ding, J.; Bao, J.; Gao, C.; Qi, Z.; Yang, X.; He, B.; Li, C. Photocatalytic degradation of gaseous toluene on Fe-TiO2 under visible light irradiation: A study on the structure, activity and deactivation mechanism. *Appl. Surf. Sci.* **2012**, *258*, 5031–5037. [CrossRef]
58. Burda, C.; Lou, Y.; Chen, X.; Samia, A.C.S.; Stout, J.; Gole, J.L. Enhanced nitrogen doping in TiO2 nanoparticles. *Nano Lett.* **2003**, *3*, 1049–1051. [CrossRef]
59. Hua, L.; Yin, Z.; Cao, S. Recent advances in synthesis and applications of carbon-doped TiO2 nanomaterials. *Catalysts* **2020**, *10*, 1431. [CrossRef]
60. Hanaor, D.A.H.; Sorrell, C.C. Review of the anatase to rutile phase transformation. *J. Mater. Sci.* **2011**, *46*, 855–874. [CrossRef]
61. Ghumro, S.S.; Lal, B.; Pirzada, T. Visible-Light-Driven Carbon-Doped TiO2-Based Nanocatalysts for Enhanced Activity toward Microbes and Removal of Dye. *ACS Omega* **2022**, *7*, 4333–4341. [CrossRef]

62. Akhter, P.; Arshad, A.; Saleem, A.; Hussain, M. Recent Development in Non-Metal-Doped Titanium Dioxide Photocatalysts for Different Dyes Degradation and the Study of Their Strategic Factors: A Review. *Catalysts* **2022**, *12*, 1331. [CrossRef]
63. Zhang, W.; Luo, N.; Huang, S.; Wu, N.L.; Wei, M. Sulfur-Doped Anatase TiO_2 as an Anode for High-Performance Sodium-Ion Batteries. *ACS Appl. Energy Mater.* **2019**, *2*, 3791–3797. [CrossRef]
64. Niu, P.; Wu, G.; Chen, P.; Zheng, H.; Cao, Q.; Jiang, H. Optimization of Boron Doped TiO_2 as an Efficient Visible Light-Driven Photocatalyst for Organic Dye Degradation With High Reusability. *Front. Chem.* **2020**, *8*, 172. [CrossRef]
65. Piątkowska, A.; Janus, M.; Szymański, K.; Mozia, S. C-, N- and S-doped TiO_2 photocatalysts: A review. *Catalysts* **2021**, *11*, 144. [CrossRef]
66. Marschall, R. Semiconductor composites: Strategies for enhancing charge carrier separation to improve photocatalytic activity. *Adv. Funct. Mater.* **2014**, *24*, 2421–2440. [CrossRef]
67. Jones, A.P. Indoor air quality and health. *Atmos. Environ.* **1999**, *33*, 4535–4564. [CrossRef]
68. Magureanu, M.; Bogdan, N.; Hu, J.; Richards, R.; Florea, M.; Parvulescu, M. Plasma-assisted catalysis total oxidation of trichloroethylene over gold nano-particles embedded in SBA-15 catalysts. *Catal. B Environ.* **2007**, *76*, 275–281. [CrossRef]
69. Xia, Y.; Xia, L.; Liu, Y.; Yang, T.; Deng, J.; Dai, H. Concurrent catalytic removal of typical volatile organic compound mixtures over Au-Pd/α-MnO_2 nanotubes. *J. Environ. Sci.* **2018**, *64*, 276–288. [CrossRef] [PubMed]
70. Bahri, M.; Haghighat, F. Plasma-based indoor air cleaning technologies: The state of the art-review. *Clean Soil Air Water* **2014**, *42*, 1667–1680. [CrossRef]
71. Shah, K.W.; Li, W. A review on catalytic nanomaterials for volatile organic compounds VOC removal and their applications for healthy buildings. *Nanomaterials* **2019**, *9*, 910. [CrossRef]
72. Tobaldi, D.M.; Dvoranová, D.; Lajaunie, L.; Rozman, N.; Figueiredo, B.; Seabra, M.P.; Škapin, A.S.; Calvino, J.J.; Brezová, V.; Labrincha, J.A. Graphene-TiO_2 hybrids for photocatalytic aided removal of VOCs and nitrogen oxides from outdoor environment. *Chem. Eng. J.* **2021**, *405*. [CrossRef]
73. Lam, S.M.; Sin, J.C.; Abdullah, A.Z.; Mohamed, A.R. Photocatalytic TiO_2/carbon nanotube nanocomposites for environmental applications: An overview and recent developments. *Fuller. Nanotub. Carbon Nanostruct.* **2014**, *22*, 471–509. [CrossRef]
74. Shayegan, Z.; Lee, C.S.; Haghighat, F. TiO_2 photocatalyst for removal of volatile organic compounds in gas phase—A review. *Chem. Eng. J.* **2018**, *334*, 2408–2439. [CrossRef]
75. Zadi, T.; Azizi, M.; Nasrallah, N.; Bouzaza, A.; Zadi, T.; Azizi, M.; Nasrallah, N.; Bouzaza, A.; Maachi, R. Indoor air treatment of refrigerated food chambers with synergetic association between cold plasma and photocatalysis: Process performance and photocatalytic poisoning. *Chem. Eng. J.* **2020**, *382*, 122951. [CrossRef]
76. Abou Saoud, W.; Kane, A.; Le Cann, P.; Gerard, A.; Lamaa, L.; Peruchon, L.; Brochier, C.; Bouzaza, A.; Wolbert, D.; Assadi, A.A. Innovative photocatalytic reactor for the degradation of VOCs and microorganism under simulated indoor air conditions: Cu-Ag/TiO_2-based optical fibers at a pilot scale. *Chem. Eng. J.* **2021**, *411*, 128622. [CrossRef]
77. Jia, Z.; Barakat, C.; Dong, B.; Rousseau, A. VOCs Destruction by Plasma Catalyst Coupling Using AL-KO PURE Air Purifier on Industrial Scale. *J. Mater. Sci. Chem. Eng.* **2015**, *3*, 19–26. [CrossRef]
78. Abidi, M.; Hajjaji, A.; Bouzaza, A.; Trablesi, K.; Makhlouf, H.; Rtimi, S.; Assadi, A.A.; Bessais, B. Simultaneous removal of bacteria and volatile organic compounds on Cu_2O-NPs decorated TiO_2 nanotubes: Competition effect and kinetic studies. *J. Photochem. Photobiol. A Chem.* **2020**, *400*, 112722. [CrossRef]
79. Vincent, G.; Queffeulou, A.; Marquaire, P.M.; Zahraa, O. Remediation of olfactory pollution by photocatalytic degradation process: Study of methyl ethyl ketone (MEK). *J. Photochem. Photobiol. A Chem.* **2007**, *191*, 42–50. [CrossRef]
80. Vincent, G.; Schaer, E.; Marquaire, P.M.; Zahraa, O. CFD modelling of an annular reactor, application to the photocatalytic degradation of acetone. *Process Saf. Environ. Prot.* **2011**, *89*, 35–40. [CrossRef]
81. Liu, H.; Li, C.; Ren, X.; Liu, K.; Yang, J. Fine platinum nanoparticles supported on a porous ceramic membrane as efficient catalysts for the removal of benzene. *Sci. Rep.* **2017**, *7*, 16589. [CrossRef]
82. Sedjame, H.J.; Fontaine, C.; Lafaye, G.; Barbier, J. On the promoting effect of the addition of ceria to platinum based alumina catalysts for VOCs oxidation. *Appl. Catal. B Environ.* **2014**, *144*, 233–242. [CrossRef]
83. Xu, Z.; Yu, J.; Jaroniec, M. Efficient catalytic removal of formaldehyde at room temperature using AlOOH nanoflakes with deposited Pt. *Appl. Catal. B Environ.* **2015**, *163*, 306–312. [CrossRef]
84. Zhang, J.; Li, Y.; Wang, L.; Zhang, C.; He, H. Catalytic oxidation of formaldehyde over manganese oxides with different crystal structures. *Catal. Sci. Technol.* **2015**, *5*, 2305–2313. [CrossRef]
85. Zhang, C.; He, H.; Tanaka, K. ichi Catalytic performance and mechanism of a Pt/TiO_2 catalyst for the oxidation of formaldehyde at room temperature. *Appl. Catal. B Environ.* **2006**, *65*, 37–43. [CrossRef]
86. Kucherov, A.V.; Tkachenko, O.P.; Kirichenko, O.A.; Kapustin, G.I.; Mishin, I.V.; Klementiev, K.V.; Ojala, S.; Kustov, L.M.; Keiski, R. Nanogold-containing catalysts for low-temperature removal of S-VOC from air. *Top. Catal.* **2009**, *52*, 351–358. [CrossRef]
87. Assadi, A.A.; Bouzaza, A.; Vallet, C.; Wolbert, D. Use of DBD plasma, photocatalysis, and combined DBD plasma/photocatalysis in a continuous annular reactor for isovaleraldehyde elimination—Synergetic effect and byproducts identification. *Chem. Eng. J.* **2014**, *254*, 124–132. [CrossRef]
88. Chen, C.Y.; Wu, L.C.; Chen, H.Y.; Chung, Y.C. Inactivation of *Staphylococcus aureus* and *Escherichia coli* in water using photocatalysis with fixed TiO_2. *Water. Air Soil Pollut.* **2010**, *212*, 231–238. [CrossRef]

89. Blanco-Galvez, J.; Fernández-Ibáñez, P.; Malato-Rodríguez, S. Solar photo catalytic detoxification and disinfection of water: Recent overview. *J. Sol. Energy Eng. Trans. ASME* **2007**, *129*, 4–15. [CrossRef]
90. Magaña-López, R.; Zaragoza-Sánchez, P.I.; Jiménez-Cisneros, B.E.; Chávez-Mejía, A.C. The use of TiO_2 as a disinfectant in water sanitation applications. *Water* **2021**, *13*, 1641. [CrossRef]
91. Anucha, C.B.; Altin, I.; Bacaksiz, E.; Stathopoulos, V.N. Titanium dioxide (TiO_2)-based photocatalyst materials activity enhancement for contaminants of emerging concern (CECs) degradation: In the light of modification strategies. *Chem. Eng. J. Adv.* **2022**, *10*, 100262. [CrossRef]
92. Rtimi, S.; Dionysiou, D.D.; Pillai, S.C.; Kiwi, J. Advances in catalytic/photocatalytic bacterial inactivation by nano Ag and Cu coated surfaces and medical devices. *Appl. Catal. B Environ.* **2019**, *240*, 291–318. [CrossRef]
93. Saoud, W.A.; Assadi, A.A.; Kane, A.; Jung, A.; Cann, P.L.; Bazantay, F.; Bouzaza, A.; Wolbert, D. Integrated process for the removal of indoor VOCs from food industry manufacturing Elimination of Butane-2,3-dione and Heptan-2-one by cold plasma-photocatalysis combination. *Photochem. Photobiol. A Chem.* **2020**, *386*, 112071. [CrossRef]
94. Venieri, D.; Gounaki, I.; Bikouvaraki, M.; Binas, V.; Zachopoulos, A.; Kiriakidis, G.; Mantzavinos, D. Solar photocatalysis as disinfection technique: Inactivation of *Klebsiella pneumoniae* in sewage and investigation of changes in antibiotic resistance profile. *J. Environ. Manag.* **2017**, *195*, 140–147. [CrossRef]
95. Ray, S.K.; Dhakal, D.; Regmi, C.; Yamaguchui, T.; Lee, S.W. Inactivation of *Staphylococcus aureus* in visible light by morphology tuned α-$NiMoO_4$. *J. Photochem. Photobiol. A Chem.* **2018**, *350*, 59–68. [CrossRef]
96. Kőrösi, L.; Pertics, B.; Schneider, G.; Bognár, B.; Kovács, J.; Meynen, V.; Scarpellini, A.; Pasquale, L.; Prato, M. Photocatalytic inactivation of plant pathogenic bacteria using TiO_2 nanoparticles prepared hydrothermally. *Nanomaterials* **2020**, *10*, 1730. [CrossRef] [PubMed]
97. Salavati-Niasari, M.; Davar, F. Synthesis of copper and copper(I) oxide nanoparticles by thermal decomposition of a new precursor. *Mater. Lett.* **2009**, *63*, 441–443. [CrossRef]
98. Yue, Y.; Zhang, P.; Wang, W.; Cai, Y.; Tan, F.; Wang, X.; Qiao, X.; Wong, P.K. Enhanced dark adsorption and visible-light-driven photocatalytic properties of narrower-band-gap Cu_2S decorated Cu_2O nanocomposites for efficient removal of organic pollutants. *J. Hazard. Mater.* **2020**, *384*, 121302. [CrossRef] [PubMed]
99. Wang, W.; Song, J.; Kang, Y.; Chai, D.; Zhao, R.; Lei, Z. Sm_2O_3 embedded in nitrogen doped carbon with mosaic structure: An effective catalyst for oxygen reduction reaction. *Energy* **2017**, *133*, 115–120. [CrossRef]
100. Gupta, S.B.; Bluhm, H. The potential of pulsed underwater streamer discharges as a disinfection technique. *IEEE Trans. Plasma Sci.* **2008**, *36*, 1621–1632. [CrossRef]
101. Tijani, J.O.; Fatoba, O.O.; Madzivire, G.; Petrik, L.F. A review of combined advanced oxidation technologies for the removal of organic pollutants from water. *Water Air Soil Pollut.* **2014**, *225*, 2102. [CrossRef]
102. Ado, A.; Tukur, A.I.; Ladan, M.; Gumel, S.M.; Muhammad, A.A.; Habibu, S.; Koki, I.B. A Review on Industrial Effluents as Major Sources of Water Pollution in Nigeria. *Chem. J.* **2015**, *1*, 159–164.
103. Weiss, C.; Carriere, M.; Fusco, L.; Fusco, L.; Capua, I.; Regla-Nava, J.A.; Pasquali, M.; Pasquali, M.; Pasquali, M.; Scott, J.A.; et al. Toward Nanotechnology-Enabled Approaches against the COVID-19 Pandemic. *ACS Nano* **2020**, *14*, 6383–6406. [CrossRef] [PubMed]
104. Karbasi, M.; Karimzadeh, F.; Raeissi, K.; Rtimi, S.; Kiwi, J.; Giannakis, S.; Pulgarin, C. Insights into the photocatalytic bacterial inactivation by flower-like Bi_2WO_6 under solar or visible light, through in situ monitoring and determination of reactive oxygen species (ROS). *Water* **2020**, *12*, 1099. [CrossRef]
105. Singh, J.; Juneja, S.; Palsaniya, S.; Manna, A.K.; Soni, R.K.; Bhattacharya, J. Evidence of oxygen defects mediated enhanced photocatalytic and antibacterial performance of ZnO nanorods. *Colloids Surf. B Biointerfaces* **2019**, *184*, 110541. [CrossRef]
106. Wang, W.; Wang, H.; Li, G.; An, T.; Zhao, H.; Wong, P.K. Catalyst-free activation of persulfate by visible light for water disinfection: Efficiency and mechanisms. *Water Res.* **2019**, *157*, 106–118. [CrossRef]
107. Hajjaji, A.; Elabidi, M.; Trabelsi, K.; Assadi, A.A.; Bessais, B.; Rtimi, S. Bacterial adhesion and inactivation on Ag decorated TiO_2-nanotubes under visible light: Effect of the nanotubes geometry on the photocatalytic activity. *Colloids Surf. B Biointerfaces* **2018**, *170*, 92–98. [CrossRef] [PubMed]
108. Ubonchonlakate, K.; Sikong, L.; Saito, F. Photocatalytic disinfection of *P. aeruginosa* bacterial Ag-doped TiO_2 film. *Procedia Eng.* **2012**, *32*, 656–662. [CrossRef]
109. Lee, M.; Shahbaz, H.M.; Kim, J.U.; Lee, H.; Lee, D.U.; Park, J. Efficacy of UV-TiO_2 photocatalysis technology for inactivation of *Escherichia coli* K12 on the surface of blueberries and a model agar matrix and the influence of surface characteristics. *Food Microbiol.* **2018**, *76*, 526–532. [CrossRef] [PubMed]

110. Gupta, K.; Singh, R.P.; Pandey, A.; Pandey, A. Photocatalytic antibacterial performance of TiO_2 and Ag-doped TiO_2 against *S. aureus, P. aeruginosa* and *E. coli*. *Beilstein J. Nanotechnol.* **2013**, *4*, 345–351. [CrossRef] [PubMed]
111. Guzman, M.; Dille, J.; Godet, S. Synthesis and antibacterial activity of silver nanoparticles against gram-positive and gram-negative bacteria. *Nanomed. Nanotechnol. Biol. Med.* **2012**, *8*, 37–45. [CrossRef] [PubMed]

Disclaimer/Publisher's Note: The statements, opinions and data contained in all publications are solely those of the individual author(s) and contributor(s) and not of MDPI and/or the editor(s). MDPI and/or the editor(s) disclaim responsibility for any injury to people or property resulting from any ideas, methods, instructions or products referred to in the content.

Article

Highly Efficient and Exceptionally Durable Photooxidation Properties on Co$_3$O$_4$/g-C$_3$N$_4$ Surfaces

Yelin Dai [1], Ziyi Feng [1], Kang Zhong [1], Jianfeng Tian [1], Guanyu Wu [1], Qing Liu [2], Zhaolong Wang [1], Yingjie Hua [3], Jinyuan Liu [1], Hui Xu [1] and Xingwang Zhu [2,*]

[1] School of the Environment and Safety Engineering, Institute for Energy Research, Jiangsu University, Zhenjiang 212013, China
[2] College of Environmental Science and Engineering, Yangzhou University, Yangzhou 225009, China
[3] The Key Laboratory of Electrochemical Energy Storage and Energy Conversion of Hainan Province, School of Chemistry and Chemical Engineering, Hainan Normal University, Haikou 571158, China
* Correspondence: zxw@yzu.edu.cn

Abstract: Water pollution is a significant social issue that endangers human health. The technology for the photocatalytic degradation of organic pollutants in water can directly utilize solar energy and has a promising future. A novel Co$_3$O$_4$/g-C$_3$N$_4$ type-II heterojunction material was prepared by hydrothermal and calcination strategies and used for the economical photocatalytic degradation of rhodamine B (RhB) in water. Benefitting the development of type-II heterojunction structure, the separation and transfer of photogenerated electrons and holes in 5% Co$_3$O$_4$/g-C$_3$N$_4$ photocatalyst was accelerated, leading to a degradation rate 5.8 times higher than that of pure g-C$_3$N$_4$. The radical capturing experiments and ESR spectra indicated that the main active species are •O$_2$$^-$ and h$^+$. This work will provide possible routes for exploring catalysts with potential for photocatalytic applications.

Keywords: photocatalyst; type-II heterojunction; carrier separation; photodegradation

Citation: Dai, Y.; Feng, Z.; Zhong, K.; Tian, J.; Wu, G.; Liu, Q.; Wang, Z.; Hua, Y.; Liu, J.; Xu, H.; et al. Highly Efficient and Exceptionally Durable Photooxidation Properties on Co$_3$O$_4$/g-C$_3$N$_4$ Surfaces. *Materials* **2023**, *16*, 3879. https://doi.org/10.3390/ma16103879

Academic Editor: Juan M. Coronado

Received: 23 March 2023
Revised: 22 April 2023
Accepted: 26 April 2023
Published: 22 May 2023

Copyright: © 2023 by the authors. Licensee MDPI, Basel, Switzerland. This article is an open access article distributed under the terms and conditions of the Creative Commons Attribution (CC BY) license (https:// creativecommons.org/licenses/by/ 4.0/).

1. Introduction

It is well known that the situation regarding water resources is linked to environmental, social, and economic risks [1,2]. However, large volumes of wastewater dyes and pharmaceutical effluents, including methylene blue, rhodamine B, tetracycline, ciprofloxacin, and so on, have been detected in our daily water bodies [3,4]. As a result, the environmental crisis over water has become one of the top risks facing the world today. Since these pollutants have become a serious threat to humans and ecosystems, there is an urgent need to clean up these colored organic dye pollutants [5]. In order to more effectively mitigate the ecological risks brought by water environment problems, environment-friendly technical methods such as adsorption, electrochemical, and photochemical methods have been proposed. The implementation of these technologies could effectively achieve the effect of purifying wastewater. Among the many technologies, photocatalysis, as a harmless technology for substance conversion, plays an important role in the field of toxic substances conversion. However, as the core of photocatalysis technology, semiconductor photocatalysts are usually limited to green, stable materials that meet the needs of industrial use [6]. To date, several types of semiconductors, such as oxides (TiO$_2$ [7], ZnO [8]), nitrides (Ta$_3$N$_5$ [9], C$_3$N$_4$ [10–14]), and sulfides (MoS$_2$ [15,16], CdS [17,18]) have been developed. In general, as a representative semiconductor material in p-type semiconductors, Co$_3$O$_4$ is highly sought after by researchers because of its excellent catalytic activity and stability in the field of photocatalysis and its high economic benefits [19–22]. However, even so, its inherent defects still greatly limit the market expansion and application of such materials, such as their low electron–hole separation rate and relatively limited optical absorption range [23,24]. Based on the above dilemma, the design idea of effectively improving the optical absorption range of Co_3O_4 and increasing the separation rate of photogenerated carriers may enable it to meet the demand gap in the field of environmental governance.

For a long time, researchers have also actively carried out a lot of research based on light absorption and carrier separation [25,26]. The implementation of many technical strategies, such as the design of morphologies, the construction of heterostructures, and the modification of precious metals, greatly optimized and improved the photocatalytic performance of Co_3O_4. Among them, the construction of semiconductor heterostructures is the most effective way to promote efficient carrier separation and migration and has shown impressive performance in many reports [24,27]. In these heterostructures, the p-type semiconductor Co_3O_4 conduction band (CB) and valence band (VB) bend towards vacuum level while the n-type semiconductor bend against vacuum level due to the formation of the built-in electric field in the catalyst and the balance of Fermi energy levels [28,29]. Moreover, the bending is only large at the region far from the depletion region. Driven by the force of the electric field, the charge is further separated efficiently, thus improving the photocatalytic efficiency. At present, the various reported n-type semiconductors that have been used to construct the p-type semiconductor Co_3O_4 include $g-C_3N_4$ [11,30], In_2O_3 [31], Bi_2O_3 [32], and MnO_2 [33]. Graphitic carbon nitride, a stable polymer semiconductor with a special 2D framework structure of heptazine rings connected via tertiary amines, could form a self-built internal electrostatic field, and the electric field and van der Waals interactions cause photogenerated separation and transport of carriers [29,34]. Moreover, due to its wide band gap, $g-C_3N_4$ exhibits efficient sunlight collection properties. And thanks to its sparse and porous structure, it is also able to easily adsorb and re-degrade pollutants [35]. Therefore, the modification of $g-C_3N_4$-based materials gives us a more practical pathway to enhance the activity of metal oxides. Based on this, we are eager to learn whether the coupling between $g-C_3N_4$ and Co_3O_4 could efficiently solve problems in the field of environmental treatment.

In our research, Co_3O_4 nanosheets and $g-C_3N_4$ were prepared by a rapid hydrothermal method and a calcination method, respectively, and then $Co_3O_4/g-C_3N_4$ nanomaterials were prepared by composing the two. The microstructure and physical and chemical properties of $Co_3O_4/g-C_3N_4$ were characterized by several methods, such as HRTEM, XPS, BET, and ESR, and the performance of different mass ratios of $Co_3O_4/g-C_3N_4$ on the catalyst photodegradation activity of RhB was investigated under simulated sunlight. The results showed that the photocatalytic activity of $Co_3O_4/g-C_3N_4$ was significantly enhanced compared with that of the pure sample, which may be due to the role of the heterojunction established between Co_3O_4 and $g-C_3N_4$, which could promote the separation of photogenerated charges and interfacial effects. Finally, a possible charge transfer pathway is proposed based on the experimental results. Our work offers new insights into the application of crystalline semiconductors for the removal of aqueous organic pollutants.

2. Materials and Methods

2.1. Materials

$CO(NH_2)_2$, $Co(NO_3)_2 \cdot 6H_2O$, C_2H_6O, NaOH, C_2H_3N, $C_6H_{15}NO_3$, and $C_{28}H_{31}ClN_2O_3$ were procured from Sinopharm Chemical Reagent Co., Ltd. (Shanghai, China). $(C_6H_9NO)_n$ (M.W. ≈ 55,000) was acquired from Aladdin Reagent Co., Ltd. (Shanghai, China). Deionized water was used throughout the experiment. All chemicals were analytically pure and required no further processing.

2.2. Synthesis of $g-C_3N_4$

Urea (20 g sample) was added to a 50 mL crucible container and transferred to a muffle furnace and calcined under an air atmosphere. The conditions were set to increase from ambient temperature to 823 K at a rate of 5 K/min for 4 h. After the sample cooled down, the sample was made into powder with a mortar and raised from the initial temperature to 773 K at a rate of 5 K/min for 2 h. The light-yellow powder obtained was $g-C_3N_4$, named CN.

2.3. Synthesis of β-Co(OH)$_2$

Co(NO$_3$)$_2$·6H$_2$O and (C$_6$H$_9$NO)$_n$ (PVP, M.W. ≈ 55,000) were thoroughly mixed in absolute ethanol and deionized water for 1 h. The mixture was then transferred to a 25 mL Teflon-lined autoclave. It was reacted for 12 h at 473 K before being cooled to room temperature. The pink product was washed several times with deionized water and anhydrous ethanol until the pH of the filtrate reached neutral, and then vacuum dried for 14 h.

2.4. Synthesis of Co$_3$O$_4$

The β-Co(OH)$_2$ precursor was heated in a tube furnace at a rate of 5 K/min and kept at 673 K for 2 h. The obtained product was the labeled Co$_3$O$_4$ nanosheet.

2.5. Synthesis of Co$_3$O$_4$/g-C$_3$N$_4$

The deionized water was added into the above-prepared Co$_3$O$_4$ and g-C3N4 and mixed with stirring, and a series of Co$_3$O$_4$/g-C$_3$N$_4$ mixture samples with different ratios were synthesized by adjusting the mass ratio between Co$_3$O$_4$ and g-C$_3$N$_4$. After being rapidly frozen with liquid nitrogen, the samples were dried in a freeze-drying oven for 72 h.

3. Results and Discussion

3.1. Microscopic Morphology and Chemical Structure Characterization

The synthesis route of the Co$_3$O$_4$/g-C$_3$N$_4$ sample is displayed in Figure 1. Here, urea was thermally oxidized to obtain g-C$_3$N$_4$ sample. At the same time, β-Co(OH)$_2$ was prepared by solvothermal reaction as a precursor of Co$_3$O$_4$. Eventually, the Co$_3$O$_4$/g-C$_3$N$_4$ heterojunction was produced by liquid nitrogen-assisted thermal oxidation.

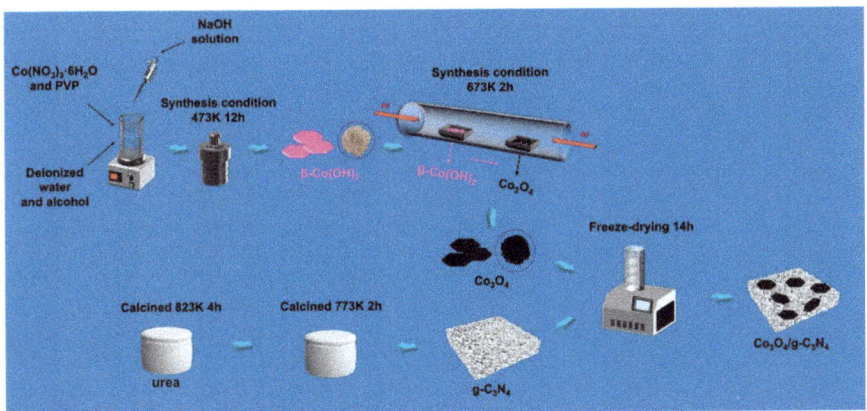

Figure 1. Schematic graph of the synthesis route of Co$_3$O$_4$/g-C$_3$N$_4$.

X-ray diffraction (XRD) was used to analyze crystallographic structures of samples in Figure 2a. It can be seen that g-C$_3$N$_4$ has broad peaks at 13.2° and 27.6°, corresponding to the (100) and (002) crystal planes of g-C$_3$N$_4$ (JCPDS No. 87-1526), and the Co$_3$O$_4$/g-C$_3$N$_4$ catalyst exhibits only very weak Co$_3$O$_4$ diffraction peaks due to the low loading percentage of the Co$_3$O$_4$ catalyst (JCPDS No. 09-0418). Five characteristic peaks were identified at 2θ = 31.3° (d = 2.86 Å), 36.85° (d = 2.44 Å), 55.64° (d = 1.65 Å), 59.35° (d = 1.56 Å), and 65.22° (d = 1.43 Å) corresponding to (220), (311), (422), (511), and (440) as cubic Co$_3$O$_4$ crystal faces [11]. The transmission electron microscopy (TEM) image of 5% Co$_3$O$_4$/g-C$_3$N$_4$ was shown in Figure 2b, where Co$_3$O$_4$ nanosheets of about 150–200 nm in size can be observed on the surface of g-C$_3$N$_4$. HRTEM and corresponding FFT studies were performed for 5% Co$_3$O$_4$/g-C$_3$N$_4$ (Figure 2c,d), and the lattice stripes with spacing of 0.285 and 0.466 nm correspond to the (220) and (111) crystal planes of Co_3O_4 (JCPDS No. 09-0418) [20]. In summary, a clear interface existed between Co_3O_4 and g-C_3N_4, and the interfacial contact facilitates the rapid transfer of photogenerated charges.

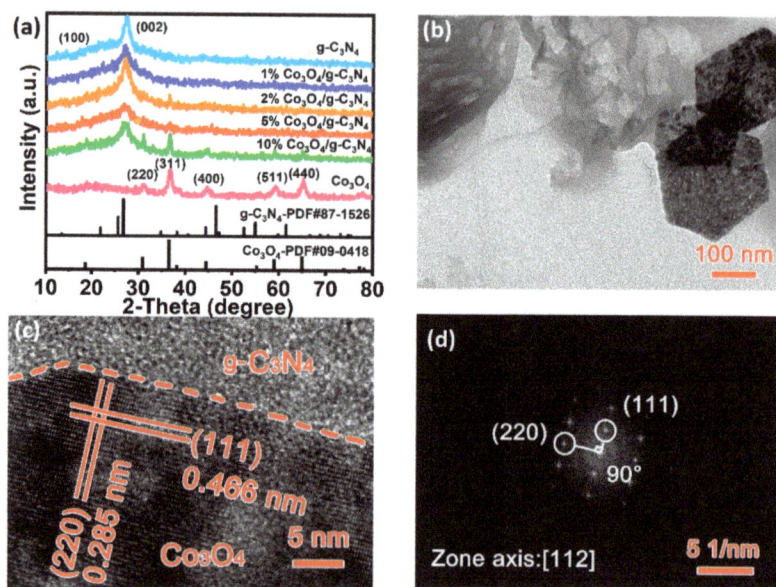

Figure 2. (a) XRD patterns; (b) TEM; (c,d) HRTEM and corresponding FFT images of 5% Co_3O_4/g-C_3N_4.

The analysis of X-ray photoelectron spectroscopy (XPS) provides insight into the surface composition and chemical changes in each sample. From the full survey spectra of samples in Figure 3a, it was found that Co, C, N, and O elements were detected in 5% Co_3O_4/g-C_3N_4, and the molar ratio of C:N:O:Co in 5% Co_3O_4/g-C_3N_4 was about 48:49.6:2:0.3, which further confirmed the complexation of Co_3O_4 with g-C_3N_4. The high-resolution XPS spectra of the C 1s spectra at energies of 288.11 eV, 286.54 eV, and 284.66 eV belong to the C-O bond and the N-C=N bond in Figures 3b and S1. The peaks of the N 1s spectra (Figure 3c) are located at 404.71 eV, 400.47 eV, 399.06 eV, and 398.43 eV, respectively, which can be attributed to sp2-hybridized nitrogen C-N=C, tertiary nitrogen N-(C)$_3$, and primary nitrogen H-N-(C)$_2$ [10,12]. The peaks of the O 1s spectra can be shown in Figure 3d, except peaks at 530.38 and 529.17 eV and at 532.35 and 531.42 eV can be found, which originate from the O-C=O and C-O groups produced at the interface of Co_3O_4 and g-C_3N_4 [33,36]. The Co 1s energy spectra of Co_3O_4 and the Co 2p energy spectra of 5% Co_3O_4/g-C_3N_4 samples (Figure 3e) showed four characteristic peaks at 795.13 eV, 794.03 eV, 780.03 eV, and 778.68 eV, which can be ascribed to the Co-O and Co=O bonds [27,37]. The effect of photocatalytic degradation is influenced by the specific surface area of the material, and the surface area of different samples was investigated by the N_2 adsorption–desorption technique (BET). The g-C_3N_4 exhibits a typical type IV isotherm with H3-type hysteresis loops, and its mesoporous structure may be due to the stacking of the g-C_3N_4 (Figure 3f). The surface area of g-C_3N_4 is about 128.5 m^2/g as calculated by the model that comes with the instrument. The higher specific surface area is attributed to the large-scale nanosheet morphology of g-C_3N_4. The specific surface area of the 5% Co_3O_4/g-C_3N_4 composite was slightly decreased after combining with Co_3O_4, probably since the decrease in specific surface area caused by the interfatial contact between Co_3O_4 and g-C_3N_4. The interfacial effect of 5% Co_3O_4/g-C_3N_4 promotes the adsorption of pollutants by the catalyst, and the abundant active sites are conducive to efficient photocatalytic reactions.

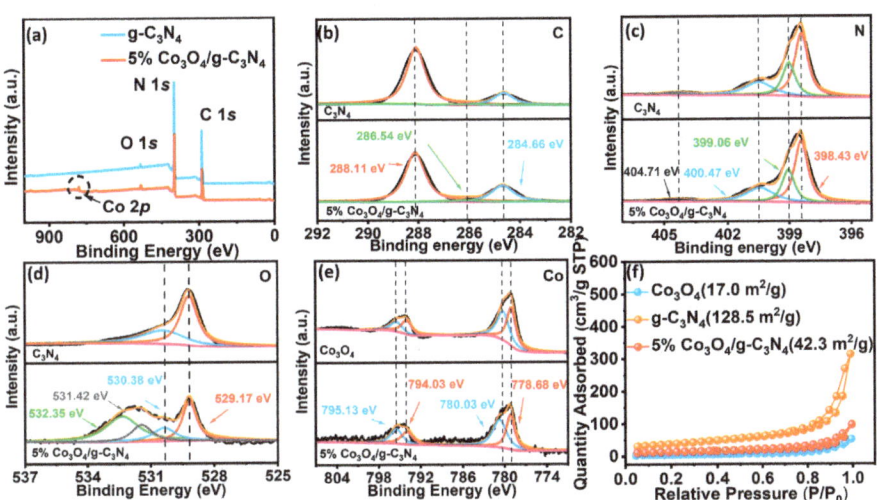

Figure 3. XPS spectra of (**a**) samples; (**b**) C 1s; (**c**) N 1s; (**d**) O 1s; (**e**) Co 2p; (**f**) N$_2$ adsorption–desorption technique.

3.2. Performance Analysis and Kinetics Study of RhB Degradation by Photocatalytic Application

The photodegradation RhB activity of different proportions of samples is usually tested under simulated sunlight conditions. As shown in Figure 4a, the degradation effect of RhB after 40 min under different sample light conditions, demonstrates that the heterogeneous combination of g-C$_3$N$_4$ and Co$_3$O$_4$ effectively promoted the photocatalytic reaction. Among them, the degradation of RhB by 5% Co$_3$O$_4$/g-C$_3$N$_4$ can reach 97.6%. At low concentrations, more Co$_3$O$_4$ facilitates the rapid carrier transfer and promotes the photocatalytic degradation reaction. However, when the concentration is high, Co$_3$O$_4$ covers the surface of g-C$_3$N$_4$, which hinders its light absorption and obscures the active site, thus causing a decrease in the reaction activity. Figure 4b shows the variation of different proportions in the samples, photocatalytic degradation of RhB over time, which more clearly confirms that the catalytic ability of 5% Co$_3$O$_4$/g-C$_3$N$_4$ is stronger than additional two monomeric catalysts. Based on the above characterization, a reaction kinetic model was established (Figure 4c), and the perfect linear relationship between $\ln(C_0/C)$ and irradiation time indicates that the photocatalytic reaction is the quasi-primary reaction; g-C$_3$N$_4$, Co$_3$O$_4$ and 5% Co$_3$O$_4$/g-C$_3$N$_4$ have rate constants k values of 0.024 min^{-1}, 0.0126 min^{-1}, and 0.0703 min^{-1}, respectively. The degradation efficiency of 5% Co$_3$O$_4$/g-C$_3$N$_4$ is approximately 3 times that of g-C$_3$N$_4$ and 5.58 times that of Co$_3$O$_4$. The Co$_3$O$_4$/g-C$_3$N$_4$ exhibited better photocatalytic activity than most of the reported photoreduction systems under similar conditions (Table S1). In addition, the repeatability of the 5% Co$_3$O$_4$/g-C$_3$N$_4$ material was tested in Figure 4d. It can be shown that the performance of 5% Co$_3$O$_4$/g-C$_3$N$_4$ did not show significant degradation after three cycles, which proved the excellent stability of the composite through interfacial compounding.

Testing the UV–vis diffuse reflectance spectroscopy (DRS) of catalysts can provide insight into their light absorption capabilities and help in studying their optical properties. It can be seen from Figure 5a, the absorption edge of g-C$_3$N$_4$ is about 450 nm, and there is almost no response in visible region beyond 450 nm. However, the absorption of 5% Co$_3$O$_4$/g-C$_3$N$_4$ is significantly stronger in visible light due to the interfacial effect formed between Co$_3$O$_4$ and g-C$_3$N$_4$, which helps to improve the photocatalytic activity of the catalyst [38–40]. The photoluminescence (PL) spectra show that the fluorescence intensities of g-C$_3$N$_4$ and Co$_3$O$_4$ were significantly higher than 5% Co$_3$O$_4$/g-C$_3$N$_4$, which indicates a higher complexation rate of photogenerated carriers on Co$_3$O$_4$ and g-C$_3$N$_4$ [41,42] (Figure 5b). To further demonstrate that 5% Co$_3$O$_4$/g-C$_3$N$_4$ has better photogenerated charge separation efficiency, the time-dependent photocurrent of samples was analyzed. The 5% Co_3O_4/g-C_3N_4 catalyst exhibited a higher photocurrent response intensity compared with single g-C_3N_4, which indicates that the composite catalyst promotes the separation and transfer of photogenerated electron–hole pairs in Figure 5c. Furthermore, the 5% Co_3O_4/g-C_3N_4 catalyst also has a smaller arc radius in the Nyquist plot of electrochemical impedance spectroscopy (EIS), which further suggests that the 5% Co_3O_4/g-C_3N_4 catalyst has better photogenerated carrier separation efficiency (Figures 5d and S2) [43,44]. Therefore, the stronger photocurrent response and the smaller charge transfer impedance suggest that the photogenerated electron–hole pairs can be effectively separated in 5% Co_3O_4/g-C_3N_4.

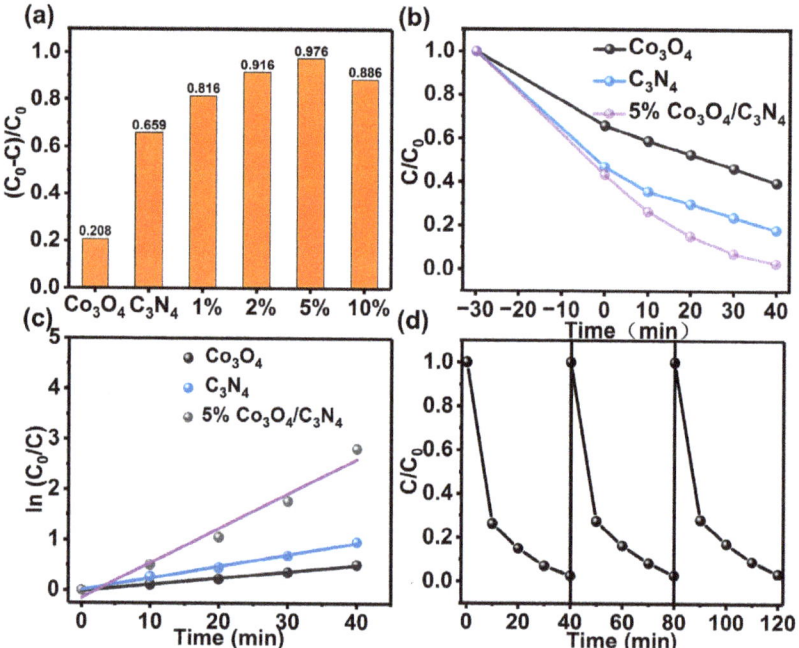

Figure 4. (a) Photodegradation rate of RhB by different samples under simulated sunlight for 40 min; (b) g-C_3N_4, Co_3O_4 and 5% Co_3O_4/g-C_3N_4 photocatalytic degradation of RhB with time; (c) photocatalytic reaction kinetics; (d) stability test of 5% Co_3O_4/g-C_3N_4.

Based on the XPS valence band (XPS-VB) spectral analysis, the VB maxima of Co_3O_4 and g-C_3N_4 can be determined to be -0.15 and 2.17 eV, respectively (Figure 6a); therefore, the conduction band (CB) minima of Co_3O_4 and g-C_3N_4 can be easily calculated as -2.92 and -3.08 eV. By analyzing the DRS spectra, the bandgaps (Eg) of Co_3O_4 and g-C_3N_4 were obtained to be 1.3 and 2.98 eV, respectively (Figure 6b). Based on the above analysis, a type-II heterojunction [2] photocatalytic mechanism is proposed in Figure 6c. The 5% Co_3O_4/g-C_3N_4 photocatalyst was excited beneath light irradiation and generates electron and hole pairs, and transferred the electrons from the CB of Co_3O_4 to g-C_3N_4. Thanks to the intrinsic force field shaped by interface contact between Co_3O_4 and g-C_3N_4, whereas the holes on the VB of g-C_3N_4 are often transferred to Co_3O_4. The Co_3O_4 can produce more photogenerated electrons on the CB of g-C_3N_4 that can promote the conversion of superoxide radicals ($\bullet O_2^-$) and accelerate the conversion of RhB to the subsequent mineralization products.

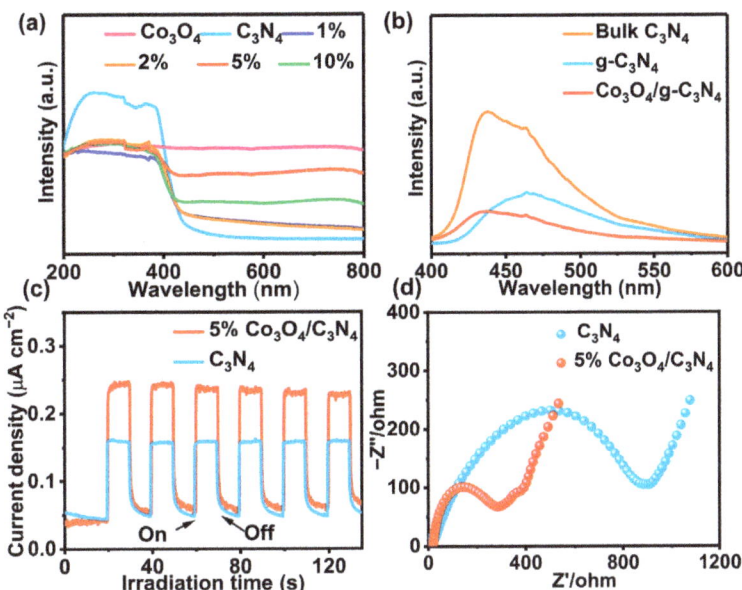

Figure 5. (**a**) DRS spectra, (**b**) PL spectra; (**c**) photocurrent responses, and (**d**) EIS measurement of samples.

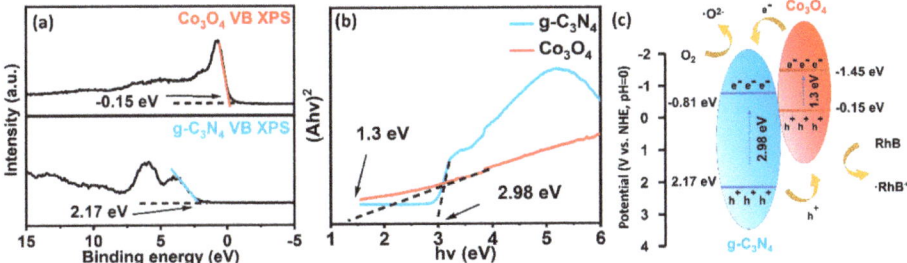

Figure 6. (**a**) XPS-VB spectra of Co_3O_4 and g-C_3N_4; (**b**) Tauc plots of Co_3O_4 and g-C_3N_4; (**c**) Co_3O_4 and g-C_3N_4 electronic band structures and schematic diagram of electrons transfer paths.

To explore the active species in the 5% Co_3O_4/g-C_3N_4 photocatalytic degradation of RhB, a series of free radical capturing experiments was performed (Figure 7a). Tertiary butanol (TBA), triethanolamine (TEOA), and benzoquinone (BQ) were used as the capture agents of •OH, h^+ and •O_2^-. After 40 min of light irradiation, it was found that the degradation efficiency of RhB by 5% Co_3O_4/g-C_3N_4 was significantly inhibited by the addition of TEOA and BQ, while the degradation effect did not change significantly after the addition of TBA. The radical capturing experiments indicated that the active species within the degradation of RhB by 5% Co_3O_4/g-C_3N_4 were in the main •O_2^- and h^+, however not •OH. In order to further verify the results, an electron spin resonance (ESR) analysis was carried out. After 10 min of irradiation with a Xe lamp (300 W), the 5% Co_3O_4/g-C_3N_4 surface produced strong •O_2^- and h^+ signals (Figure 7b,c), while almost no signal of •OH appeared (Figure 7d), proving that •O_2^- and h^+ are the main reactive groups in the reaction system, which is in line with the results of the radical capturing experiments. In addition, it was often found that the signals of •O_2^- and h^+ on the surface of 5% Co_3O_4/g-C_3N_4 significantly exceeded those of g-C_3N_4, indicating that the created heterojunction will higher separate the photogenerated carriers, confirming the results of the previous analysis.

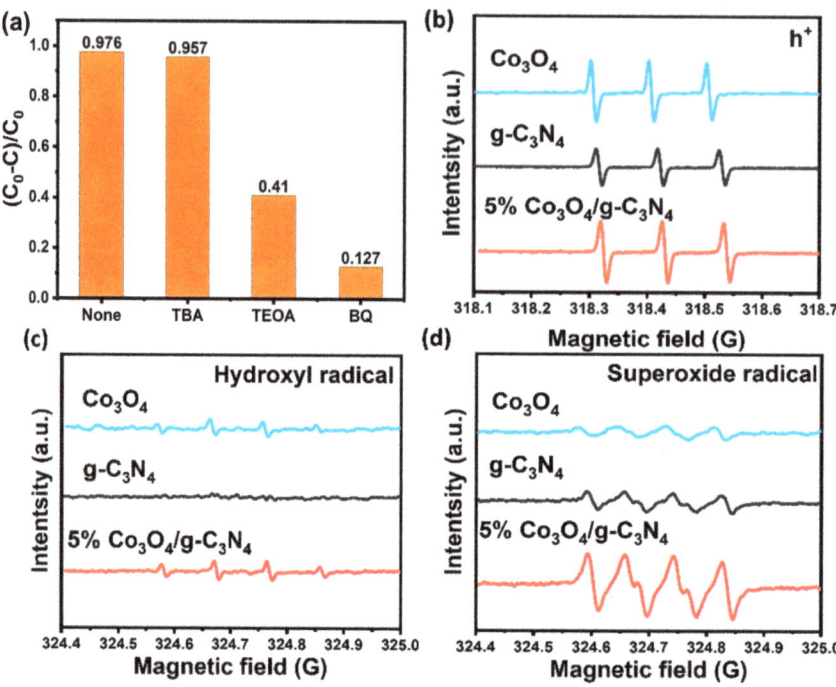

Figure 7. (**a**) Photodegradation rate of RhB by 5% Co_3O_4/g-C_3N_4 under different scavengers within 40 min; ESR spectra of g-C_3N_4, Co_3O_4 and 5% Co_3O_4/g-C_3N_4 (**b**) holes; (**c**) hydroxyl radicals; (**d**) superoxide radicals.

4. Conclusions

In summary, a novel type-II heterojunction photocatalyst (Co_3O_4/g-C_3N_4) was prepared by simple hydrothermal and calcination methods and used to efficiently degrade RhB in water. The experimental results showed that the Co_3O_4/g-C_3N_4 photocatalyst had robust photocatalytic degradation activity toward RhB under light irradiation. The DRS, PL, time-dependent photocurrent, and EIS analyses revealed that the type-II heterojunction structure effectively reduced the composite rate of photogenerated electrons and holes, and therefore the holes in VB of Co_3O_4 and therefore the electrons in CB of g-C_3N_4 were utilized to reinforce the oxidation–reduction ability of the photocatalyst, which resulted in the speedy degradation of pollutants. Among them, the best degradation potency was achieved by a 5% Co_3O_4/g-C_3N_4 photocatalyst, and the RhB degradation potency was increased by 48% compared with the g-C_3N_4 photocatalyst. This study provides some reference data for the development of different heterojunction photocatalysts for the degradation of organic pollutants.

Supplementary Materials: The following supporting information can be downloaded at: https://www.mdpi.com/article/10.3390/ma16103879/s1. Characterization of the photocatalysts. Photoelectrochemical test. Photocatalytic degradation activity test. Figure S1. The high-resolution O 1s spectra of Co_3O_4. Figure S2. EIS spectra under light conditions. Table S1. Common photocatalyst and the effect of degrading organic pollutants in water. Refs [45–54] were cited in the Supplementary Materials.

Author Contributions: Y.D.: Methodology, Conceptualization, Investigation, Data curation, Writing—original draft. Z.F.: Investigation, Data curation, Writing—review and editing. K.Z.: Writing—review and editing. J.T.: Writing—review and editing, Investigation. G.W.: Writing—review and editing. Q.L.: Writing—review and editing. Z.W.: Writing—review and editing. Y.H.: Investigation. J.L.:

Project administration, Methodology. H.X.: Resources, Funding acquisition, Methodology, Writing—review and editing. X.Z.: Supervision, Project administration, Writing—review and editing, Funding acquisition. All authors have read and agreed to the published version of the manuscript.

Funding: This study was funded by the Natural Science Foundation of Jiangsu Province (BK20220598), Key Laboratory of Electrochemical Energy Storage and Energy Conversion of Hainan Province (KFKT2022001).

Institutional Review Board Statement: Not applicable.

Informed Consent Statement: Not applicable.

Data Availability Statement: Raw data are available upon request.

Conflicts of Interest: The authors declare that they have no known competing financial interest or personal relationships that could have appeared to influence the work reported in this paper.

References

1. Chen, C.; Ma, W.; Zhao, J. Semiconductor-mediated photodegradation of pollutants under visible-light irradiation. *Chem. Soc. Rev.* **2010**, *39*, 4206–4219. [CrossRef]
2. Li, X.; Yu, J.; Jaroniec, M. Hierarchical photocatalysts. *Chem. Soc. Rev.* **2016**, *45*, 2603–2636. [CrossRef]
3. Jing, L.; Xu, Y.; Liu, J.; Zhou, M.; Xu, H.; Xie, M.; Li, H.; Xie, J. Direct Z-scheme red carbon nitride/rod-like lanthanum vanadate composites with enhanced photodegradation of antibiotic contaminants. *Appl. Catal. B Environ.* **2020**, *277*, 119245. [CrossRef]
4. Jing, L.; He, M.; Xie, M.; Song, Y.; Wei, W.; Xu, Y.; Xu, H.; Li, H. Realizing the synergistic effect of electronic modulation over graphitic carbon nitride for highly efficient photodegradation of bisphenol A and 2-mercaptobenzothiazole: Mechanism, degradation pathway and density functional theory calculation. *J. Colloid Interface Sci.* **2021**, *583*, 113–127. [CrossRef] [PubMed]
5. Jing, L.; Xu, Y.; Zhou, M.; Deng, J.; Wei, W.; Xie, M.; Song, Y.; Xu, H.; Li, H. Novel broad-spectrum-driven oxygen-linked band and porous defect co-modified orange carbon nitride for photodegradation of Bisphenol A and 2-Mercaptobenzothiazole. *J. Hazard. Mater.* **2020**, *396*, 122659. [CrossRef]
6. Kubacka, A.; Fernández-García, M.; Colón, G. Advanced nanoarchitectures for solar photocatalytic applications. *Chem. Rev.* **2012**, *112*, 1555–1614. [CrossRef]
7. Schneider, J.; Matsuoka, M.; Takeuchi, M.; Zhang, J.; Horiuchi, Y.; Anpo, M.; Bahnemann, D.W. Understanding TiO_2 photocatalysis: Mechanisms and materials. *Chem. Rev.* **2014**, *114*, 9919–9986. [CrossRef] [PubMed]
8. Wang, L.; Li, D.-B.; Li, K.; Chen, C.; Deng, H.-X.; Gao, L.; Zhao, Y.; Jiang, F.; Li, L.; Huang, F.; et al. Stable 6%-efficient Sb2Se3 solar cells with a ZnO buffer layer. *Nat. Energy* **2017**, *2*, 17046. [CrossRef]
9. Fu, J.; Jiang, K.; Qiu, X.; Yu, J.; Liu, M. Product Cocatalysts of photocatalytic CO_2 reduction reactions. *Mater. Today* **2020**, *32*, 222–243. [CrossRef]
10. Zhu, X.; Yang, J.; She, X.; Song, Y.; Qian, J.; Wang, Y.; Xu, H.; Li, H.; Yan, Q. Rapid synthesis of ultrathin 2D materials through liquid-nitrogen and microwave treatments. *J. Mater. Chem. A* **2019**, *7*, 5209–5213. [CrossRef]
11. Zhu, X.; Ji, H.; Yi, J.; Yang, J.; She, X.; Ding, P.; Li, L.; Deng, J.; Qian, J.; Xu, H.; et al. A Specifically Exposed Cobalt Oxide/Carbon Nitride 2D Heterostructure for Carbon Dioxide Photoreduction. *Ind. Eng. Chem. Res.* **2018**, *57*, 17394–17400. [CrossRef]
12. Mo, Z.; Zhu, X.; Jiang, Z.; Song, Y.; Liu, D.; Li, H.; Yang, X.; She, Y.; Lei, Y.; Yuan, S.; et al. Porous nitrogen-rich g-C_3N_4 nanotubes for efficient photocatalytic CO_2 reduction. *Appl. Catal. B Environ.* **2019**, *256*, 117854. [CrossRef]
13. Zhou, X.; Zhang, X.; Ding, P.; Zhong, K.; Yang, J.; Yi, J.; Hu, Q.; Zhou, G.; Wang, X.; Xu, H.; et al. Simultaneous manipulation of scalable absorbance and the electronic bridge for efficient CO_2 photoreduction. *J. Mater. Chem. A* **2022**, *10*, 25661–25670. [CrossRef]
14. Yang, J.; Jing, L.; Zhu, X.; Zhang, W.; Deng, J.; She, Y.; Nie, K.; Wei, Y.; Li, H.; Xu, H. Modulating electronic structure of lattice O-modified orange polymeric carbon nitrogen to promote photocatalytic CO_2 conversion. *Appl. Catal. B Environ.* **2023**, *320*, 122005. [CrossRef]
15. Zhu, C.; Xian, Q.; He, Q.; Chen, C.; Zou, W.; Sun, C.; Wang, S.; Duan, X. Edge-Rich Bicrystalline 1T/2H-MoS_2 Cocatalyst-Decorated {110} Terminated CeO_2 Nanorods for Photocatalytic Hydrogen Evolution. *ACS Appl. Mater. Interfaces* **2021**, *13*, 35818–35827. [CrossRef] [PubMed]
16. Xu, H.; Yi, J.; She, X.; Liu, Q.; Song, L.; Chen, S.; Yang, Y.; Song, Y.; Vajtai, R.; Lou, J.; et al. 2D heterostructure comprised of metallic 1T-MoS_2/Monolayer O-g-C_3N_4 towards efficient photocatalytic hydrogen evolution. *Appl. Catal. B Environ.* **2018**, *220*, 379–385. [CrossRef]
17. Zhu, Z.; Li, X.; Qu, Y.; Zhou, F.; Wang, Z.; Wang, W.; Zhao, C.; Wang, H.; Li, L.; Yao, Y.; et al. A hierarchical heterostructure of CdS QDs confined on 3D $ZnIn_2S_4$ with boosted charge transfer for photocatalytic CO_2 reduction. *Nano Res.* **2021**, *14*, 81–90. [CrossRef]
18. Naya, S.-I.; Kume, T.; Akashi, R.; Fujishima, M.; Tada, H. Red-Light-Driven Water Splitting by Au(Core)-CdS(Shell) Half-Cut Nanoegg with Heteroepitaxial Junction. *J. Am. Chem. Soc.* **2018**, *140*, 1251–1254. [CrossRef]
19. Li, J.-X.; Ye, C.; Li, X.-B.; Li, Z.-J.; Gao, X.-W.; Chen, B.; Tung, C.-H.; Wu, L.-Z. A Redox Shuttle Accelerates O_2 Evolution of Photocatalysts Formed In Situ under Visible Light. *Adv. Mater.* **2017**, *29*, 1606009. [CrossRef]

20. Hu, L.; Peng, Q.; Li, Y. Selective Synthesis of Co_3O_4 Nanocrystal with Different Shape and Crystal Plane Effect on Catalytic Property for Methane Combustion. *J. Am. Chem. Soc.* **2008**, *130*, 16136–16137. [CrossRef]
21. Liu, G.; Wang, L.; Chen, X.; Zhu, X.; Wang, B.; Xu, X.; Chen, Z.; Zhu, W.; Li, H.; Xia, J. Crafting of plasmonic Au nanoparticles coupled ultrathin BiOBr nanosheets heterostructure: Steering charge transfer for efficient CO_2 photoreduction. *Green Chem. Eng.* **2022**, *3*, 157–164. [CrossRef]
22. Liu, G.; Wang, L.; Wang, B.; Zhu, X.; Yang, J.; Liu, P.; Zhu, W.; Chen, Z.; Xia, J. Synchronous activation of Ag nanoparticles and BiOBr for boosting solar-driven CO_2 reduction. *Chin. Chem. Lett.* **2023**, *34*, 157–164. [CrossRef]
23. Wu, H.; Li, C.; Che, H.; Hu, H.; Hu, W.; Liu, C.; Ai, J.; Dong, H. Decoration of mesoporous Co_3O_4 nanospheres assembled by monocrystal nanodots on g-C_3N_4 to construct Z-scheme system for improving photocatalytic performance. *Appl. Surf. Sci.* **2018**, *440*, 308–319. [CrossRef]
24. Guo, S.; Zhao, S.; Wu, X.; Li, H.; Zhou, Y.; Zhu, C.; Yang, N.; Jiang, X.; Gao, J.; Bai, L.; et al. A Co_3O_4-CDots-C_3N_4 three component electrocatalyst design concept for efficient and tunable CO_2 reduction to syngas. *Nat. Commun.* **2017**, *8*, 1828. [CrossRef] [PubMed]
25. Tian, J.; Zhong, K.; Zhu, X.; Yang, J.; Mo, Z.; Liu, J.; Dai, J.; She, Y.; Song, Y.; Li, H.; et al. Highly exposed active sites of Au nanoclusters for photocatalytic CO_2 reduction. *Chem. Eng. J.* **2023**, *451*, 138392. [CrossRef]
26. Zhu, X.; Zhou, G.; Yi, J.; Ding, P.; Yang, J.; Zhong, K.; Song, Y.; Hua, Y.; Zhu, X.; Yuan, J.; et al. Accelerated Photoreduction of CO_2 to CO over a Stable Heterostructure with a Seamless Interface. *ACS Appl. Mater. Interfaces* **2021**, *13*, 39523–39532. [CrossRef]
27. Si, J.; Xiao, S.; Wang, Y.; Zhu, L.; Xia, X.; Huang, Z.; Gao, Y. Sub-nanometer Co_3O_4 clusters anchored on TiO_2(B) nano-sheets: Pt replaceable Co-catalysts for H_2 evolution. *Nanoscale* **2018**, *10*, 2596–2602. [CrossRef]
28. Zhu, X.; Yang, J.; Zhu, X.; Yuan, J.; Zhou, M.; She, X.; Yu, Q.; Song, Y.; She, Y.; Hua, Y.; et al. Exploring deep effects of atomic vacancies on activating CO_2 photoreduction via rationally designing indium oxide photocatalysts. *Chem. Eng. J.* **2021**, *422*, 129888. [CrossRef]
29. She, X.; Xu, H.; Li, L.; Mo, Z.; Zhu, X.; Yu, Y.; Song, Y.; Wu, J.; Qian, J.; Yuan, S.; et al. Steering charge transfer for boosting photocatalytic H_2 evolution: Integration of two-dimensional semiconductor superiorities and noble-metal-free Schottky junction effect. *Appl. Catal. B Environ.* **2019**, *245*, 477–485. [CrossRef]
30. Han, Q.; Ge, L.; Chen, C.; Li, Y.; Xiao, X.; Zhang, Y.; Guo, L. Novel visible light induced Co_3O_4-g-C_3N_4 heterojunction photocatalysts for efficient degradation of methyl orange. *Appl. Catal. B Environ.* **2014**, *147*, 546–553. [CrossRef]
31. Cao, J.; Zhang, N.; Wang, S.; Zhang, H. Electronic structure-dependent formaldehyde gas sensing performance of the In_2O_3/Co_3O_4 core/shell hierarchical heterostructure sensors. *J. Colloid Interface Sci.* **2020**, *577*, 19–28. [CrossRef] [PubMed]
32. Hu, L.; Zhang, G.; Liu, M.; Wang, Q.; Wang, P. Enhanced degradation of Bisphenol A (BPA) by peroxymonosulfate with Co_3O_4-Bi_2O_3 catalyst activation: Effects of pH, inorganic anions, and water matrix. *Chem. Eng. J.* **2018**, *338*, 300–310. [CrossRef]
33. Mo, Z.; Xu, H.; Chen, Z.; She, X.; Song, Y.; Lian, J.; Zhu, X.; Yan, P.; Lei, Y.; Yuan, S.; et al. Construction of MnO_2/Monolayer g-C_3N_4 with Mn vacancies for Z-scheme overall water splitting. *Appl. Catal. B Environ.* **2019**, *241*, 452–460. [CrossRef]
34. Zhu, X.; Liu, J.; Zhao, Y.; Yan, J.; Xu, Y.; Song, Y.; Ji, H.; Xu, H.; Li, H. Hydrothermal synthesis of mpg-C_3N_4 and Bi_2WO_6 nest-like structure nanohybrids with enhanced visible light photocatalytic activities. *RSC Adv.* **2017**, *7*, 38682–38690. [CrossRef]
35. Jing, L.; Xu, Y.; Chen, Z.; He, M.; Xie, M.; Liu, J.; Xu, H.; Huang, S.; Li, H. Different Morphologies of SnS_2 Supported on 2D g-C_3N_4 for Excellent and Stable Visible Light Photocatalytic Hydrogen Generation. *ACS Sustain. Chem. Eng.* **2018**, *6*, 5132–5141. [CrossRef]
36. She, X.; Wu, J.; Zhong, J.; Xu, H.; Yang, Y.; Vajtai, R.; Lou, J.; Liu, Y.; Du, D.; Li, H.; et al. Ajayan, Oxygenated monolayer carbon nitride for excellent photocatalytic hydrogen evolution and external quantum efficiency. *Nano Energy* **2016**, *27*, 138–146. [CrossRef]
37. Dong, G.; Zhang, Y.; Bi, Y. The synergistic effect of Bi_2WO_6 nanoplates and Co_3O_4 cocatalysts for enhanced photoelectrochemical properties. *J. Mater. Chem. A* **2017**, *5*, 20594–20597. [CrossRef]
38. Yang, J.; Zhu, X.; Yu, Q.; He, M.; Zhang, W.; Mo, Z.; Yuan, J.; She, Y.; Xu, H.; Li, H. Multidimensional In_2O_3/In_2S_3 heterojunction with lattice distortion for CO_2 photoconversion. *Chin. J. Catal.* **2022**, *43*, 1286–1294. [CrossRef]
39. Yang, J.; Zhu, X.; Mo, Z.; Yi, J.; Yan, J.; Deng, J.; She, Y.; Qian, J.; Xu, H.; et al. A multidimensional In_2S_3–$CuInS_2$ heterostructure for photocatalytic carbon dioxide reduction. *Inorg. Chem. Front.* **2018**, *5*, 3163–3169. [CrossRef]
40. Mo, Z.; Xu, H.; She, X.; Song, Y.; Yan, P.; Yi, J.; Zhu, X.; Lei, Y.; Yuan, S.; Li, H. Constructing Pd/2D-C_3N_4 composites for efficient photocatalytic H_2 evolution through nonplasmon-induced bound electrons. *Appl. Surf. Sci.* **2019**, *467–468*, 151–157. [CrossRef]
41. Zhang, G.; Zhu, X.; Chen, D.; Li, N.; Xu, Q.; Li, H.; He, J.; Xu, H.; Lu, J. Hierarchical Z-scheme g-C_3N_4/Au/$ZnIn_2S_4$ photocatalyst for highly enhanced visible-light photocatalytic nitric oxide removal and carbon dioxide conversion. *Environ. Sci. Nano* **2020**, *7*, 676–687. [CrossRef]
42. Yan, P.; She, X.; Zhen, X.; Xu, L.; Qian, J.; Xia, J.; Zhang, Y.; Xu, H.; Li, H. Efficient photocatalytic hydrogen evolution by engineering amino groups into ultrathin 2D graphitic carbon nitride. *Appl. Surf. Sci.* **2020**, *507*, 145085. [CrossRef]
43. Li, L.; Yi, J.; Zhu, X.; Zhou, M.; Zhang, S.; She, X.; Chen, Z.; Li, H.-M.; Xu, H. Nitriding Nickel-Based Cocatalyst: A Strategy To Maneuver Hydrogen Evolution Capacity for Enhanced Photocatalysis. *ACS Sustain. Chem. Eng.* **2019**, *8*, 884–892. [CrossRef]
44. Li, Q.; Zhu, X.; Yang, J.; Yu, Q.; Zhu, X.; Chu, J.; Du, Y.; Wang, C.; Hua, Y.; Li, H.; et al. Plasma treated Bi_2WO_6 ultrathin nanosheets with oxygen vacancies for improved photocatalytic CO_2 reduction. *Inorg. Chem. Front.* **2020**, *7*, 597–602. [CrossRef]
45. Feng, J.; Zhang, D.; Zhou, H.; Pi, M.; Wang, X.; Chen, S. Coupling P Nanostructures with P-Doped g-C_3N_4 As Efficient Visible Light Photocatalysts for H_2 Evolution and RhB Degradation. *ACS Sustain. Chem. Eng.* **2018**, *6*, 6342–6349. [CrossRef]

46. Liu, G.; Liao, M.; Zhang, Z.; Wang, H.; Chen, D.; Feng, Y. Enhanced photodegradation performance of Rhodamine B with g-C_3N_4 modified by carbon nanotubes. *Sep. Purif. Technol.* **2020**, *244*, 116618. [CrossRef]
47. Li, W.; Wang, Z.; Li, Y.; Ghasemi, J.B.; Li, J.; Zhang, G. Visible-NIR light-responsive 0D/2D CQDs/Sb_2WO_6 nanosheets with enhanced photocatalytic degradation performance of RhB: Unveiling the dual roles of CQDs and mechanism study. *J. Hazard. Mater.* **2022**, *424*, 127595. [CrossRef] [PubMed]
48. Hu, Q.; Dong, J.; Chen, Y.; Yi, J.; Xia, J.; Yin, S.; Li, H. In-situ construction of bifunctional MIL-125(Ti)/BiOI reactive adsorbent/photocatalyst with enhanced removal efficiency of organic contaminants. *Appl. Surf. Sci.* **2022**, *583*, 152423. [CrossRef]
49. Nandigana, P.; Mahato, S.; Dhandapani, M.; Pradhan, B.; Subramanian, B.; Panda, S.K. Lyophilized tin-doped MoS_2 as an efficient photocatalyst for overall degradation of Rhodamine B dye. *J. Alloys Compd.* **2022**, *907*, 164470. [CrossRef]
50. Preetha, R.; Govinda raj, M.; Vijayakumar, E.; Narendran, M.G.; Varathan, E.; Neppolian, B.; Jeyapaul, U.; John Bosco, A. Promoting photocatalytic interaction of boron doped reduced graphene oxide supported $BiFeO_3$ nanocomposite for visible-light-induced organic pollutant degradation. *J. Alloys Compd.* **2022**, *904*, 164038. [CrossRef]
51. Chen, Y.; Su, X.; Ma, M.; Hou, Y.; Lu, C.; Wan, F.; Ma, Y.; Xu, Z.; Liu, Q.; Hao, M.; et al. One-dimensional magnetic flower-like $CoFe_2O_4$@Bi_2WO_6@BiOBr composites for visible-light catalytic degradation of Rhodamine B. *J. Alloys Compd.* **2022**, *929*, 167297. [CrossRef]
52. Naing, H.H.; Li, Y.; Ghasemi, J.B.; Wang, J.; Zhang, G. Enhanced visible-light-driven photocatalysis of in-situ reduced of bismuth on BiOCl nanosheets and montmorillonite loading: Synergistic effect and mechanism insight. *Chemosphere* **2022**, *304*, 135354. [CrossRef] [PubMed]
53. Li, C.; Zhao, Y.; Fan, J.; Hu, X.; Liu, E.; Yu, Q. Nanoarchitectonics of S-scheme 0D/2D $SbVO_4$/g-C_3N_4 photocatalyst for enhanced pollution degradation and H_2 generation. *J. Alloys Compd.* **2022**, *919*, 165752. [CrossRef]
54. Chen, Y.; Jiang, Y.; Chen, B.; Tang, H.; Li, L.; Ding, Y.; Duan, H.; Wu, D. Insights into the enhanced photocatalytic activity of O-g-C_3N_4 coupled with SnO_2 composites under visible light irradiation. *J. Alloys Compd.* **2022**, *903*, 163739. [CrossRef]

Disclaimer/Publisher's Note: The statements, opinions and data contained in all publications are solely those of the individual author(s) and contributor(s) and not of MDPI and/or the editor(s). MDPI and/or the editor(s) disclaim responsibility for any injury to people or property resulting from any ideas, methods, instructions or products referred to in the content.

Article

Theoretical and Experimental Study of the Photocatalytic Properties of ZnO Semiconductor Nanoparticles Synthesized by *Prosopis laevigata*

Mizael Luque Morales [1], Priscy Alfredo Luque Morales [1], Manuel de Jesús Chinchillas Chinchillas [2], Víctor Manuel Orozco Carmona [3,*], Claudia Mariana Gómez Gutiérrez [1], Alfredo Rafael Vilchis Nestor [4] and Rubén César Villarreal Sánchez [1,*]

1. Facultad de Ingeniería Arquitectura y Diseño, Universidad Autónoma de Baja California, Ensenada 22860, Mexico; mizael.luque@uabc.edu.mx (M.L.M.); pluque@uabc.edu.mx (P.A.L.M.); cmgomezg@uabc.edu.mx (C.M.G.G.)
2. Departamento de Ingeniería y Tecnología, Universidad Autónoma de Occidente, Guasave 81048, Mexico; manuel.chinchillas@uadeo.mx
3. Departamento de Metalurgia e Integridad Estructural, Centro de Investigación en Materiales Avanzados, Chihuahua 31136, Mexico
4. Centro Conjunto de Investigación en Química Sustentable, UAEM-UNAM, Toluca 50200, Mexico; arvilchisn@uaemex.mx
* Correspondence: victor.orozco@cimav.edu.mx (V.M.O.C.); ruben.villarreal@uabc.edu.mx (R.C.V.S.)

Abstract: In this work, the photocatalytic activity of nanoparticles (NPs) of zinc oxide synthetized by *Prosopis laevigata* as a stabilizing agent was evaluated in the degradation of methylene blue (MB) dye under UV radiation. The theoretical study of the photocatalytic degradation process was carried out by a Langmuir–Hinshelwood–Hougen–Watson (LHHW) model. Zinc oxide nanoparticles were synthesized by varying the concentration of natural extract of Prosopis laevigata from 1, 2, and 4% (weight/volume), identifying the samples as ZnO_PL1%, ZnO_PL2%, and ZnO_PL4%, respectively. The characterization of the nanoparticles was carried out by Fourier transform infrared spectroscopy (FT-IR), where the absorption band for the Zn-O vibration at 400 cm^{-1} was presented; by ultraviolet–visible spectroscopy (UV–vis) the value of the band gap was calculated, resulting in 2.80, 2.74 and 2.63 eV for the samples ZnO_PL1%, ZnO_PL2%, and ZnO_PL4%, respectively; XRD analysis indicated that the nanoparticles have a hexagonal zincite crystal structure with an average crystal size of 55, 50, and 49 in the sample ZnO_PL1%, ZnO_PL2%, and ZnO_PL4%, respectively. The morphology observed by TEM showed that the nanoparticles had a hemispherical shape, and the ZnO_PL4% sample presented sizes ranging between 29 and 45 nm. The photocatalytic study showed a total degradation of the MB in 150, 120, and 60 min for the samples ZnO_PL1%, ZnO_PL2%, and ZnO_PL4%, respectively. Also, the model explains the experimental observation of the first-order kinetic model in the limit of low concentrations of dye, indicating the influence of the mass transfer processes.

Keywords: zinc oxide; Langmuir–Hinshelwood–Hougen–Watson model; photocatalysis; methylene blue

1. Introduction

Water is essential for the survival of all living organisms. There is a wide range of harmful organic contaminants present in water organisms, such as pesticides, phenols, detergents, pharmaceuticals, carbohydrates, and organic dyes, among others. In general, water pollution causes eutrophication of plants, damage to aquatic systems, and severe threats to human health, causing diseases [1–3]. Of all the pollutants mentioned above, dyes are an important class of pollutants in the effluents of various industries, such as textile, plastics, food, and paper [4]. The presence of dyes in water, even in minute quantities, is undesirable since most are toxic, mutagenic, and carcinogenic [5].

There are many methods for removing organic pollutants from wastewater, each with its characteristics [6]. However, photocatalytic degradation has been reported to be an environmentally friendly option with much potential. In recent years, photocatalysis has gained significant interest in treating waters with industrial pollutants due to its advantages over other methods, such as waste disposal, low cost, and low environmental impact, among others [7–9]. Among the materials that have been used as photocatalysts, some of the most common are semiconductors, as they have band gap values ranging from 1.5 eV to 3.5 eV, allowing them to conduct electrons in the presence of light [10].

Zinc oxide (ZnO) is a semiconductor material that has attracted significant attention and has become well known for its various advantages over other semiconductor nanoparticle oxides, including low toxicity, ease of availability, chemical stability, and low cost [11–13]. Among the applications that stand out for zinc oxide are: anticancer, antibacterial, and antidiabetic activities, photocatalysis, cosmetics, solar energy harvesting, sensors, pigments, optical filters, sunscreens, dye-sensitized solar cells, etc. [14–18]. Currently, there are various routes for the synthesis of ZnO nanoparticles, such as physical methods (mechanical milling, laser ablation, sputtering, among others) and chemical methods (pyrolysis, sol gel, photochemical reduction, among others) [19,20]. However, most of these methods are expensive, require sophisticated equipment, require high energy consumption, use reagents that are harmful to the environment, generate secondary waste, etc. Another method that has been used in recent years is green synthesis. This process is cheap, fast, and uses natural materials such as leaves, fruit shells, flowers, or stems of plants and trees to reduce metallic salts and obtain nanoparticles with desired characteristics [21]. Using plant extracts brings advantages such as high availability, safety, non-toxic (in most cases), and various phytochemicals that help reduce metal salts and provide stability to nanoparticles [22].

The synthesis of this material by using green chemistry methods has gained importance in recent years. Kamarajan et al., 2022, reported the green synthesis of nanoparticles using *Acalypha indica* showing promising results in the degradation of methylene blue [23]. Sasi et al., 2022, worked on the synthesis of nanoparticles using *Garcinia cambogia* demonstrating promising results in photocatalytic activity degrading methylene blue, crystal violet and phenol red [24]. Finally Vu Nu et al., 2022, reported the use of *Cordia myxa* for the synthesis of nanoparticles and good results were obtained in the photocatalytic degradation of methylene blue [25], among other related works. The mesquite tree (*Prosopis laevigata*) is one of the primary natural resources of the United States and Mexico. It is widely used as animal feed, to obtain wood and coal, and it is used in the construction industry and medicine, among other applications. It has been reported that *Prosopis laevigata* contains sugars, proteins, carbohydrates, fats, crude fiber, phenolic compounds, and polyphenols, among others [26–28]. These phytochemicals actively contribute to the formation of nanoparticles through green synthesis.

What has not yet to be reported in the literature is the use of *Prosopis laevigata* extract in the synthesis of ZnO nanoparticles, nor the evaluation of a theoretical model on the photocatalytic behavior of these nanomaterials. Therefore, in this work, *Prosopis laevigata* extract has been used as a reducing agent in the green synthesis of ZnO nanoparticles, which proved their photocatalytic effectiveness in the degradation of methylene blue (MB) by ultraviolet radiation. Also, the theoretical modeling of the photocatalytic process by a Langmuir–Hinshelwood–Hougen–Watson (LHHW) approach was carried out, and variations in the dye concentration were performed to study the effect of mass transfer phenomena on the degradation kinetics.

2. Materials and Methods

2.1. Materials

For the green synthesis of ZnO NPs, *Prosopis laevigata* (mesquite) leaves were used as a reducing agent (purchased locally), zinc nitrate, $Zn(NO_3)_2 \cdot 6H_2O$, as precursor salt (purchased from Faga Lab, Los Mochis, Mexico), and deionized water as solvent.

2.2. Methodology

2.2.1. Preparation of Extract

For the extraction process of the reducing agent, *Prosopis laevigata* extracts were prepared at concentrations of 1, 2, and 4% (w/v) in 50 mL of deionized water. Previously, *P. laevigata* leaves were cleaned, crushed, and added in deionized water and kept under magnetic agitation for 2 h. Subsequently, they were placed in a thermal bath at 60 °C for 1 h. Finally, they were vacuum filtered using a Watman #4 filter. The extract was stored for later use.

2.2.2. Synthesis of ZnO Nanoparticles

To synthesize ZnO NPs, 2 g of zinc nitrate was added to 42 mL of *P. laevigata* extract (amount of extract obtained after the filtration process). The solutions were magnetically stirred for 1 h at room temperature. Subsequently, the samples were placed in a thermal bath at 60 °C for 13 h until most of the water evaporated and a high viscosity consistency was obtained. Finally, the samples were calcined at 400 °C for 1 h. Figure 1 shows the process carried out in the green synthesis of ZnO nanoparticles using *Prosopis laevigata* extract.

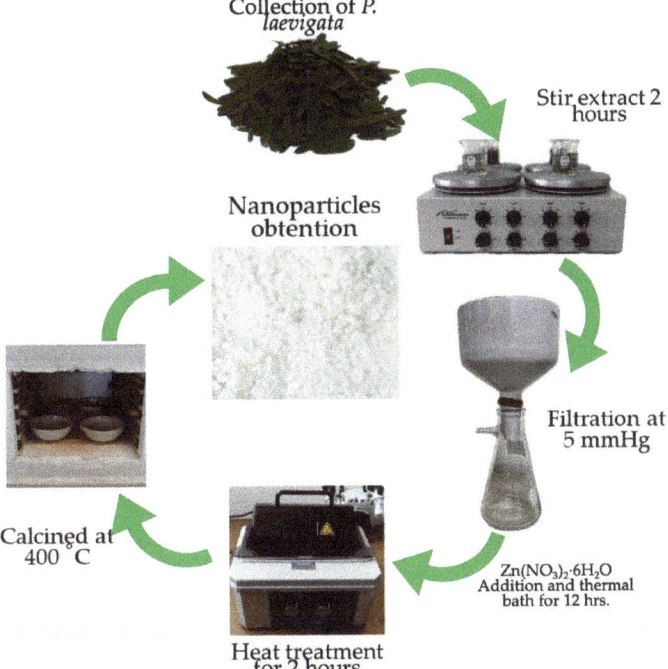

Figure 1. Green synthesis process of ZnO nanoparticles using *Prosopis laevigata* extract.

2.2.3. Characterization

Different techniques were used to study the physical, chemical, and optical properties of the synthesized ZnO NPs to characterize the material. In order to know the functional groups presents in the material, Fourier transform infrared spectroscopy (FT-IR) ATR mode was used, and the spectra were obtained with Perkin Elmer brand equipment (Waltham, MA, USA) at 0.5 cm^{-1} resolution and 400 to 3500 cm^{-1} measurement range. For the study of the optical properties and the band gap of the materials ultraviolet–visible (UV–vis) spectroscopy was used at a wavelength of 200 to 800 nm, using a Perkin Elmer UV/VIS Lambda 365 spectrometer. Crystal structure analysis was determined by X-ray diffractometry (XRD) using a Bruker D2-phase diffractometer (Billerica, MA, USA). The nanoparticles' size, shape, and structure were determined using transmission electron microscopy (TEM) with a JEOL microscope (Tokyo, Japan).

2.2.4. Photocatalytic Activity

The photocatalytic activity of ZnO NPs was analyzed to determine the degradation of methylene blue (MB) under ultraviolet (UV) radiation. The degradation process was carried out inside a stainless-steel reactor, with a length of 26.6 cm and a width of 5 cm, equipped with a UV lamp. To analyze the photocatalytic activity of the different samples, 50 mg of ZnO nanoparticles were added to 50 mL of the MB-contaminated solution (at 15 ppm); consecutively, the samples were shaken for 30 min in the dark to carry out the adsorption–desorption process. Subsequently, the suspensions were irradiated with UV light for 3 h. Aliquots were taken every 10 min during the first 30 min and every 30 min during the following 2 h 30 min. The percentage of MB degradation was determined by UV–vis spectroscopy.

2.2.5. Model

The LHHW model for the degradation kinetic process is based on the following assumptions: (1) The number of adsorption sites on the surface is limited. (2) Each site can adsorb only one molecule, and a maximum of one layer can cover the catalyst surface. (3) The adsorption reaction can be reversible. (4) The catalyst surface is homogeneous. (5) There is no interaction between the adsorbed molecules [29–32]. For this purpose, two stages are taken into account during the degradation of the chemical species, the adsorption stage and the reaction stage, which are defined below:

Adsorption stage:

$$A + l \underset{K_d}{\overset{K_A}{\rightleftarrows}} A.l \tag{1}$$

where A represents a dye molecule, l represents an active site on the catalyst surface, K_A and K_d are kinetic constants.

Reaction stage:

$$A.l \underset{K'_{sr}}{\overset{K_{sr}}{\rightleftarrows}} R.l \tag{2}$$

where A.l represents a molecule A adsorbed on a site l, and R.l represents the degraded product adsorbed on the catalyst, with K_{sr} and K'_{sr} as kinetic constants of the reaction.

The model states that the degradation rate $(-r_A)$ for the adsorbed chemical species A is [33]:

$$-r_A = \frac{B\,C_A}{1 + AC_A} \tag{3}$$

If the desorption process is compared to the adsorption process, then $A = K_A/K_{sr}$ and $B = K_A C_t$; C_A is the concentration of the dye around the catalyst; K_A the adsorption constant; and C_t the concentration of active sites on the catalyst surface.

Integrating Equation (1) and defining the variable $X = C/C_{A0}$, where C_{A0} is the initial concentration, we have for the degradation time:

$$t = -\frac{1}{B}\ln\left(\frac{1}{X}\right) + \frac{AC_{A0}}{B}(1 - X) \tag{4}$$

For the calculation of the constants A and B, an objective function (3) is defined to minimize the difference between the value of X calculated by the model and the value of X measured experimentally.

$$Fobj = \sum_{i=0}^{n}(X_{calculated} - X_{measured})^2 \tag{5}$$

where n is the number of experimental data and $X_{calculated}$ is obtained from numerical solve of Equation (2). The minimization of the objective function is performed by means of the conjugate gradient method [34].

3. Results and Discussion

3.1. FTIR

Figure 2 shows the FTIR spectra, analyzed in the range from 4000 cm^{-1} to 400 cm^{-1}, for the ZnO_PL1%, ZnO_PL2%, and ZnO_PL4% samples. The materials show vibrations at 3410, 1385, 1120, and 400 cm^{-1}. The band at 3410 cm^{-1} represents the vibration of the hydroxyl group (O-H) [35]. The bands at 1384 and 1120 cm^{-1} are attributed to the C-H and C-O vibrations of the carboxylic groups present in the organic molecules of the *Prosopis laevigata* extract [36,37]. Finally, the band observed at 400 cm^{-1} in the three samples analyzed is attributed to the stretching of metal–oxygen (Zn-O), which confirms the obtaining of ZnO nanoparticles [38].

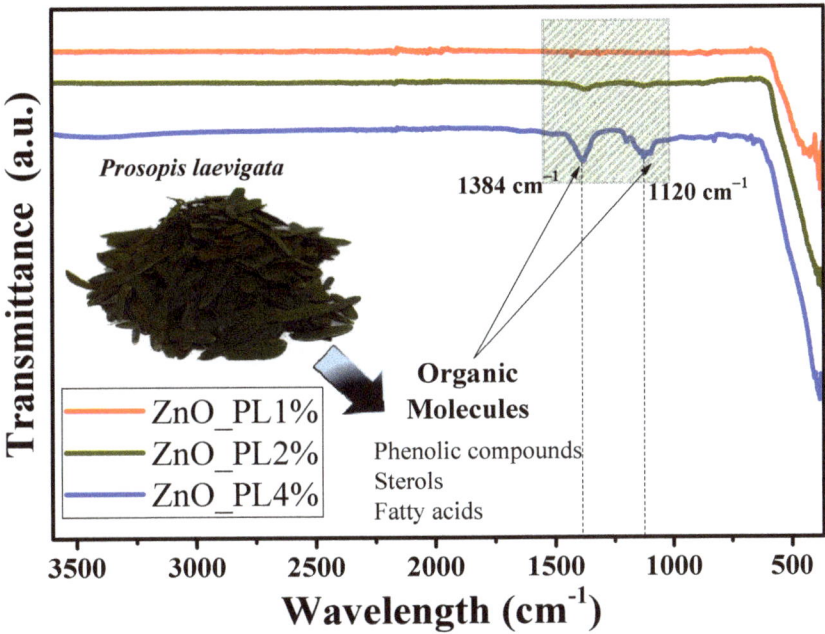

Figure 2. Study of functional groups with FT-IR of ZnO nanoparticles synthesized with different percentages (1, 2, and 4%) of *Prosopis laevigata*.

3.2. Optical Properties

The determination of the optical properties of the material was carried out using UV–vis spectroscopy. Figure 3 shows the spectrum obtained for ZnO NPs dispersed in water, in a wavelength range from 200 nm to 600 nm, with a maximum absorption peak located around 350 nm for the 3 samples. The Tauc model was used to calculate the band gap.

$$(\alpha h\nu)^{1/n} = B(h\nu - Eg) \qquad (6)$$

where $\alpha(\nu)$ is the absorption coefficient (Lambert-Beer), $h\nu$ is the photon energy, B is a proportionality constant (it is determined by the refractive index, the effective electron, and hole masses), Eg is the energy of the band gap and the value of n corresponds directly to the band gap semiconductor ($n = 1/2$) [39,40]. The calculated values of the band gap by the method of Tauc resulted in 2.80, 2.74, and 2.63 eV for ZnO_PL1%, ZnO_PL 2%, and

ZnO_PL 4% of ZnO NPs, respectively. These values are very similar to those reported in the literature for ZnO NPs [41]. It should be noted that as the percentage of extract used for material biosynthesis increases, the band gap decreases. This decrease in the energy band gap is due to the size of the NPs; the greater the amount of extract used in biosynthesis, the more organic molecules remain on the surface of the nanoparticles (residual carbon) acting as photosensitizers [35,42,43].

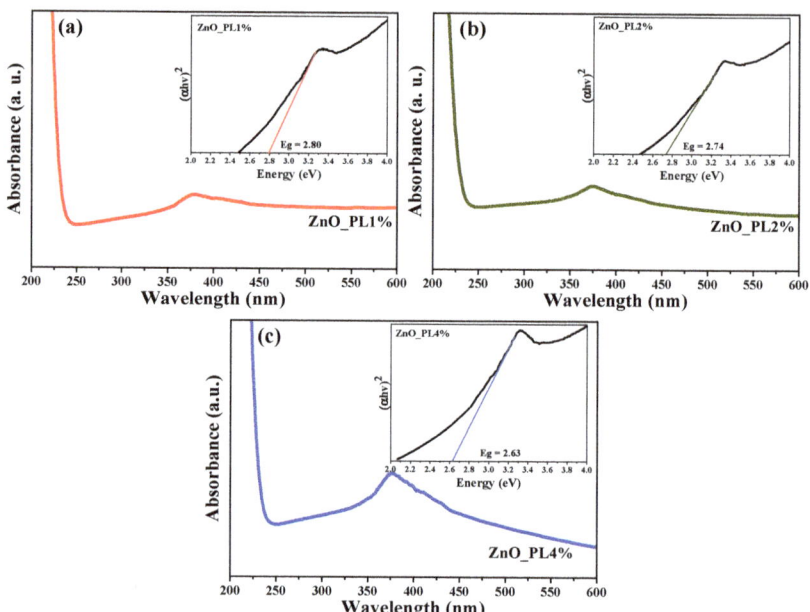

Figure 3. Study of optical properties and band gap ZnO. (**a**) ZnO_PL1%. (**b**) ZnO_PL2% and (**c**) ZnO_PL4%.

3.3. XRD

Figure 4 shows the XRD analysis of the ZnO NPs. The characteristic peaks of this material are observed in the diffraction pattern, being identified at 31.73, 34.40, 36.23, 47.51, 56.56, 62.82, 66.42, 67.92°, and 69.04 2θ, corresponding to Miller indices of (100), (002), (101), (102), (110), (103), (200), (112), and (210), respectively. These peaks coincide with those of the JCPDS Card No. 76-0704, which describes the ZnO NPs as hexagonal structures type zincite [35]. To calculate the crystallite size, the Debye–Scherrer formula (5) was used for the three most intense peaks (100), (002), and (101):

$$L = \frac{0.9\,\lambda}{\beta \cos \theta} \qquad (7)$$

where L is the crystallite size (nm), λ the incident wavelength, β the full width at half maximum of the peak, θ the Bragg angle [44]. The results showed that the size of the crystallites was 55, 50, and 49 nm for samples ZnO_PL1%, ZnO_PL2%, and ZnO_PL4%, respectively. The size obtained in this study is within the range of previous reports in the literature for ZnO nanoparticles [38]. Research has demonstrated that using more extract in green synthesis results in smaller crystallites. This is due to the organic molecules acting as a barrier and preventing the nanomaterial from agglomerating and growing, according to a study [30].

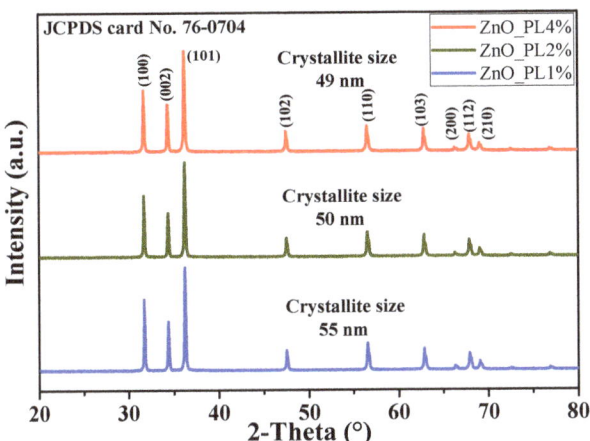

Figure 4. Study of the crystal structure of ZnO NPs.

3.4. TEM

Figure 5 shows the morphology of ZnO NPs by TEM. TEM micrographs (a), (b), and (c) show a semi-spherical shape of different sizes for the three ZnO samples, with small agglomerations. The imageJ (version 1.6.0) software was used to measure nanoparticles observed in the micrographs. The ZnO_PL1% NPs showed sizes between 64 and 114 nm. The ZnO_PL2% sample presented particle sizes between 63 and 83 nm, and finally, ZnO_PL4% presented measurements between 29 and 45 nm. The effect caused by the concentration of *P. laevigata* extract during the synthesis of ZnO on the nanoparticle size is evident; it is observed that the higher the concentration of extract the nanoparticle size is lower, as previously reported in the literature [45,46]. Furthermore, the crystalline nature of the ZnO nanoparticles was confirmed with the selected area electron diffraction (SAED) pattern. The ring-shaped diffraction pattern demonstrates that the nanoparticles are nanocrystalline [47,48].

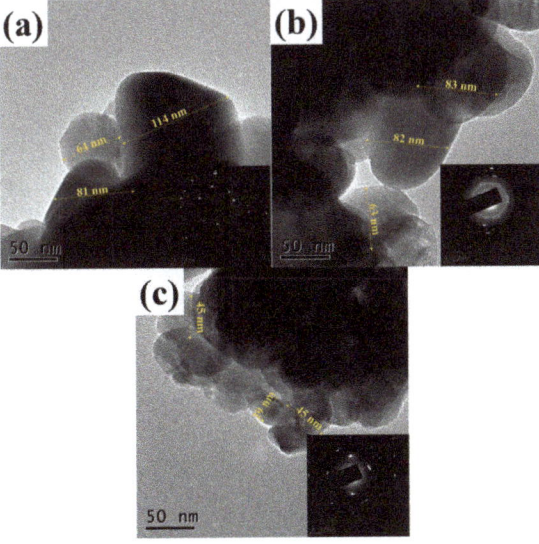

Figure 5. Morphology of the nanoparticles by TEM, (**a**) ZnO_PL1%, (**b**) ZnO_PL2%, and (**c**) ZnO_PL4%.

3.5. Formation Mechanism of ZnO Nanoparticles

Figure 6 presents a proposed mechanism for forming ZnO nanoparticles biosynthesized with *P. laevigata* extract. The nanoparticle formation reaction begins with the hydrolysis of the metal precursor (zinc nitrate) when in contact with the *P. laevigata* extract. Subsequently, the reaction continues with nucleation and later agglomeration, leading to the formation of ZnO nanoparticles [49]. In the formation of the nanomaterial, the organic molecules present in the extract are functionalized on the surface of the nanoparticles with interactions with the OH groups of the extract [50]. These functionalized molecules help as stabilizing agents and prevent excessive growth of the nanoparticles [51].

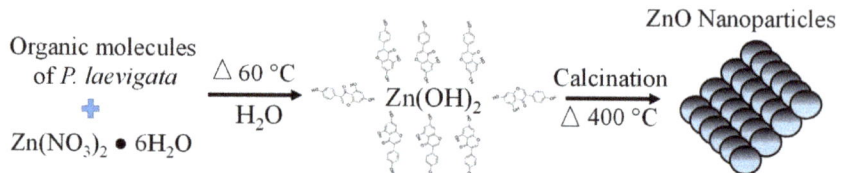

Figure 6. Proposal for a mechanism for the formation of biosynthesized ZnO nanoparticles with *P. laeveigata*.

3.6. Model

Figure 7 shows degradation graphs, where a nice fitting of the LHHW model with the experimental data can be observed. According to the model, the value of the constant B is related to the concentration of active sites on the surface of the NPs. The ZnO_PL1% sample, which has the largest size of NPs, has a value of the constant equal to 0.0328, while the samples of ZnO_PL2% and ZnO_PL4% have a value of B of 0.0527 and 0.0757, respectively. This increase in the value of constant B highlights an increase in the number of active sites when the size of the NPs decreases, which increases their reactivity, as can be seen with the decrease in degradation time. Figure 7a shows that the time to degrade 90% of the dye is 70 min for the sample of ZnO_PL1%, while in Figure 7b a time of 45 min is obtained for the sample of 2%. Finally, Figure 7c shows a time of 31 min for the sample of ZnO_PL4%. This increase in reactivity is also proven with the value of constant A, which according to the model is related to the adsorption and reaction constants. By decreasing the value of the constant A, from a value of 0.0216 for the sample of ZnO_PL1% to a value of 0.01 for the samples of ZnO_PL2% and ZnO_PL4%, we observe an increase in the kinetic constant of the reaction carried out on the surface of the catalyst, as the size of the NPs decreases. On the other hand, the speed of the photocatalytic reaction of the nanoparticles was 0.0305, 0.0387, and 0.0771 for the ZnO_PL1%, ZnO_PL2%, and ZnO_PL4% samples, respectively. Similar values have been found in the literature, for example, the work of Ludmila Motelica et al. in 2022 where they synthesized ZnO nanoparticles using various alcohols, reporting reaction rates of the nanoparticles between 0.0569 and 0.0935 [52]. In the present study, the ZnO_PL4% sample was the one that reached a higher reaction rate, which is attributed to the percentage of extract used in the biosynthesis. Some authors mention that the sharp edges in the nanoparticles lead to an improvement in the photocatalytic activity. In addition, the smaller the size of the nanoparticles, the greater the number of defects on the nanoparticle surface because there is a greater surface area. These defects represent catalytic centers for the photodegradation of polluting molecules, leading to improved photocatalytic activity [40,52,53].

In order to study the effect of mass transfer resistance of the dye to the catalytic surface, a variation in the amount of MB was made with the largest nanoparticles (1%), which is important for the control of the pressure drop in the design of packed bed reactors. Figure 8a shows the degradation curve for 15 ppm of dye concentration. It can be seen how the LHHW model fits the experimental data, but so does the first-order kinetic model $(-r_A = BC_A)$, since at low dye concentrations (at the limit when it tends to zero), the LHHW model tends to the first-order kinetic model, explaining the observed convergence and the apparent kinetic order shown. The degradation curve for 20 ppm of dye is observed in Figure 8b. The graph shows how the LHHW model adjusts to the experimental data, but in these conditions of dye concentration, a kinetic of the first order is no longer observed, due to the increase in dye concentration in the solution [54]. The adjustment of the constant A under these conditions generates an A value of 0.01 for the sample of 15 ppm and a value of 0.0276 for the sample of 20 ppm. The increase in the value of constant A shows an increase in the kinetic constant of adsorption, due to the mass transfer resistance of dye molecules towards the catalytic surface.

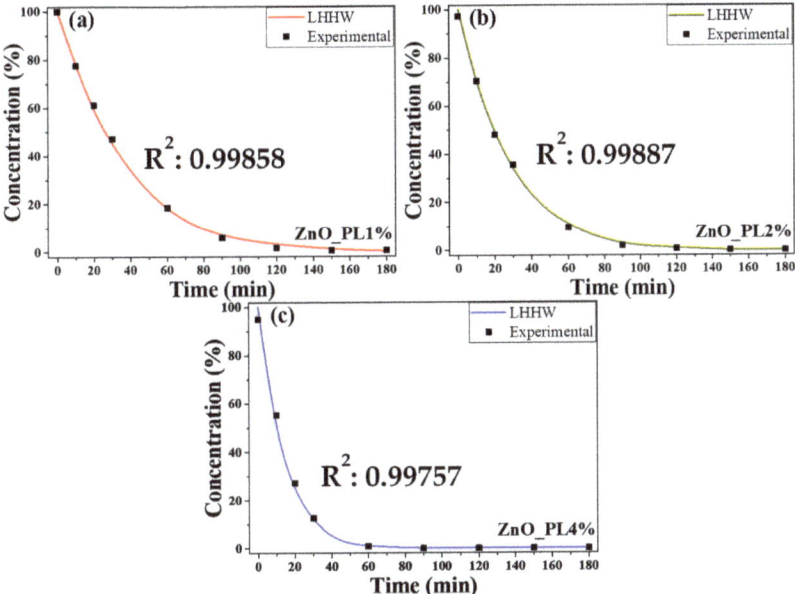

Figure 7. Comparison of degradation capacity of nanoparticles from different extract concentrations in the synthesis, (**a**) ZnO_PL1%, (**b**) ZnO_PL2%, and (**c**) ZnO_PL4%.

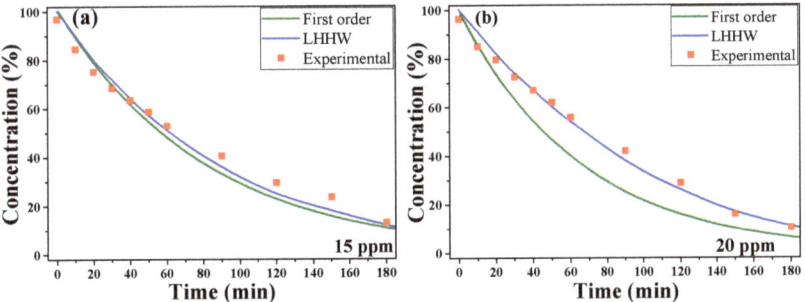

Figure 8. Comparison of degradation capacity of nanoparticles from variation in dye concentration, (**a**) 15 ppm and (**b**) 20 ppm.

3.7. Reaction Mechanism

Free radicals are essential for degrading dyes due to their high oxidizing power that can destroy harmful organic contaminants [55]. The reaction mechanism for the photocatalytic reaction is shown below in Figure 9.

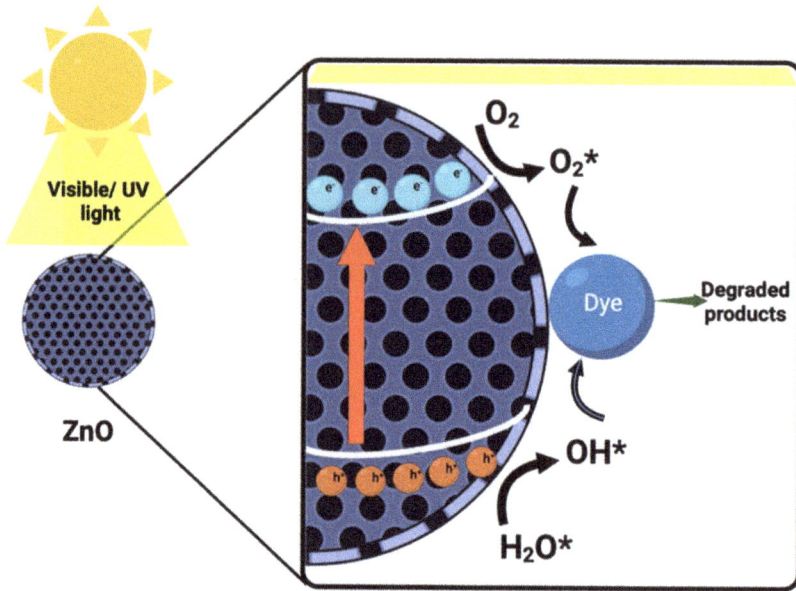

Figure 9. Reaction mechanism of nanoparticles in pollutant degradation.

The photocatalysis reaction of the organic dyes happens in a homogenous way; this implies that the dye molecules remain adsorbed on the catalyst surface (active sites). When UV light is irradiated over the system, electrons (e^-) are generated in the conduction band and holes (h^+) in the valence band [56]. The reactions in the surface are described as follows [55]:

Photon absorption

$$ZnO + h\upsilon(UV) \rightarrow h_{VB}^+ + e_{CB}^- \, (ZnO) \tag{8}$$

Reduction of O_2 to superoxide radicals O_2^-

$$O_2 + e_{CB}^- \rightarrow O_2^{-*} \tag{9}$$

Oxidation of hydroxyl ions and water molecules to generate hydroxyl radicals

$$h_{VB}^+ + OH^- \rightarrow OH^* \tag{10}$$

The holes (h^+) react with water and, upon oxidation, generate hydroxyl radicals

$$h_{VB}^+ + H_2O \rightarrow OH^* + H^+ \tag{11}$$

Radicals OH^* and O_2^{-*} are responsible for the photodegradation of dyes to degraded products.

$$Dye + OH^* \rightarrow intermediates \rightarrow degraded\ products \tag{12}$$

$$Dye + O_2^{-*} \rightarrow intermediates \rightarrow degraded\ products \tag{13}$$

4. Conclusions

In this work, we report the green synthesis of ZnO NPs using *Prosopis laevigata* extract with a simple and environmentally friendly process, as well as the photocatalytic degradation of methylene blue on the NPs. In addition, the degradation kinetics are described by the LHHW model, in order to study the photocatalytic and mass transfer processes. The LHHW model nicely fits the experimental data, evidencing the stages assumed in the photocatalytic process. The size of the NPs was varied using different concentrations of *P. laevigata* as a reducing and stabilizing agent. The LHHW model demonstrates that by decreasing the nanoparticle size, the number of active sites increases, generating greater reactivity. In the limit of low concentrations, first-order kinetics and the LHHW model coincide, showing that the resistance to the mass transfer of molecules generates an apparent degradation kinetic order of one. This research shows that it is possible to use an extract of *P. laevigata* as a reducing and stabilizing agent in forming ZnO nanoparticles. In addition, this study presents a theoretical model that adapts to a high percentage of the experimental. This work can help develop industrial catalytic reactors that help reduce pollution in bodies of water around the world.

Author Contributions: M.L.M.: investigation, conceptualization, methodology, and writing—original draft. P.A.L.M.: investigation, supervision, software, and validation. M.d.J.C.C.: investigation, data curation, methodology, and writing—review and editing. V.M.O.C.: methodology, resources, conceptualization, funding acquisition, and visualization. C.M.G.G.: conceptualization, formal analysis, and data curation. A.R.V.N.: methodology, writing—review and supervision. R.C.V.S.: investigation, formal analysis, writing—review and editing, supervision and validation. All authors have read and agreed to the published version of the manuscript.

Funding: This research was funded by the Frontier Science Project with Conahcyt number 1805.

Institutional Review Board Statement: Not applicable.

Informed Consent Statement: Not applicable.

Data Availability Statement: Data sharing is not applicable to this article.

Acknowledgments: The authors thank the Universidad Autónoma de Baja California (UABC) project number 402/3391. The authors acknowledge Uvaldo Hernández and María Citlalit Martínez Soto for technical assistance during this work.

Conflicts of Interest: The authors declare that they have no known competing financial interest or personal relationships that could have appeared to influence the work reported in this paper.

References

1. Cui, X.; Li, P.; Lei, H.; Tu, C.; Wang, D.; Wang, Z.; Chen, W. Greatly enhanced tribocatalytic degradation of organic pollutants by TiO$_2$ nanoparticles through efficiently harvesting mechanical energy. *Sep. Purif. Technol.* **2022**, *289*, 120814. [CrossRef]
2. Liu, Z.; Demeestere, K.; Van Hulle, S. Comparison and performance assessment of ozone-based AOPs in view of trace organic contaminants abatement in water and wastewater: A review. *J. Environ. Chem. Eng.* **2021**, *9*, 105599. [CrossRef]
3. Wei, Z.; Van Le, Q.; Peng, W.; Yang, Y.; Yang, H.; Gu, H.; Lam, S.S.; Sonne, C. A review on phytoremediation of contaminants in air, water and soil. *J. Hazard. Mater.* **2021**, *403*, 123658. [CrossRef]
4. Aljeboree, A.M.; Radia, N.D.; Jasim, L.S.; Alwarthan, A.A.; Khadhim, M.M.; Salman, A.W.; Alkaim, A.F. Synthesis of a new nanocomposite with the core TiO$_2$/hydrogel: Brilliant green dye adsorption, isotherms, kinetics, and DFT studies. *J. Ind. Eng. Chem.* **2022**, *109*, 475–485. [CrossRef]
5. Rajput, R.B.; Jamble, S.N.; Kale, R.B. A review on TiO$_2$/SnO$_2$ heterostructures as a photocatalyst for the degradation of dyes and organic pollutants. *J. Environ. Manag.* **2022**, *307*, 114533. [CrossRef]
6. Kumar, S.; Tewari, C.; Sahoo, N.G.; Philip, L. Mechanistic insights into carbo-catalyzed persulfate treatment for simultaneous degradation of cationic and anionic dye in multicomponent mixture using plastic waste–derived carbon. *J. Hazard. Mater.* **2022**, *435*, 128956. [CrossRef]
7. Wu, S.; Wang, F.; Li, Q.; Wang, J.; Zhou, Y.; Duan, N.; Niazi, S.; Wang, Z. Photocatalysis and degradation products identification of deoxynivalenol in wheat using upconversion nanoparticles@ TiO$_2$ composite. *Food Chem.* **2020**, *323*, 126823. [CrossRef] [PubMed]
8. Rajendrachari, S.; Taslimi, P.; Karaoglanli, A.C.; Uzun, O.; Alp, E.; Jayaprakash, G.K. Photocatalytic degradation of Rhodamine B (RhB) dye in waste water and enzymatic inhibition study using cauliflower shaped ZnO nanoparticles synthesized by a novel One-pot green synthesis method. *Arab. J. Chem.* **2021**, *14*, 103180. [CrossRef]
9. Yadav, R.; Chundawat, T.S.; Rawat, P.; Rao, G.K.; Vaya, D. Photocatalytic degradation of malachite green dye by ZnO and ZnO–β-cyclodextrin nanocomposite. *Bull. Mater. Sci.* **2021**, *44*, 250. [CrossRef]

10. Silva-Osuna, E.R.; Vilchis-Nestor, A.R.; Villarreal-Sanchez, R.C.; Castro-Beltran, A.; Luque, P.A. Study of the optical properties of TiO$_2$ semiconductor nanoparticles synthesized using *Salvia rosmarinus* and its effect on photocatalytic activity. *Opt. Mater.* **2022**, *124*, 112039. [CrossRef]
11. Mugundan, S.; Praveen, P.; Sridhar, S.; Prabu, S.; Mary, K.L.; Ubaidullah, M.; Shaikh, S.F.; Kanagesan, S. Sol-gel synthesized barium doped TiO$_2$ nanoparticles for solar photocatalytic application. *Inorg. Chem. Commun.* **2022**, *139*, 109340. [CrossRef]
12. Chaudhari, K.B.; Rane, Y.N.; Shende, D.A.; Gosavi, N.M.; Gosavi, S.R. Effect of annealing on the photocatalytic activity of chemically prepared TiO$_2$ thin films under visible light. *Optik* **2019**, *193*, 163006. [CrossRef]
13. Esakki, E.S.; Deepa, G.; Vivek, P.; Devi, L.R.; Sheeba, N.L.; Sundar, S.M. Investigation on electrochemical analysis of ZnO nanoparticles and its performance for dye-sensitized solar cells using various natural dyes. *J. Indian Chem. Soc.* **2023**, *100*, 100889. [CrossRef]
14. Ahmad, I.; Aslam, M.; Jabeen, U.; Zafar, M.N.; Malghani, M.N.K.; Alwadai, N.; Alshammari, F.H.; Almuslem, A.S.; Ullah, Z. ZnO and Ni-doped ZnO photocatalysts: Synthesis, characterization and improved visible light driven photocatalytic degradation of methylene blue. *Inorganica Chim. Acta* **2022**, *543*, 121167. [CrossRef]
15. Shubha, J.P.; Roopashree, B.; Patil, R.C.; Khan, M.; Shaik, M.R.; Alaqarbeh, M.; Alwarthan, A.; Karami, A.M.; Adil, S.F. Facile synthesis of ZnO/CuO/Eu heterostructure photocatalyst for the degradation of industrial effluent. *Arab. J. Chem.* **2023**, *16*, 104547. [CrossRef]
16. Zhou, R.; Zhang, W.; Tang, N. The study of ZnO/InGaAs solar cells and three methods to enhance the performances. *Opt. Mater.* **2022**, *127*, 112095. [CrossRef]
17. Agarwal, L.; Singh, R.; Varshney, G.; SambasivaRao, K.; Tripathi, S. Design and analysis of Yb doped ZnO (YZO) and P–Si bilayer nano-stacked reflector for optical filter applications. *Superlattices Microstruct.* **2020**, *146*, 106670. [CrossRef]
18. Mahalakshmi, S.; Hema, N.; Vijaya, P.P. In vitro biocompatibility and antimicrobial activities of zinc oxide nanoparticles (ZnO NPs) prepared by chemical and green synthetic route—A comparative study. *Bionanoscience* **2020**, *10*, 112–121.
19. Ijaz, I.; Gilani, E.; Nazir, A.; Bukhari, A. Detail review on chemical, physical and green synthesis, classification, characterizations and applications of nanoparticles. *Green Chem. Lett. Rev.* **2020**, *13*, 223–245. [CrossRef]
20. Salem, S.S.; Fouda, A. Green synthesis of metallic nanoparticles and their prospective biotechnological applications: An overview. *Biol. Trace Elem. Res.* **2021**, *199*, 344–370. [CrossRef]
21. Bandeira, M.; Giovanela, M.; Roesch-Ely, M.; Devine, D.M.; da Silva Crespo, J. Green synthesis of zinc oxide nanoparticles: A review of the synthesis methodology and mechanism of formation. *Sustain. Chem. Pharm.* **2020**, *15*, 100223. [CrossRef]
22. Pal, G.; Rai, P.; Pandey, A. Green synthesis of nanoparticles: A greener approach for a cleaner future. In *Green Synthesis, Characterization and Applications of Nanoparticles*; Elsevier: Amsterdam, The Netherlands, 2019; pp. 1–26.
23. Kamarajan, G.; Anburaj, D.B.; Porkalai, V.; Muthuvel, A.; Nedunchezhian, G. Green synthesis of ZnO nanoparticles using Acalypha indica leaf extract and their photocatalyst degradation and antibacterial activity. *J. Indian Chem. Soc.* **2022**, *99*, 100695. [CrossRef]
24. Sasi, S.; Fasna, P.H.F.; Sharmila, T.K.B.; Chandra, C.S.J.; Antony, J.V.; Raman, V.; Nair, A.B.; Ramanathan, H.N. Green synthesis of ZnO nanoparticles with enhanced photocatalytic and antibacterial activity. *J. Alloys Compd.* **2022**, *924*, 166431. [CrossRef]
25. Batool, S.; Hasan, M.; Dilshad, M.; Zafar, A.; Tariq, T.; Wu, Z.; Chen, R.; Hassan, S.G.; Munawar, T.; Iqbal, F. Green synthesis of Cordia myxa incubated ZnO, Fe$_2$O$_3$, and Co$_3$O$_4$ nanoparticle: Characterization, and their response as biological and photocatalytic agent. *Adv. Powder Technol.* **2022**, *33*, 103780. [CrossRef]
26. Peña-Avelino, L.Y.; Pinos-Rodríguez, J.M.; Yáñez-Estrada, L.; Juárez-Flores, B.I.; Mejia, R.; Andrade-Zaldivar, H. Chemical composition and in vitro degradation of red and white mesquite (*Prosopis laevigata*) pods. *S. Afr. J. Anim. Sci.* **2014**, *44*, 298–306. [CrossRef]
27. García-López, J.C.; Durán-García, H.M.; José, A.; Álvarez-Fuentes, G.; Pinos-Rodríguez, J.M.; Lee-Rangel, H.A.; López-Aguirre, S.; Ruiz-Tavares, D.; Rendón-Huerta, J.A.; Vicente-Martínez, J.G. Producción y contenido nutrimental de vainas de tres variantes de mezquite (*Prosopis laevigata*) en el altiplano potosino, México. *Agrociencia* **2019**, *53*, 821–831.
28. Díaz-Batalla, L.; Hernández-Uribe, J.P.; Gutiérrez-Dorado, R.; Téllez-Jurado, A.; Castro-Rosas, J.; Pérez-Cadena, R.; Gómez-Aldapa, C.A. Nutritional Characterization of *Prosopis laevigata* Legume Tree (Mesquite) Seed Flour and the Effect of Extrusion Cooking on its Bioactive Components. *Foods* **2018**, *7*, 124. [CrossRef]
29. Konstantinou, I.K.; Albanis, T.A. TiO$_2$-assisted photocatalytic degradation of azo dyes in aqueous solution: Kinetic and mechanistic investigations: A review. *Appl. Catal. B Environ.* **2004**, *49*, 1–14. [CrossRef]
30. Vautier, M.; Guillard, C.; Herrmann, J.-M. Photocatalytic degradation of dyes in water: Case study of indigo and of indigo carmine. *J. Catal.* **2001**, *201*, 46–59. [CrossRef]
31. Cunningham, J.; Al-Sayyed, G.; Srijaranai, S. Adsorption of model pollutants onto TiO$_2$ particles in relation to photoremediation of contaminated water. In *Aquatic and Surface Photochemistry*; CRC Press: Boca Raton, FL, USA, 2018; pp. 317–348.
32. Vulliet, E.; Chovelon, J.-M.; Guillard, C.; Herrmann, J.-M. Factors influencing the photocatalytic degradation of sulfonylurea herbicides by TiO$_2$ aqueous suspension. *J. Photochem. Photobiol. A Chem.* **2003**, *159*, 71–79. [CrossRef]
33. Kumar, K.V.; Porkodi, K.; Rocha, F. Langmuir–Hinshelwood kinetics–a theoretical study. *Catal. Commun.* **2008**, *9*, 82–84. [CrossRef]
34. Li, X.; Zhao, X. A hybrid conjugate gradient method for optimization problems. *Nat. Sci.* **2011**, *3*, 85–90. [CrossRef]
35. Yang, L.; Ren, L.; Zhao, Y.; Liu, S.; Wang, H.; Gao, X.; Niu, B.; Li, W. Preparation and characterization of PVA/arginine chitosan/ZnO NPs composite films. *Int. J. Biol. Macromol.* **2023**, *226*, 184–193. [CrossRef] [PubMed]

36. Ananthalakshmi, R.; Rajarathinam, S.R.; Sadiq, A.M. Antioxidant activity of ZnO Nanoparticles synthesized using Luffa acutangula peel extract. *Res. J. Pharm. Technol.* **2019**, *12*, 1569–1572. [CrossRef]
37. Mansour, A.T.; Alprol, A.E.; Khedawy, M.; Abualnaja, K.M.; Shalaby, T.A.; Rayan, G.; Ramadan, K.M.A.; Ashour, M. Green Synthesis of Zinc Oxide Nanoparticles Using Red Seaweed for the Elimination of Organic Toxic Dye from an Aqueous Solution. *Materials* **2022**, *15*, 5169. [CrossRef] [PubMed]
38. Pham, Q.P.; Le Nguyen, Q.N.; Nguyen, N.H.; Doan, U.T.T.; Ung, T.D.T.; Tran, V.C.; Phan, T.B.; Pham, A.T.T.; Pham, N.K. Calcination-dependent microstructural and optical characteristics of eco-friendly synthesized ZnO nanoparticles and their implementation in analog memristor application. *Ceram. Int.* **2023**, *49*, 20742–20755. [CrossRef]
39. Singh, G.P.; Aman, A.K.; Singh, R.K.; Roy, M.K. Effect of low Co-doping on structural, optical, and magnetic performance of ZnO nanoparticles. *Optik* **2020**, *203*, 163966. [CrossRef]
40. Hendrix, Y.; Rauwel, E.; Nagpal, K.; Haddad, R.; Estephan, E.; Boissière, C.; Rauwel, P. Revealing the Dependency of Dye Adsorption and Photocatalytic Activity of ZnO Nanoparticles on Their Morphology and Defect States. *Nanomaterials* **2023**, *13*, 1998. [CrossRef]
41. Luque-Morales, P.A.; Lopez-Peraza, A.; Nava-Olivas, O.J.; Amaya-Parra, G.; Baez-Lopez, Y.A.; Orozco-Carmona, V.M.; Garrafa-Galvez, H.E.; Chinchillas-Chinchillas, M.d.J. ZnO Semiconductor Nanoparticles and Their Application in Photocatalytic Degradation of Various Organic Dyes. *Materials* **2021**, *14*, 7535. [CrossRef]
42. Pratomo, U.; Pratama, R.A.; Irkham, I.; Sulaeman, A.P.; Mulyana, J.Y.; Primadona, I. 3D-ZnO Superstructure Decorated with Carbon-Based Material for Efficient Photoelectrochemical Water-Splitting under Visible-Light Irradiation. *Nanomaterials* **2023**, *13*, 1380. [CrossRef]
43. Tammina, S.K.; Mandal, B.K.; Ranjan, S.; Dasgupta, N. Cytotoxicity study of *Piper nigrum* seed mediated synthesized SnO$_2$ nanoparticles towards colorectal (HCT116) and lung cancer (A549) cell lines. *J. Photochem. Photobiol. B Biol.* **2017**, *166*, 158–168. [CrossRef] [PubMed]
44. Rajarajeswari, P.; Shaikh, R.S.; Ravangave, L.S. Effect of temperature reaction on chemically synthesized ZnO nanoparticles change in particle size. *Mater. Today Proc.* **2021**, *45*, 3997–4001. [CrossRef]
45. Soto-Robles, C.A.; Nava, O.; Cornejo, L.; Lugo-Medina, E.; Vilchis-Nestor, A.R.; Castro-Beltrán, A.; Luque, P.A. Biosynthesis, characterization and photocatalytic activity of ZnO nanoparticles using extracts of *Justicia spicigera* for the degradation of methylene blue. *J. Mol. Struct.* **2021**, *1225*, 129101. [CrossRef]
46. Falih, A.; Ahmed, N.M.; Rashid, M. Green synthesis of zinc oxide nanoparticles by fresh and dry alhagi plant. *Mater. Today Proc.* **2022**, *49*, 3624–3629. [CrossRef]
47. Gawade, V.V.; Gavade, N.L.; Shinde, H.M.; Babar, S.B.; Kadam, A.N.; Garadkar, K.M. Green synthesis of ZnO nanoparticles by using *Calotropis procera* leaves for the photodegradation of methyl orange. *J. Mater. Sci. Mater. Electron.* **2017**, *28*, 14033–14039. [CrossRef]
48. Donga, S.; Chanda, S. Caesalpinia crista seeds mediated green synthesis of zinc oxide nanoparticles for antibacterial, antioxidant, and anticancer activities. *Bionanoscience* **2022**, *12*, 451–462. [CrossRef]
49. Prasad, A.R.; Williams, L.; Garvasis, J.; Shamsheera, K.O.; Basheer, S.M.; Kuruvilla, M.; Joseph, A. Applications of phytogenic ZnO nanoparticles: A review on recent advancements. *J. Mol. Liq.* **2021**, *331*, 115805. [CrossRef]
50. Aldeen, T.S.; Mohamed, H.E.A.; Maaza, M. ZnO nanoparticles prepared via a green synthesis approach: Physical properties, photocatalytic and antibacterial activity. *J. Phys. Chem. Solids* **2022**, *160*, 110313. [CrossRef]
51. Chinchillas-Chinchillas, M.J.; Garrafa-Gálvez, H.E.; Orozco-Carmona, V.M.; Luque-Morales, P.A. Comparative Study of SnO$_2$ and ZnO Semiconductor Nanoparticles (Synthesized Using Randia echinocarpa) in the Photocatalytic Degradation of Organic Dyes. *Symmetry* **2022**, *14*, 1970. [CrossRef]
52. Motelica, L.; Oprea, O.-C.; Vasile, B.-S.; Ficai, A.; Ficai, D.; Andronescu, E.; Holban, A.M. Antibacterial Activity of Solvothermal Obtained ZnO Nanoparticles with Different Morphology and Photocatalytic Activity against a Dye Mixture: Methylene Blue, Rhodamine B and Methyl Orange. *Int. J. Mol. Sci.* **2023**, *24*, 5677. [CrossRef]
53. Motelica, L.; Vasile, B.-S.; Ficai, A.; Surdu, A.-V.; Ficai, D.; Oprea, O.-C.; Andronescu, E.; Jinga, D.C.; Holban, A.M. Influence of the Alcohols on the ZnO Synthesis and Its Properties: The Photocatalytic and Antimicrobial Activities. *Pharmaceutics* **2022**, *14*, 2842. [CrossRef] [PubMed]
54. Kumar, A.; Pandey, G. A review on the factors affecting the photocatalytic degradation of hazardous materials. *Mater. Sci. Eng. Int. J.* **2017**, *1*, 1–10. [CrossRef]
55. Singh, A.; Goyal, V.; Singh, J.; Rawat, M. Structural, morphological, optical and photocatalytic properties of green synthesized TiO$_2$ NPs. *Curr. Res. Green Sustain. Chem.* **2020**, *3*, 100033. [CrossRef]
56. Sahu, K.; kuriakose, S.; Singh, J.; Satpati, B.; Mohapatra, S. Facile synthesis of ZnO nanoplates and nanoparticle aggregates for highly efficient photocatalytic degradation of organic dyes. *J. Phys. Chem. Solids* **2018**, *121*, 186–195. [CrossRef]

Disclaimer/Publisher's Note: The statements, opinions and data contained in all publications are solely those of the individual author(s) and contributor(s) and not of MDPI and/or the editor(s). MDPI and/or the editor(s) disclaim responsibility for any injury to people or property resulting from any ideas, methods, instructions or products referred to in the content.

Article

Phosphorus-Doped Hollow Tubular g-C$_3$N$_4$ for Enhanced Photocatalytic CO$_2$ Reduction

Manying Sun [1], Chuanwei Zhu [1], Su Wei [1], Liuyun Chen [1], Hongbing Ji [1,2], Tongming Su [1,*] and Zuzeng Qin [1,*]

[1] Guangxi Key Laboratory of Petrochemical Resource Processing and Process Intensification Technology, School of Chemistry and Chemical Engineering, Guangxi University, Nanning 530004, China; sunmy0723@163.com (M.S.); cnpp0924@163.com (C.Z.); 18378045306@163.com (S.W.); 18811313166@163.com (L.C.); jihb@mail.sysu.edu.cn (H.J.)
[2] Fine Chemical Industry Research Institute, Sun Yat-sen University, Guangzhou 510275, China
* Correspondence: sutm@gxu.edu.cn (T.S.); qinzuzeng@gxu.edu.cn (Z.Q.)

Abstract: Photocatalytic CO$_2$ reduction is a tactic for solving the environmental pollution caused by greenhouse gases. Herein, NH$_4$H$_2$PO$_4$ was added as a phosphorus source in the process of the hydrothermal treatment of melamine for the first time, and phosphorus-doped hollow tubular g-C$_3$N$_4$ (x-P-HCN) was fabricated and used for photocatalytic CO$_2$ reduction. Here, 1.0-P-HCN exhibited the largest CO production rate of 9.00 μmol·g^{-1}·h^{-1}, which was 10.22 times higher than that of bulk g-C$_3$N$_4$. After doping with phosphorus, the light absorption range, the CO$_2$ adsorption capacity, and the specific surface area of the 1.0-P-HCN sample were greatly improved. In addition, the separation of photogenerated electron–hole pairs was enhanced. Furthermore, the phosphorus-doped g-C$_3$N$_4$ effectively activated the CO$_2$ adsorbed on the surface of phosphorus-doped g-C$_3$N$_4$ photocatalysts, which greatly enhanced the CO production rate of photocatalytic CO$_2$ reduction over that of g-C$_3$N$_4$.

Keywords: phosphorus; doped; g-C$_3$N$_4$; photocatalytic; CO$_2$ reduction

1. Introduction

Human beings are facing two major challenges today: a huge energy demand and serious environmental problems [1,2]. As an inexhaustible clean energy source, solar energy has been widely studied and utilized for decades, and the high-efficiency utilization of solar energy through photocatalysts has become a research hotspot in recent decades [3,4]. The photocatalytic reduction of CO$_2$ into valuable products such as CO and CH$_4$ is considered a promising technology for alleviating the greenhouse effect [5,6]. The discovery of single-layer graphene brought extensive attention to 2D materials, and its unique properties are favored in different research fields [7]. For example, as a two-dimensional nonmetal semiconductor material, graphitic carbon nitride (g-C$_3$N$_4$) has attracted increasing attention from researchers due to its inexpensive raw materials, simple preparation methods, excellent photoelectric physical structure properties, and chemical stability [8]. At present, g-C$_3$N$_4$ has been extensively studied in photocatalytic CO$_2$ reduction and has shown excellent performance and considerable application prospects [9]. However, the large block microstructure and low specific surface area of bulk g-C$_3$N$_4$ prepared directly through traditional thermal polymerization are not conducive to the full exposure of active sites, the migration of photogenerated charges, and the mass transfer process of reactants [10]. In addition, bulk g-C$_3$N$_4$ tends to exhibit a narrow light absorption range and severe recombination of photogenerated electrons, resulting in lower catalytic activity [11].

The preparation of g-C$_3$N$_4$ with a specific microstructure has become an available strategy for augmenting the specific surface area and accelerating the transfer of photogenerated electrons and holes. In addition, 3D g-C$_3$N$_4$ with special microstructures of hollow spheres and a 3D network structure can be prepared by using SiO$_2$ microspheres

Citation: Sun, M.; Zhu, C.; Wei, S.; Chen, L.; Ji, H.; Su, T.; Qin, Z. Phosphorus-Doped Hollow Tubular g-C$_3$N$_4$ for Enhanced Photocatalytic CO$_2$ Reduction. *Materials* **2023**, *16*, 6665. https://doi.org/10.3390/ma16206665

Academic Editor: Klára Hernádi

Received: 22 September 2023
Revised: 5 October 2023
Accepted: 10 October 2023
Published: 12 October 2023

Copyright: © 2023 by the authors. Licensee MDPI, Basel, Switzerland. This article is an open access article distributed under the terms and conditions of the Creative Commons Attribution (CC BY) license (https://creativecommons.org/licenses/by/4.0/).

as hard templates [12,13]. However, the complex preparation and removal processes for hard templates such as SiO_2 microspheres increase the cost of catalyst preparation, which is not conducive to practical applications. The preparation of g-C_3N_4 precursors by treating melamine and other raw materials through the hydrothermal method is a feasible way of controlling the microstructure of g-C_3N_4. Melamine and other raw materials can be self-assembled by using the hydrothermal method, and a hard template is unnecessary [14]. In addition, strategies such as heteroatom doping, structural modification, heterojunction construction, and combination with cocatalysts are favorable for the efficient separation of photogenerated electrons and holes in g-C_3N_4 [15–18]. Furthermore, the introduction of P, S, B, O, halogen atoms, and other heteroatom dopants can be used to adjust the electrical characteristics of g-C_3N_4 and, thus, enhance its light absorption ability and inhibit the recombination of photogenerated electrons and holes [19–22]. P doping can be used to replace C or N in g-C_3N_4 and form chemical bonds with contiguous N or C, and a P-containing lone pair of electrons can serve as active sites for trapping holes, which helps to improve the conductivity and charge transfer capabilities [23,24].

In this work, precursors were obtained through the hydrothermal treatment of melamine and $NH_4H_2PO_4$ for the first time, and they were subsequently calcined to obtain P-doped hollow tubular g-C_3N_4; the obtained photocatalyst was used for the photocatalytic reduction of CO_2 with H_2O. The physicochemical properties of P-doped tubular g-C_3N_4 were investigated. In addition, the impacts of P doping on the separation and transfer of photogenerated charge carriers and the CO production rate in photocatalytic CO_2 reduction were revealed. This work provides an in-depth study of the role of P doping in boosting photocatalytic CO_2 reduction.

2. Materials and Methods

2.1. Materials

All chemicals were of analytical grade and used without further purification. Melamine and Nafion® solutions were purchased from Aladdin Industries. $NH_4H_2PO_4$ was purchased from Xilong Technology Co., Ltd. (Shantou, China).

2.2. Synthesis of the Photocatalyst

Synthesis of bulk g-C_3N_4: 5 g of melamine was placed in a crucible with a lid and transferred to a muffle furnace, which was heated to 550 °C with a heating rate of 5 °C/min, and then kept for 2 h in an air atmosphere. The bulk g-C_3N_4 was ground and collected after cooling to room temperature.

Synthesis of phosphorus-doped hollow tubular g-C_3N_4 and hollow tubular g-C_3N_4: First, the precursors were synthesized with a hydrothermal method; 5 g of melamine and x g of $NH_4H_2PO_4$ (x = 0.5, 1.0, 1.5, 2.0) were dispersed in 60 mL of deionized water and stirred for 30 min. Then, the mixture was transferred to a stainless steel autoclave with Teflon lining and kept at 180 °C for 12 h. The solid product was separated through centrifugal filtration, washed with deionized water, and then dried at 60 °C for 12 h. Finally, the obtained solid was placed in a covered crucible, and the temperature was raised to 550 °C at a rate of 2.5 °C/min and kept for 2 h. The phosphorus-doped hollow tubular g-C_3N_4 was ground and collected after cooling to room temperature; it was designated as x-P-HCN (x = 0.5, 1.0, 1.5, 2.0), and the synthesis process is shown in Scheme 1. In addition, the hollow tubular g-C_3N_4 (HCN) was prepared with the same method without adding the $NH_4H_2PO_4$.

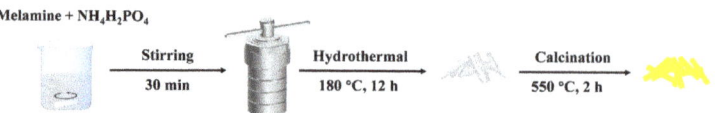

Scheme 1. Schematic illustration of the synthesis of x-P-HCN.

2.3. Characterization

X-ray powder diffraction (XRD) was performed on a SMARTLAB3KW X-ray powder diffractometer (Akishima, Japan) equipped with a Cu–Kα radiation source. Raman spectra were obtained on a LabRam HR 800 laser confocal Raman spectrometer (Paris, Franch). SEM images were recorded on a ZEISS Gemini 300 (Oberkochen, Germany) field-emission scanning electron microscope (FE-SEM). Transmission electron microscopy (TEM) and high-resolution transmission electron microscopy (HR-TEM) images and energy-dispersive spectroscopy (EDS) mappings were acquired on an FEI Talos F200S transmission electron microscope (Waltham, MA, USA). X-ray photoelectron spectroscopy (XPS) was performed on a Thermo Scientific K-Alpha system (Waltham, MA, USA). Time-resolved fluorescence spectra were acquired on an FLS 1000 steady-state/transient fluorescence spectrometer (Livingston, UK). Photoluminescence (PL) spectra were recorded on a Thermo Scientific Lumina fluorescence spectrometer (Waltham, MA, USA). Ultraviolet–visible diffuse reflectance spectra (UV–vis DRS) were recorded using a TU-19 ultraviolet–visible spectrophotometer (Beijing, China). The N_2 adsorption and desorption curves, Brunauer–Emmett–Teller (BET) specific surface area, and pore size distribution of the samples were obtained on a TriStar II system (Norcross, GA, USA). CO_2 temperature-programmed desorption (CO_2-TPD) was carried out on an Altamira AMI 300 system (Pittsburgh, PA, USA). The samples were treated in a He atmosphere at 300 °C for 1 h to eliminate surface substances contributing to physical adsorption, cooled to room temperature, injected with pure CO_2 for 1 h, and then heated from 50 °C to 450 °C at a rate of 10 °C/min. Fourier transform infrared (FT-IR) spectra and in situ diffuse reflectance infrared Fourier transform spectroscopy (DRIFTS) experiments were performed on a Bruker TENSOR II infrared spectrometer (Karlsruhe, Germany). For in situ DRIFTS, the samples were treated in Ar (30 mL/min) at 300 °C for 1 h to eliminate impurities on the sample surface and then cooled to room temperature. After that, CO_2 (20 mL/min) was passed through deionized water and introduced into the infrared cell for 60 min in the dark. The reaction temperature was maintained at 25 °C using a cooling circulating water system. Then, argon gas was injected into the infrared cell for 20 min to discharge free CO_2 and water vapor, and a 300 W xenon lamp (CEL-HXF300, Beijing China Education Au-light Co., Ltd., Beijing, China) was used as the light source. The IR spectra were collected in the dark for 60 min and under light irradiation for 60 min.

2.4. Photocatalytic CO_2 Reduction

First, 30 mg of the photocatalyst was dispersed at the bottom of a 220 mL quartz reactor containing 3 mL of deionized water, and then the photocatalyst at the bottom of the reactor was dried at 60 °C for 8 h. Before the photocatalytic reaction, CO_2 and water vapor were allowed to enter the reactor at a rate of 40 mL min^{-1} for 0.5 h by bubbling CO_2 gas through deionized water. A 300 W Xe lamp with a 400 nm cutoff filter was used as the light source. Gas products were analyzed with a Shimadzu GC-2030 gas chromatograph (Kyoto, Japan) with a barrier discharge ionization detector (BID). When carrying out the cycle experiment, the quartz reactor containing the photocatalyst was vacuum-dried after each cycle reaction; then, CO_2 and water vapor were reintroduced into the reactor, and the other reaction conditions remained unchanged.

2.5. Photoelectrochemical Measurements

Photoelectrochemical measurements were performed on an electrochemical workstation (CHI 760E, Shanghai Chenhua, Shanghai, China) with a three-electrode system. In this system, the Ag/AgCl electrode was used as the reference electrode, the platinum electrode was used as the counter-electrode, and an FTO substrate loaded with the photocatalyst was used as the working electrode. Briefly, the working electrode was prepared as follows: 20 mg of photocatalyst and 20 μL of Nafion® solution were mixed with 400 μL of absolute ethanol and sonicated for 0.5 h. Then, 20 μL of the above suspension was evenly spread on the FTO to cover an area of 1 cm × 1 cm and dried at room temperature to obtain a working electrode. In addition, 0.5 M Na_2SO_4 solution was used as the electrolyte in all photoelectrochemical measurements. The transient photocurrent response was measured with a 300 W Xe lamp (equipped with a 400 nm cutoff filter, CEL-HXF300, Beijing China Education Au-light Co., Ltd., Hangzhou, China) as the light source. Electrochemical impedance spectroscopy (EIS) was performed in the frequency range from 0.01 to 1,000,000 Hz with an amplitude of 5 mV. The Mott–Schottky plot was tested using three frequencies of 1000, 1500, and 2000 Hz.

3. Results and Discussion

The X-ray diffraction (XRD) patterns of the synthesized samples are presented in Figure 1A. Two peaks at 13° and 27.3° were found in all of the samples; these corresponded to the typical (110) and (002) of g-C_3N_4, and (100) and (002) were ascribed to the in-plane repeat unit of heptazine and the characteristic interlayer structure of g-C_3N_4, respectively [25]. Both of the (002) peaks of HCN and x-P-HCN were weaker than those of g-C_3N_4, indicating their poor in-plane periodicity, which is in agreement with other carbon nitrides with tubular structures [26].

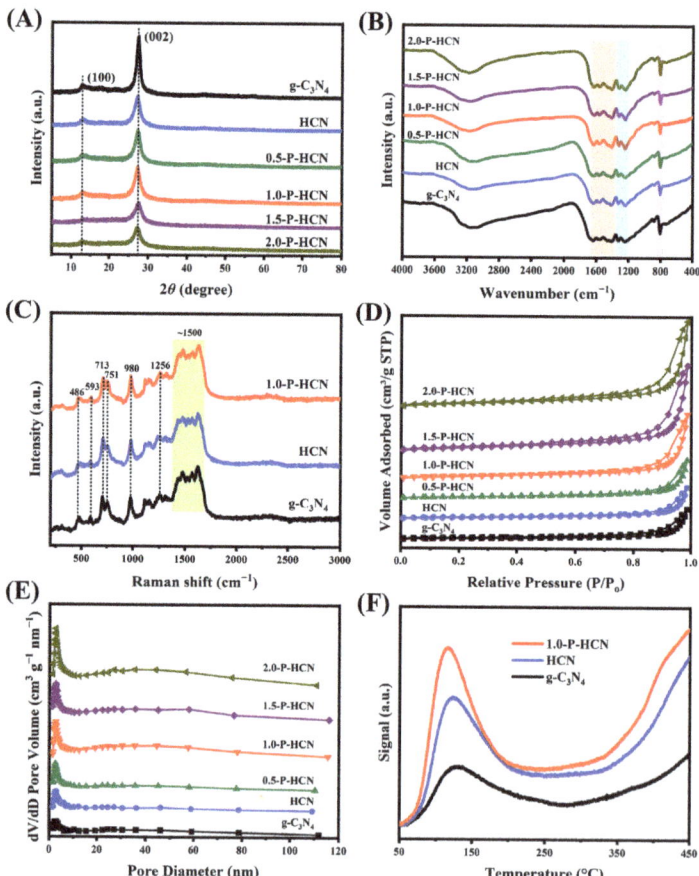

Figure 1. XRD patterns (**A**), FT-IR spectra (**B**), N_2 adsorption–desorption isotherms (**D**), and pore size distributions (**E**) of g-C_3N_4, HCN, and x-P-HCN. Raman spectra (**C**) and CO_2-TPD (**F**) of g-C_3N_4, HCN, and 1.0-P-HCN.

The chemical groups of the catalysts were determined with FT-IR spectra. As shown in Figure 1B, the peak at 808 cm^{-1} (pink area) of the different catalysts was ascribed to the out-of-plane bending vibration of the heptazine ring, and the peaks around 1408–1638 cm^{-1} (orange area) corresponded to the stretching vibration of the heptazine-derived repeating unit. Moreover, the peaks around 1240–1320 cm^{-1} (blue area) were attributed to the stretching vibration of C-N(-C)-C or C-NH-C, and the broad peak at 3000–3500 cm^{-1} was attributed to terminal uncondensed -NH or -NH$_2$ [27]. However, the absorption peaks of P-C or P-N functional groups were not observed, which may have been due to the low doping amount of P [28]. The peaks at 400–1300 cm^{-1} in the Raman spectra (Figure 1C) were attributed to typical heptazine units, and the peaks at 486, 593, 713, 751, 980, and 1256 cm^{-1} were related to the vibrational modes of CN heterocycles, among which the peaks at 713 and 1256 cm^{-1} were attributed to the ring breathing mode of s-triazine, and the broad peak at ~1500 cm^{-1} (green area) was attributed to the D and G bands of the typical graphitic structure [29,30]. Notably, the P-doped g-C$_3$N$_4$ obtained with the hydrothermal treatment had the same Raman peaks as those of g-C$_3$N$_4$, indicating that the hydrothermal pretreatment and incorporation of P did not change the framework of g-C$_3$N$_4$, which was consistent with the results of the XRD patterns and FT-IR spectra. Furthermore, no new peaks of P species were found because of the low doping amount of P.

The textural properties of different catalysts were revealed by the N$_2$ adsorption–desorption isotherms and pore size distributions. As shown in Figure 1D,E and Table S1, the adsorption isotherms of g-C$_3$N$_4$, HCN, and x-P-HCN were all type II, and the hysteresis loops all belonged to type H3, indicating that the hydrothermal treatment and P doping did not significantly change the mesoporous structure of g-C$_3$N$_4$. In addition, the BET specific surface area of HCN (7.91 m$^2 \cdot$g^{-1}) obtained after the hydrothermal pretreatment slightly increased compared with that of bulk g-C$_3$N$_4$ (6.81 m$^2 \cdot$g^{-1}). With the increase in NH$_4$H$_2$PO$_4$ addition, the specific surface area significantly increased, and the BET specific surface areas of 0.5-P-HCN, 1.0-P-HCN, 1.5-P-HCN, and 2.0-P-HCN reached 9.61, 13.85, 17.55, and 20.01 m$^2 \cdot$g^{-1}, respectively. It is worth noting that the specific surface area of g-C$_3$N$_4$ was increased by doping with P, which might have been due to the change in morphology after doping with P. It can be seen from the SEM images (Figures 2 and S1) that a hollow tubular morphology was formed and a large number of nanopores were generated in the P-doped g-C$_3$N$_4$, and this was able to greatly increase its specific surface area g-C$_3$N$_4$. This larger surface area can expose more active sites to improve the photocatalytic performance of the reduction of CO$_2$ into CO.

The CO$_2$ adsorption capacities of the three catalysts were studied with CO$_2$ temperature-programmed desorption (CO$_2$-TPD). The CO$_2$ adsorption capacity of HCN was higher than that of g-C$_3$N$_4$, indicating that the hydrothermal pretreatment of the precursor was beneficial for CO$_2$ adsorption, as shown in Figure 1F. In addition, 1.0-P-HCN exhibited the largest CO$_2$ adsorption capacity in comparison with those of g-C$_3$N$_4$ and HCN, determining that the CO$_2$ adsorption of HCN was further enhanced by doping with P, which promoted the photocatalytic reduction of CO$_2$ into CO [31].

The microstructure and morphology of the photocatalyst are exhibited in SEM images (Figures 2 and S1). g-C$_3$N$_4$ exhibited an irregular structure, while HCN showed a rodlike and nanosheet structure, indicating that the hydrothermal pretreatment of the precursor was able to impact the morphology of g-C$_3$N$_4$. After doping with P, the x-P-HCN samples displayed a hollow tubular structure, indicating that the morphology of HCN could be regulated by doping with P. The hollow tubular structure was conducive to exposing abundant active sites for CO$_2$ reduction on the photocatalyst surface and decreased the transfer distance of the photogenerated electrons and holes [32].

TEM and HRTEM images (Figures 3, S2 and S3) were used for a further investigation of the morphology and microstructure of the photocatalyst. An amorphous structure was found in the samples of g-C$_3$N$_4$, HCN, and 1.0-P-HCN. Notably, the hollow tubular morphology was not observed in the TEM images of 1.0-P-HCN, which might have been because the TEM images were obtained from some local positions of 1.0-P-HCN. In addition, based on the HADDF image and the corresponding EDS elemental mapping of 1.0-P-HCN (Figure 3D), the P was evenly distributed in the catalyst, which confirmed the incorporation of P in g-C_3N_4.

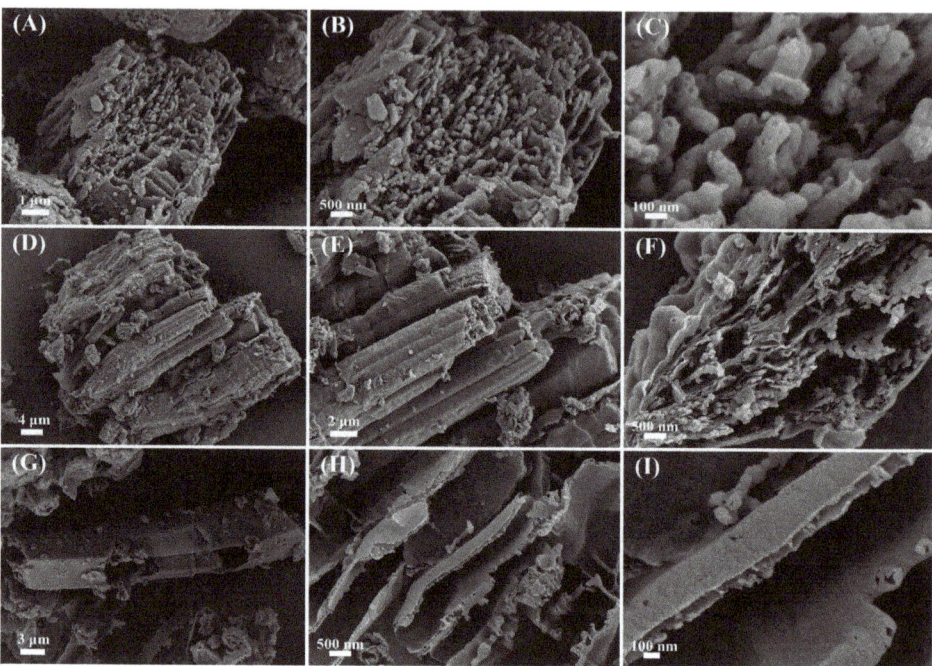

Figure 2. SEM images of g-C_3N_4 (**A–C**), HCN (**D–F**), and 1.0-P-HCN (**G–I**).

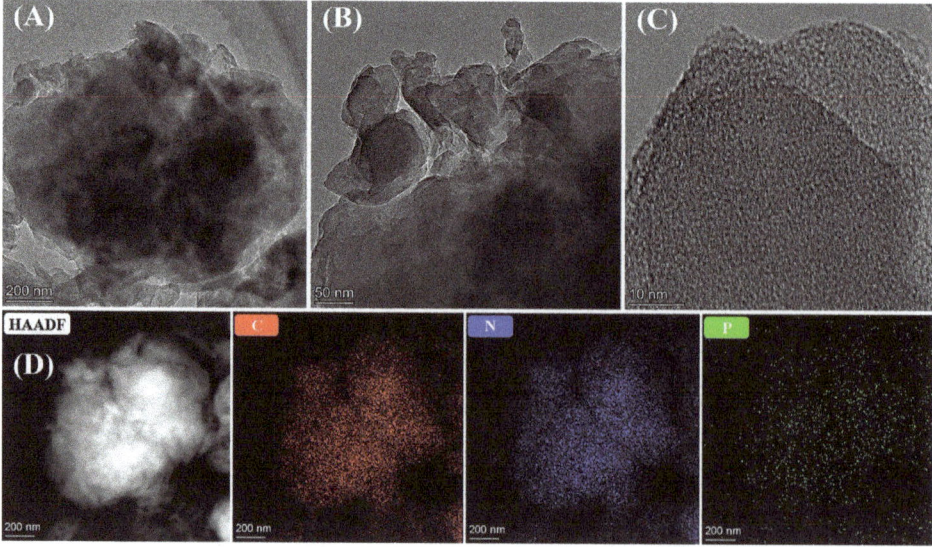

Figure 3. TEM images (**A–C**) and EDS elemental mapping (**D**) of 1.0-P-HCN.

The surface chemical states of the photocatalyst were revealed with X-ray photoelectron spectroscopy (XPS). As shown in the XPS spectra of C 1s (Figure 4A), the three peaks at 288.5, 286.6, and 284.8 eV were assigned to N-C=N of sp^3, C-N-C of sp^2, and C-C of graphite carbon, respectively, and the broad peak at 294.0 eV was attributed to π–π* excitation of interlayer [33]. In the N 1s XPS spectra (Figure 4B), three peaks could be observed at 401.6, 400.0, and 399.0 eV, corresponding to C-N-H, sp^2 hybridized N bonded to three atoms (C-N(-C)-C or C-N(-H)-C), and the aromatic N (C-N=C) in the triazine ring, respectively; the broad peak from 403 eV to 406 eV was attributed to the interlayer π–π* excitation [26]. Notably, the signal of P was not detected in the XPS spectra of 1.0-P-HCN due to the low doping amount.

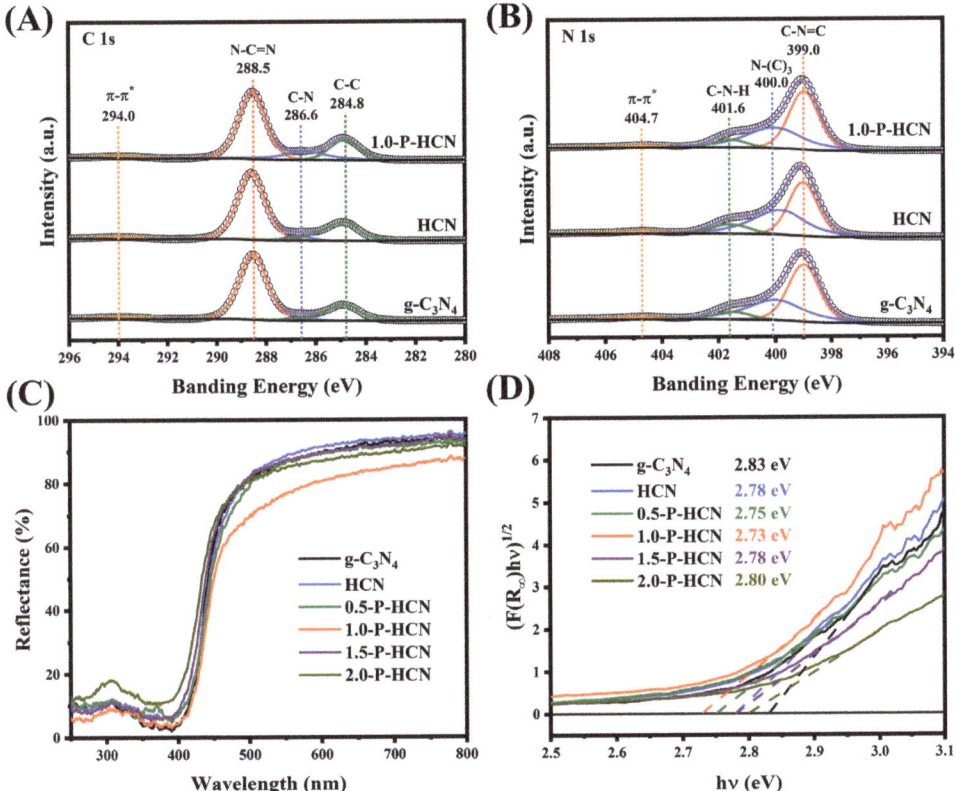

Figure 4. XPS spectra of C 1s (**A**) and N 1s (**B**) for g-C$_3$N$_4$, HCN, and 1.0-P-HCN. UV–Vis diffuse reflectance spectra (**C**) and corresponding band gaps (**D**) for g-C$_3$N$_4$, HCN, and x-P-HCN.

The light absorption abilities of g-C$_3$N$_4$, HCN, and x-P-HCN were illustrated with UV–Vis diffuse reflection spectra (Figure 4C,D). The light absorption ranges and the corresponding band gaps of g-C$_3$N$_4$ were slightly affected by doping with P. Notably, the 1.0-P-HCN sample exhibited the strongest light absorption ability compared to the other samples, which might have been because the electronic structure of g-C$_3$N$_4$ was changed by doping with moderate P [26]. Moreover, the band gaps of g-C$_3$N$_4$, HCN, 0.5-P-HCN, 1.0-P-HCN, 1.5-P-HCN, and 2.0-P-HCN were calculated to be 2.83, 2.78, 2.75, 2.73, 2.78, and 2.80 eV.

The separation and transfer of photogenerated charge carriers in the photocatalyst were revealed with steady-state and time-resolved photoluminescence spectra (Figure 5A,B).

Bulk g-C$_3$N$_4$ exhibited a strong fluorescence peak, indicating that it had a high photogenerated electron–hole recombination rate, which is unfavorable for the photocatalytic reduction of CO$_2$ [34]. However, the fluorescence peak of HCN was lower than that of bulk g-C$_3$N$_4$, indicating that the hydrothermal treatment had a beneficial effect by allowing the avoidance of the recombination of photogenerated electrons and holes. Meanwhile, after doping with P, the fluorescence intensity of g-C$_3$N$_4$ was further decreased compare with that of HCN, indicating that the separation and transfer of the photogenerated charge carrier in HCN could be enhanced by doping with P. Noteworthily, the fluorescence intensity of the 1.0-P-HCN sample was the lowest, demonstrating that the photogenerated electron–hole pairs can be greatly separated after doping with moderate P, which can boost the reaction rate of photocatalytic CO$_2$ reduction [35].

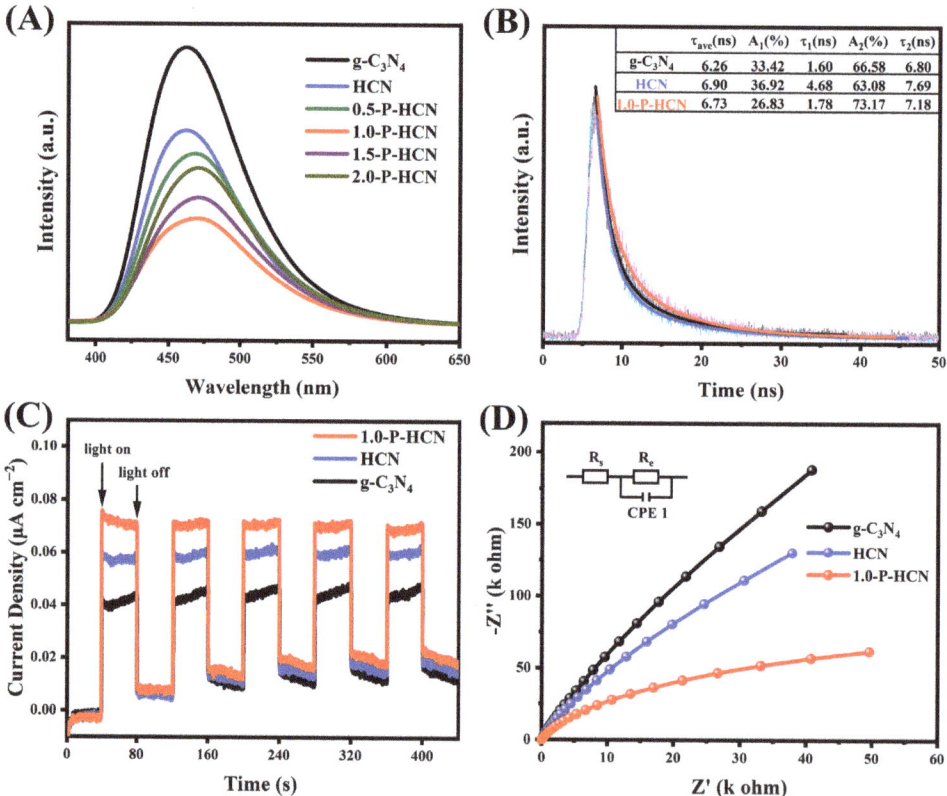

Figure 5. Steady-state PL spectra (**A**) of g-C$_3$N$_4$, HCN, and x-P-HCN. Time-resolved photoluminescence spectra (**B**), transient photocurrent density (**C**), and EIS Nyquist plots (**D**) of g-C$_3$N$_4$, HCN, and 1.0-P-HCN.

In addition, the kinetics of photoinduced electrons and holes in the g-C$_3$N$_4$, HCN, and 1.0-P-HCN samples were further investigated with their TRPL spectra, and the fluorescence lifetime was calculated [36]. From Figure 5B, the gray, cyan and pink lines are the curves of the original data of g-C$_3$N$_4$, HCN, and 1.0-P-HCN respectively, and the black, blue and red lines are the fitted curves of g-C$_3$N$_4$, HCN, and 1.0-P-HCN respectively. The average photocarrier lifetimes of g-C$_3$N$_4$, HCN, and 1.0-P-HCN were 6.26, 6.90, and 6.73 ns, respectively, indicating that the recombination rates of photoinduced electrons and holes in HCN and 1.0-P-HCN were significantly lower than that in g-C$_3$N$_4$, which further confirmed that the hydrothermal treatment of the precursor and doping with P are available ways to enhance the separation of photogenerated electron–hole pairs [37]. However, the average photocarrier lifetime of 1.0-P-HCN was shorter than that of HCN, which might have been because the surface defects were adjusted by P doping to affect the transfer of photogenerated electrons [28].

The separation and transfer of photogenerated charge carriers of g-C_3N_4, HCN, and 1.0-P-HCN were further revealed with photoelectrochemical measurements. As shown by the transient photocurrent densities of g-C_3N_4, HCN, and 1.0-P-HCN (Figure 5C), the 1.0-P-HCN sample exhibited the highest photocurrent density, showing that the incorporation of P enhanced the separation of photoinduced electron–hole pairs in g-C_3N_4, which can enhance the photocatalytic CO_2 reduction performance [38]. Furthermore, as shown by the EIS Nyquist plots of g-C_3N_4, HCN, and 1.0-P-HCN (Figure 5D), the smallest arc radius of 1.0-P-HCN indicated that it had the smallest charge transfer resistance, which accelerated the transfer of the charge carriers and enhanced the performance of photocatalytic CO_2 reduction [39].

The photocatalytic CO_2 reduction performance of different catalysts under visible light (>400 nm) was investigated. As shown in Figure 6A,B, the CO production rate of HCN was 2.78 $\mu mol \cdot g^{-1} \cdot h^{-1}$, which was 3.16 times that of g-C_3N_4 (0.88 $\mu mol \cdot g^{-1} \cdot h^{-1}$). Notably, the 1.0-P-HCN sample exhibited the best CO production rate (9.00 $\mu mol \cdot g^{-1} \cdot h^{-1}$), which was 3.24 times and 10.23 times those of HCN and g-C_3N_4. However, the CO production rates of 1.5-P-HCN and 2.0-P-HCN decreased with the increase in P doping, which might have been because the increased P could be used as a site for the recombination of photogenerated electrons and holes. To show the advantage of 1.0-P-HCN, the photocatalytic CO_2 reduction activity of some reported g-C_3N_4 and phosphorus-doped g-C_3N_4 photocatalysts is summarized in Table S2. Compared with the other photocatalysts in Table S2, 1.0-P-HCN showed excellent photocatalytic CO_2 reduction performance without cocatalysts or sacrificial agents under visible light irradiation. This was attributed to the enhanced CO_2 adsorption capacity and enhanced photogenerated carrier separation ability caused by the hollow tubular morphology and phosphorus doping.

The carbon source of the CO product was ascertained by using control experiments. As shown in Figure 6C, without CO_2, light, or a catalyst, no CO products were detected, indicating that CO was generated through the photocatalytic reduction of CO_2 by 1.0-P-HCN. Notably, when water vapor was not added to the photocatalytic reactor, the rate of photocatalytic CO production was 0.49 $\mu mol \cdot g^{-1} \cdot h^{-1}$, indicating that water vapor played a vital role in the photocatalytic CO_2 reduction reaction.

The stability of photocatalytic CO_2 reduction with 1.0-P-HCN was investigated through cycling experiments. As shown in Figure 6D, the performance of the photocatalytic reduction of CO_2 into CO did not significantly decrease after three cycles, and the average CO generation rate in the third cycle remained at 90% of that in the first cycle, which indicated that 1.0-P-HCN is stable for the photocatalytic reduction of CO_2 into CO. After three cycles of the reaction, 1.0-P-HCN was characterized via FT-IR, XRD, SEM, TEM, and XPS to clarify the its stability. As shown in Figures S4 and S5, no obvious changes were observed in the XRD patterns and FT-IR spectra of 1.0-P-HCN before and after the reaction, demonstrating that the structure of 1.0-P-HCN did not change after the reaction. In addition, the SEM images (Figure S6), TEM images, and EDS elemental mapping (Figure S7) showed that the morphology of 1.0-P-HCN before and after the reaction was unaltered. Moreover, according to the XPS spectra in Figure S8, the chemical composition and chemical state of 1.0-P-HCN did not change significantly before and after the reaction. The above results indicate that 1.0-P-HCN was stable in the photocatalytic CO_2 reduction reaction.

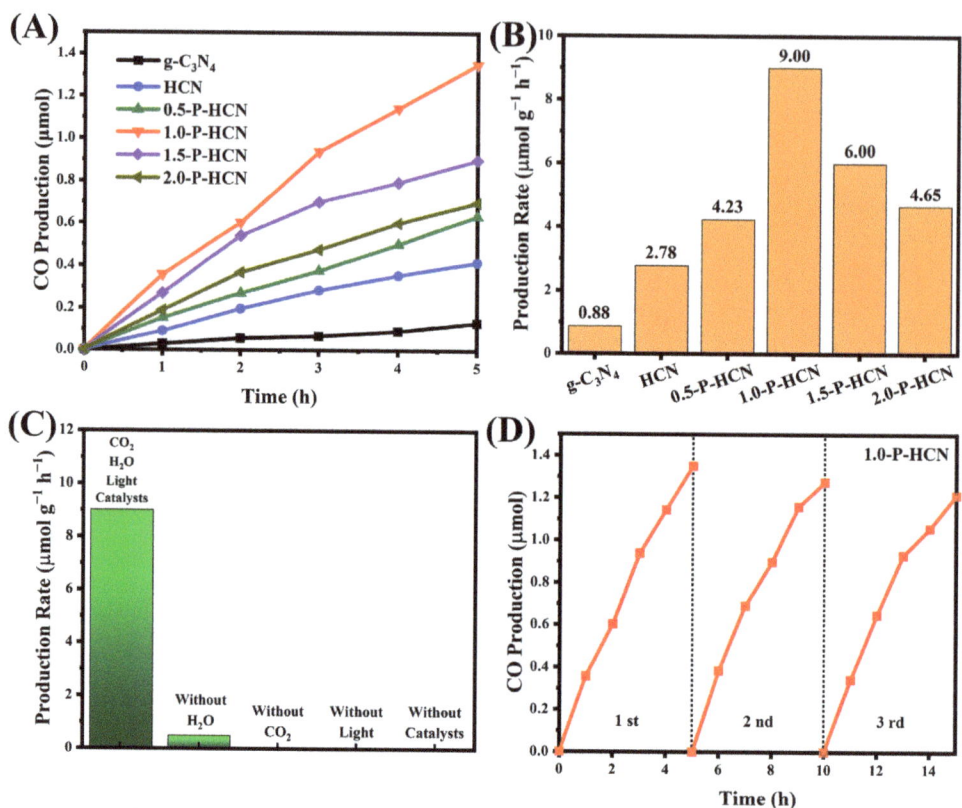

Figure 6. Yield (**A**) and production rate (**B**) of the photocatalytic reduction of CO_2 into CO for g-C_3N_4, HCN, and x-P-HCN. The CO production rate of photocatalytic CO_2 reduction by 1.0-P-HCN under different conditions (**C**). Cyclic test of photocatalytic reduction of CO_2 into CO by 1.0-P-HCN (**D**).

The in situ DRIFTS spectra were used to investigate the reaction pathway and mechanism of photocatalytic CO_2 reduction. As shown in Figure 7A, several peaks could be observed after the co-adsorption of CO_2 and H_2O vapor onto g-C_3N_4, HCN, and 1.0-P-HCN in the dark for 60 min. The four peaks at 1356, 1419, 1497, and 1509 cm^{-1} corresponded to monodentate carbonate (m-CO_3^{2-}) [40–42], and the peaks at 1396, 1436, 1457, 1473, 1647, and 1653 cm^{-1} could be attributed to HCO_3^- [12,42–44]. The peak at 1489 cm^{-1} was attributed to methoxy (CH_3O) [41], and the peaks at 1521 and 1576 cm^{-1} were attributable to bidentate carbonate (b-CO_3^{2-}) [41,43]. In addition, the peak of COOH* (The * represents an adsorbed intermediate species.) was also found at 1541 cm^{-1}; COOH* is generally suggested to be a crucial intermediate in the formation of CO [40,44]. In addition, $HCOO^-$ (1558, 1636, and 1793 cm^{-1}) [12,41,45], CO_2^- (1670, 1684, and 1698 cm^{-1}) [37,43,44], and chelating-bridged carbonate (c-CO_3^{2-}, 1715, 1733, 1748, and 1773 cm^{-1}) [42,46] can be observed in Figure 7A. Furthermore, the peaks at 1868, 2017, and 2179 cm^{-1} ascribed to CO* [42,44,46] can be observed in Figure 7B. These carbon species are important intermediates for CO_2 conversion. Notably, the peak intensity of intermediates on 1.0-P-HCN is much stronger than those of g-C_3N_4 and HCN, indicating that more CO_2 can be adsorbed and activated on the surface of 1.0-P-HCN.

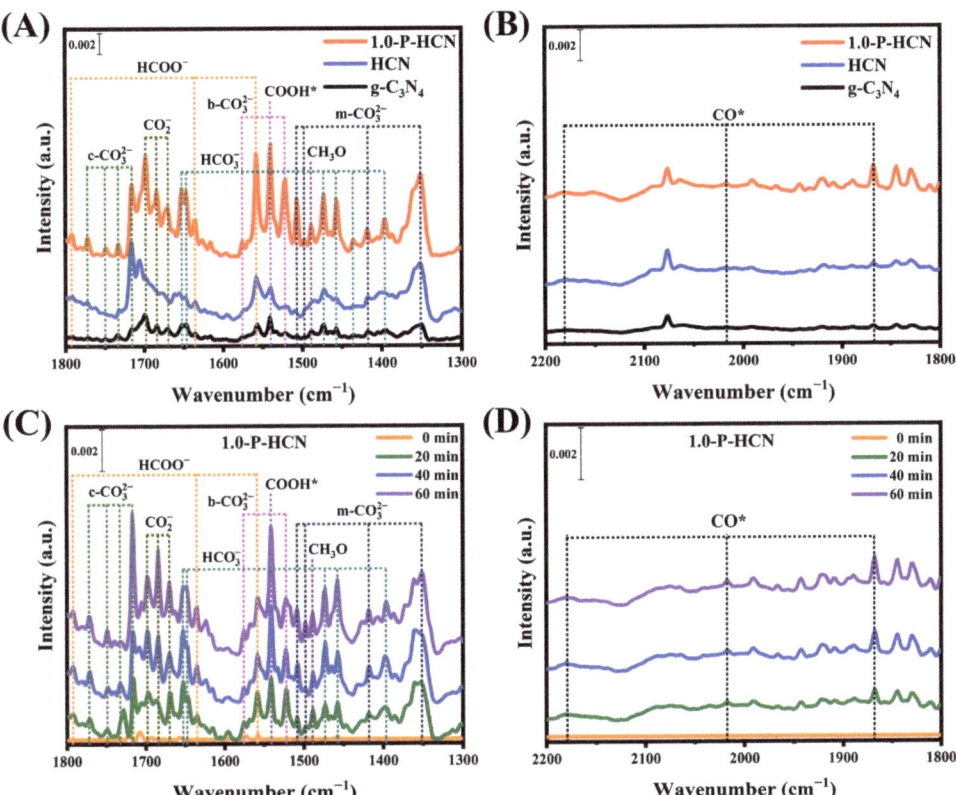

Figure 7. In situ DRIFTS spectra of g-C$_3$N$_4$, HCN, and 1.0-P-HCN with CO$_2$ and H$_2$O vapor for 60 min in the dark (**A**,**B**). In situ DRIFTS spectra of 1.0-P-HCN with CO$_2$ and H$_2$O vapor under illumination (**C**,**D**).

The in situ DRIFTS spectra of 1.0-P-HCN with CO$_2$ and H$_2$O vapor under illumination were also recorded to reveal the impact of light on the activation of adsorbed CO$_2$ with 1.0-P-HCN. As shown in Figure 7C,D, the peak intensity of the adsorbed carbon species was greatly enhanced after the light was turned on, indicating that the adsorption capacity for CO$_2$ was increased under light irradiation. In addition, the peak intensity of various CO$_2$ intermediates was increased with the extension of the illumination time. Compared to the in situ DRIFTS spectra in the dark, no new characteristic peaks were detected. However, the ratio of the COOH* peak was greatly enhanced under light irradiation, and the CO* was also increased with the increase in the illumination time. Therefore, the photocatalytic CO$_2$ reduction reaction with 1.0-P-HCN might follow the pathway of CO$_2$→COOH*→CO*→CO.

Mott–Schottky plots were obtained to determine the Fermi level of the photocatalyst. As shown in Figure S9, the positive slope of the Mott–Schottky plots suggests that the photocatalysts were n-type semiconductors. The flat band potentials (E_{fb}) of g-C$_3$N$_4$, HCN, and 1.0-P-HCN were ascertained to be −1.03, −0.97, and −1.08 V (vs. Ag/AgCl, pH = 7). According to Equation (1), where E^θ(Ag/AgCl) = 0.197 V at 25 °C [47], the E_{fb} values of g-C$_3$N$_4$, HCN, and 1.0-P-HCN were calculated to be −0.42, −0.36, and −0.47 V (vs. NHE, pH = 0). Theoretically, the value of the Fermi level of the photocatalyst was equal to the flat band potential. Therefore, the Fermi levels of g-C$_3$N$_4$, HCN, and 1.0-P-HCN were considered to be −0.42, −0.36, and −0.47 V (vs. NHE, pH = 0) [12].

$$E(NHE) = E(Ag/AgCl) + E^\theta(Ag/AgCl) + 0.059 pH \tag{1}$$

In addition, the valence band position relative to the Fermi level was measured with the XPS-VB spectra (Figure S10). The valence bands of g-C$_3$N$_4$, HCN, and 1.0-P-HCN relative to the Fermi level were 2.88, 2.73, and 2.70 eV. Therefore, the valence band positions of g-C$_3$N$_4$, HCN, and 1.0-P-HCN were calculated to be 2.46, 2.37, and 2.23 V, respectively (vs. NHE, pH = 0). Moreover, as shown in Figure 4D, the band gaps of g-C$_3$N$_4$, HCN, and 1.0-P-HCN were 2.83, 2.78, and 2.73 eV. Hence, the conduction band positions of g-C$_3$N$_4$, HCN, and 1.0-P-HCN were calculated to be −0.37, −0.41, and −0.50 V (vs. NHE, pH = 0). The positions of CB were more negative than that of the reduction of CO$_2$ into CO (−0.12 V vs. NHE, pH = 0), indicating that g-C$_3$N$_4$, HCN, and 1.0-P-HCN were able to reduce CO$_2$ into CO. The band arrangements of g-C$_3$N$_4$, HCN, and 1.0-P-HCN are shown in Figure 8A. In Figure 8A, we can find that 1.0-P-HCN had the narrowest band gap, which enhanced the photocatalytic CO$_2$ reaction under visible light. Moreover, 1.0-P-HCN had the most negative conduction band, which indicated that it had the strongest reducibility among these samples.

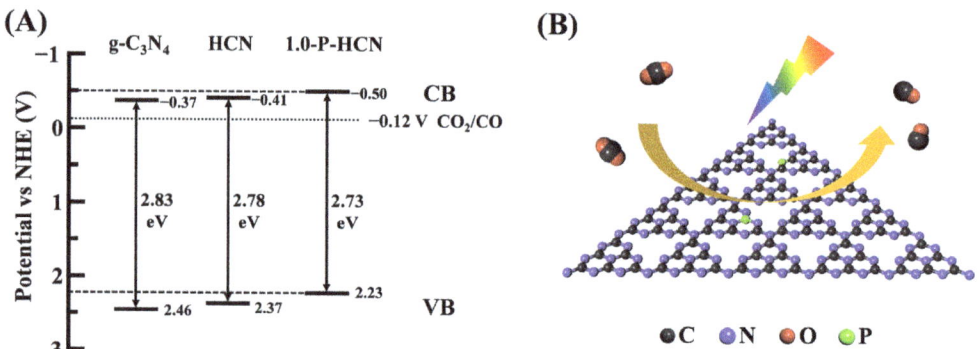

Figure 8. Energy band alignment of g-C$_3$N$_4$, HCN, and 1.0-P-HCN (**A**). Schematic illustration of photocatalytic CO$_2$ reduction to generate CO on 1.0-P-HCN (**B**).

According to the above results of the photocatalytic CO$_2$ reduction test, characterization, and in situ DRIFTS, a mechanism of photocatalytic CO$_2$ reduction with 1.0-P-HCN is proposed (Figure 8B). Under visible light illumination, photogenerated electrons were excited and transitioned from VB to CB of 1.0-P-HCN and were then transferred to the surface of 1.0-P-HCN for the photocatalytic CO$_2$ reduction reaction. The CO$_2$ was first adsorbed and activated on the surface of 1.0-P-HCN and then converted into the adsorbed carbon intermediates; then, the generation of CO followed the pathway of CO$_2$→COOH*→CO*→CO. Furthermore, the P doping of g-C$_3$N$_4$ enhanced its light absorption properties, reduced its charge transfer resistance, improved its CO$_2$ adsorption capacity, and improved the separation efficiency of the photogenerated electron–hole pairs, all of which is conducive to boosting photocatalytic CO$_2$ reduction.

4. Conclusions

In summary, NH$_4$H$_2$PO$_4$ was used as a phosphorus source in the hydrothermal treatment of melamine for the first time, and phosphorus-doped hollow tubular g-C$_3$N$_4$ (x-P-HCN) photocatalysts were finally prepared for photocatalytic CO$_2$ reduction. x-P-HCN presented much higher photocatalytic CO$_2$ reduction performance than that of g-C$_3$N$_4$. The production rate of CO with 1.0-P-HCN reached 9.00 μmol·g^{-1}·h^{-1}, which was 10.22

times that of bulk g-C_3N_4. Moreover, the enhanced performance of 1.0-P-HCN might have been due to the increased specific surface area, excellent CO_2 adsorption ability, enhanced light absorption capacity, lower charge transfer resistance, and more efficient separation of photogenerated electrons and holes of 1.0-P-HCN, which significantly boosted its photocatalytic CO_2 reduction. This work provides an in-depth research perspective for accelerating the photocatalytic CO_2 reduction rate of g-C_3N_4.

Supplementary Materials: The following supporting information can be downloaded at: https://www.mdpi.com/article/10.3390/ma16206665/s1, Figure S1. SEM images of 0.5-P-HCN (A, B, C), 1.5-P-HCN (D, E, F), and 2.0-P-HCN (G, H, I). Figure S2. TEM images (A, B, C) and EDS elemental mapping (D) of g-C_3N_4. Figure S3. TEM images (A, B, C) and EDS elemental mapping (D) of HCN. Figure S4. FT-IR spectra of 1.0-P-HCN before and after reaction for three cycles. Figure S5. XRD patterns of 1.0-P-HCN before and after reaction for three cycles. Figure S6. SEM images of 1.0-P-HCN after reaction for three cycles. Figure S7. TEM images (A, B, C) and EDS elemental mapping (D) of 1.0-P-HCN after reaction for three cycles. Figure S8. XPS spectra of C 1s (A), N 1s (B) of 1.0-P-HCN before and after reaction for three cycles. Figure S9. Mott-Schottky plots of g-C_3N_4 (A), HCN (B), and 1.0-P-HCN (C) at the frequency of 1000 Hz, 1500 Hz, and 2000 Hz. Figure S10. XPS valence band spectra of g-C_3N_4 (A), HCN (B) and 1.0-P-HCN (C). Table S1. Specific surface area and average pore diameter of g-C_3N_4, HCN, and x-P-HCN. Table S2. Summary of the photocatalytic CO_2 reduction performance over g-C_3N_4 and phosphorus doped g-C_3N_4 photocatalysts. References [48–52] are cited in the supplementary Materials

Author Contributions: Methodology, T.S.; Formal analysis, L.C.; Investigation, M.S., C.Z. and S.W.; Writing—original draft, M.S.; Writing—review & editing, H.J., T.S. and Z.Q.; Supervision, H.J., T.S. and Z.Q.; Project administration, T.S.; Funding acquisition, T.S. and Z.Q. All authors have read and agreed to the published version of the manuscript.

Funding: This work was supported by the National Natural Science Foundation of China (22208065), Guangxi Natural Science Foundation (2022GXNSFBA035483, 2020GXNSFDA297007), Opening Project of Guangxi Key Laboratory of Petrochemical Resource Processing and Process Intensification Technology (2021K009), Special Funding for 'Guangxi Bagui Scholars', and National College Students' Innovation and Entrepreneurship Training Programs (202210593011).

Institutional Review Board Statement: Not applicable.

Informed Consent Statement: Not applicable.

Data Availability Statement: Raw data are available upon request.

Conflicts of Interest: The authors declare no conflict of interest.

References

1. Vu, N.N.; Kaliaguine, S.; Do, T.O. Critical Aspects and Recent Advances in Structural Engineering of Photocatalysts for Sunlight-Driven Photocatalytic Reduction of CO_2 into Fuels. *Adv. Funct. Mater.* **2019**, *29*, 1901825. [CrossRef]
2. Xu, Q.L.; Xia, Z.H.; Zhang, J.M.; Wei, Z.Y.; Guo, Q.; Jin, H.L.; Tang, H.; Li, S.Z.; Pan, X.C.; Su, Z.; et al. Recent advances in solar-driven CO_2 reduction over g-C_3N_4-based photocatalysts. *Carbon Energy* **2023**, *5*, e205. [CrossRef]
3. Liu, X.J.; Chen, T.Q.; Xue, Y.H.; Fan, J.C.; Shen, S.L.; Hossain, M.S.A.; Amin, M.A.; Pan, L.K.; Xu, X.T.; Yamauchi, Y. Nanoarchitectonics of MXene/semiconductor heterojunctions toward artificial photosynthesis via photocatalytic CO_2 reduction. *Coord. Chem. Rev.* **2022**, *459*, 214440. [CrossRef]
4. Dong, M.; Li, W.; Zhou, J.; You, S.Q.; Sun, C.Y.; Yao, X.H.; Qin, C.; Wang, X.L.; Su, Z.M. Microenvironment Modulation of Imine-Based Covalent Organic Frameworks for CO_2 Photoreduction. *Chin. J. Chem.* **2022**, *40*, 2678–2684. [CrossRef]
5. Zhao, G.Q.; Hu, J.; Long, X.; Zou, J.; Yu, J.G.; Jiao, F.P. A Critical Review on Black Phosphorus-Based Photocatalytic CO_2 Reduction Application. *Small* **2021**, *17*, 2102155. [CrossRef]
6. Chen, L.Y.; Huang, K.L.; Xie, Q.R.; Lam, S.M.; Sin, J.C.; Su, T.M.; Ji, H.B.; Qin, Z.Z. The enhancement of photocatalytic CO_2 reduction by the in situ growth of TiO_2 on Ti_3C_2 MXene. *Catal. Sci. Technol.* **2021**, *11*, 1602–1614. [CrossRef]
7. Su, T.M.; Ma, X.H.; Tong, J.H.; Ji, H.B.; Qin, Z.Z.; Wu, Z.L. Surface engineering of MXenes for energy and environmental applications. *J. Mater. Chem. A* **2022**, *10*, 10265–10296. [CrossRef]
8. Catherine, H.N.; Liu, Z.T.; Lin, C.Y.; Chung, P.W.; Tsunekawa, S.; Lin, S.D.; Yoshida, M.; Hu, C. Understanding the intermediates and carbon dioxide adsorption of potassium chloride-incorporated graphitic carbon nitride with tailoring melamine and urea as precursors. *J. Colloid Interface Sci.* **2023**, *633*, 598–607. [CrossRef] [PubMed]

9. Alshamkhani, M.T.; Putri, L.K.; Lahijani, P.; Lee, K.T.; Mohamed, A.R. A metal-free electrochemically exfoliated graphene/graphitic carbon nitride nanocomposite for CO_2 photoreduction to methane under visible light irradiation. *J. Environ. Chem. Eng.* **2023**, *11*, 109086. [CrossRef]
10. Pei, X.Q.; An, W.X.; Zhao, H.L.; He, H.; Fu, Y.C.; Shen, X.M. Enhancing visible-light degradation performance of g-C_3N_4 on organic pollutants by constructing heterojunctions via combining tubular g-C_3N_4 with Bi_2O_3 nanosheets. *J. Alloys Compd.* **2023**, *934*, 167928. [CrossRef]
11. Yu, Y.T.; Huang, H.W. Coupled adsorption and photocatalysis of g-C_3N_4 based composites: Material synthesis, mechanism, and environmental applications. *Chem. Eng. J.* **2023**, *453*, 139755. [CrossRef]
12. Xiao, Y.; Men, C.Z.; Chu, B.X.; Qin, Z.Z.; Ji, H.B.; Chen, J.H.; Su, T.M. Spontaneous reduction of copper on $Ti_3C_2T_x$ as fast electron transport channels and active sites for enhanced photocatalytic CO_2 reduction. *Chem. Eng. J.* **2022**, *446*, 137028. [CrossRef]
13. Cui, L.F.; Song, J.L.; McGuire, A.F.; Kang, S.F.; Fang, X.Y.; Wang, J.J.; Yin, C.C.; Li, X.; Wang, Y.G.; Cui, B.X. Constructing Highly Uniform Onion-Ring-like Graphitic Carbon Nitride for Efficient Visible-Light-Driven Photocatalytic Hydrogen Evolution. *ACS Nano* **2018**, *12*, 5551–5558. [CrossRef]
14. Wang, Z.Y.; Chen, M.J.; Huang, Y.; Shi, X.J.; Zhang, Y.F.; Huang, T.T.; Cao, J.J.; Ho, W.K.; Lee, S.C. Self-assembly synthesis of boron-doped graphitic carbon nitride hollow tubes for enhanced photocatalytic NO_x removal under visible light. *Appl. Catal. B* **2018**, *239*, 352–361. [CrossRef]
15. Liang, Y.J.; Wu, X.; Liu, X.Y.; Li, C.H.; Liu, S.W. Recovering solar fuels from photocatalytic CO_2 reduction over W^{6+}-incorporated crystalline g-C_3N_4 nanorods by synergetic modulation of active centers. *Appl. Catal. B* **2022**, *304*, 120978. [CrossRef]
16. Li, Y.H.; Sun, Y.J.; Ho, W.K.; Zhang, Y.X.; Huang, H.W.; Cai, Q.; Dong, F. Highly enhanced visible-light photocatalytic NO_x purification and conversion pathway on self-structurally modified g-C_3N_4 nanosheets. *Sci. Bull.* **2018**, *63*, 609–620. [CrossRef] [PubMed]
17. Cao, H.; Yan, Y.M.; Wang, Y.; Chen, F.F.; Yu, Y. Dual role of g-C_3N_4 microtubes in enhancing photocatalytic CO_2 reduction of Co_3O_4 nanoparticles. *Carbon* **2023**, *201*, 415–424. [CrossRef]
18. Tan, X.; Wu, S.W.; Li, Y.Z.; Zhang, Q.; Hu, Q.Q.; Wu, J.C.; Zhang, A.; Zhang, Y.D. Highly Efficient Photothermocatalytic CO_2 Reduction in Ni/Mg-Doped Al_2O_3 with High Fuel Production Rate, Large Light-to-Fuel Efficiency, and Good Durability. *Energy Environ. Mater.* **2022**, *5*, 582–591. [CrossRef]
19. Jiang, J.Z.; Xiong, Z.G.; Wang, H.T.; Liao, G.D.; Bai, S.S.; Zou, J.; Wu, P.X.; Zhang, P.; Li, X. Sulfur-doped g-C_3N_4/g-C_3N_4 isotype step-scheme heterojunction for photocatalytic H_2 evolution. *J. Mater. Sci. Technol.* **2022**, *118*, 15–24. [CrossRef]
20. Liu, Q.Q.; Shen, J.Y.; Yu, X.H.; Yang, X.F.; Liu, W.; Yang, J.; Tang, H.; Xu, H.; Li, H.M.; Li, Y.Y.; et al. Unveiling the origin of boosted photocatalytic hydrogen evolution in simultaneously (S, P, O)-Codoped and exfoliated ultrathin g-C_3N_4 nanosheets. *Appl. Catal. B* **2019**, *248*, 84–94. [CrossRef]
21. Kim, D.; Yong, K. Boron doping induced charge transfer switching of a C_3N_4/ZnO photocatalyst from Z-scheme to type II to enhance photocatalytic hydrogen production. *Appl. Catal. B* **2021**, *282*, 119538. [CrossRef]
22. Fu, J.W.; Zhu, B.C.; Jiang, C.J.; Cheng, B.; You, W.; Yu, J.G. Hierarchical Porous O-Doped g-C_3N_4 with Enhanced Photocatalytic CO_2 Reduction Activity. *Small* **2017**, *13*, 1603938. [CrossRef] [PubMed]
23. Patnaik, S.; Sahoo, D.P.; Parida, K. Recent advances in anion doped g-C_3N_4 photocatalysts: A review. *Carbon* **2021**, *172*, 682–711. [CrossRef]
24. Wang, G.; Chen, Z.; Wang, T.; Wang, D.S.; Mao, J.J. P and Cu Dual Sites on Graphitic Carbon Nitride for Photocatalytic CO_2 Reduction to Hydrocarbon Fuels with High C_2H_6 Evolution. *Angew. Chem. Int. Ed.* **2022**, *61*, e202210789. [CrossRef]
25. Wang, X.C.; Maeda, K.; Thomas, A.; Takanabe, K.; Xin, G.; Carlsson, J.M.; Domen, K.; Antonietti, M. A metal-free polymeric photocatalyst for hydrogen production from water under visible light. *Nat. Mater.* **2009**, *8*, 76–80. [CrossRef]
26. Wu, M.; Zhang, J.; He, B.B.; Wang, H.W.; Wang, R.; Gong, Y.S. In-situ construction of coral-like porous P-doped g-C_3N_4 tubes with hybrid 1D/2D architecture and high efficient photocatalytic hydrogen evolution. *Appl. Catal. B* **2019**, *241*, 159–166. [CrossRef]
27. Liu, J.H.; Zhang, T.K.; Wang, Z.C.; Dawson, G.; Chen, W. Simple pyrolysis of urea into graphitic carbon nitride with recyclable adsorption and photocatalytic activity. *J. Mater. Chem.* **2011**, *21*, 14398–14401. [CrossRef]
28. Zhou, Y.J.; Zhang, L.X.; Liu, J.J.; Fan, X.Q.; Wang, B.Z.; Wang, M.; Ren, W.C.; Wang, J.; Li, M.L.; Shi, J.L. Brand new P-doped g-C_3N_4: Enhanced photocatalytic activity for H_2 evolution and Rhodamine B degradation under visible light. *J. Mater. Chem. A* **2015**, *3*, 3862–3867. [CrossRef]
29. Dong, H.; Guo, X.T.; Yang, C.; Ouyang, Z.Z. Synthesis of g-C_3N_4 by different precursors under burning explosion effect and its photocatalytic degradation for tylosin. *Appl. Catal. B* **2018**, *230*, 65–76. [CrossRef]
30. Lan, Z.A.; Zhang, G.G.; Wang, X.C. A facile synthesis of Br-modified g-C_3N_4 semiconductors for photoredox water splitting. *Appl. Catal. B* **2016**, *192*, 116–125. [CrossRef]
31. Yu, H.; Sun, C.; Xuan, Y.; Zhang, K.; Chang, K. Full solar spectrum driven plasmonic-assisted efficient photocatalytic CO_2 reduction to ethanol. *Chem. Eng. J.* **2022**, *430*, 132940. [CrossRef]
32. Guo, S.E.; Deng, Z.P.; Li, M.X.; Jiang, B.J.; Tian, C.G.; Pan, Q.J.; Fu, H.G. Phosphorus-Doped Carbon Nitride Tubes with a Layered Micro-nanostructure for Enhanced Visible-Light Photocatalytic Hydrogen Evolution. *Angew. Chem. Int. Ed.* **2016**, *55*, 1830–1834. [CrossRef] [PubMed]

33. Mohamed, M.A.; Zain, M.F.M.; Jeffery Minggu, L.; Kassim, M.B.; Saidina Amin, N.A.; Salleh, W.N.W.; Salehmin, M.N.I.; Md Nasir, M.F.; Mohd Hir, Z.A. Constructing bio-templated 3D porous microtubular C-doped g-C_3N_4 with tunable band structure and enhanced charge carrier separation. *Appl. Catal. B* **2018**, *236*, 265–279. [CrossRef]
34. Jiang, J.; Cao, S.W.; Hu, C.L.; Chen, C.H. A comparison study of alkali metal-doped g-C_3N_4 for visible-light photocatalytic hydrogen evolution. *Chin. J. Catal.* **2017**, *38*, 1981–1989. [CrossRef]
35. Hu, S.Z.; Ma, L.; You, J.G.; Li, F.Y.; Fan, Z.P.; Wang, F.; Liu, D.; Gui, J.Z. A simple and efficient method to prepare a phosphorus modified g-C_3N_4 visible light photocatalyst. *RSC Adv.* **2014**, *4*, 21657–21663. [CrossRef]
36. Su, T.M.; Men, C.Z.; Chen, L.Y.; Chu, B.X.; Luo, X.; Ji, H.B.; Chen, J.H.; Qin, Z.Z. Sulfur Vacancy and $Ti_3C_2T_x$ Cocatalyst Synergistically Boosting Interfacial Charge Transfer in 2D/2D $Ti_3C_2T_x$/$ZnIn_2S_4$ Heterostructure for Enhanced Photocatalytic Hydrogen Evolution. *Adv. Sci.* **2022**, *9*, 2103715. [CrossRef]
37. Wang, R.N.; Wang, Z.; Qiu, Z.Y.; Wan, S.P.; Ding, J.; Zhong, Q. Nanoscale 2D g-C_3N_4 decorating 3D hierarchical architecture LDH for artificial photosynthesis and mechanism insight. *Chem. Eng. J.* **2022**, *448*, 137338. [CrossRef]
38. Liu, Y.Z.; Zhao, L.; Zeng, X.H.; Xiao, F.; Fang, W.; Du, X.; He, X.; Wang, D.H.; Li, W.X.; Chen, H. Efficient photocatalytic reduction of CO_2 by improving adsorption activation and carrier utilization rate through N-vacancy g-C_3N_4 hollow microtubule. *Mater. Today Energy* **2023**, *31*, 101211. [CrossRef]
39. Song, X.H.; Zhang, X.Y.; Li, X.; Che, H.N.; Huo, P.W.; Ma, C.C.; Yan, Y.S.; Yang, G.Y. Enhanced light utilization efficiency and fast charge transfer for excellent CO_2 photoreduction activity by constructing defect structures in carbon nitride. *J. Colloid Interface Sci.* **2020**, *578*, 574–583. [CrossRef]
40. Ojha, N.; Bajpai, A.; Kumar, S. Visible light-driven enhanced CO_2 reduction by water over Cu modified S-doped g-C_3N_4. *Catal. Sci. Technol.* **2019**, *9*, 4598–4613. [CrossRef]
41. Fang, R.M.; Yang, Z.Q.; Kadirova, Z.C.; He, Z.Q.; Wang, Z.Q.; Ran, J.Y.; Zhang, L. High-efficiency photoreduction of CO_2 to solar fuel on alkali intercalated Ultra-thin g-C_3N_4 nanosheets and enhancement mechanism investigation. *Appl. Surf. Sci.* **2022**, *598*, 153848. [CrossRef]
42. Ran, L.; Li, Z.W.; Ran, B.; Cao, J.Q.; Zhao, Y.; Shao, T.; Song, Y.R.; Leung, M.K.H.; Sun, L.C.; Hou, J.G. Engineering Single-Atom Active Sites on Covalent Organic Frameworks for Boosting CO_2 Photoreduction. *J. Am. Chem. Soc.* **2022**, *144*, 17097–17109. [CrossRef] [PubMed]
43. Qin, H.; Guo, R.T.; Liu, X.Y.; Shi, X.; Wang, Z.Y.; Tang, J.Y.; Pan, W.G. 0D NiS_2 quantum dots modified 2D g-C_3N_4 for efficient photocatalytic CO_2 reduction. *Colloids Surf. A* **2020**, *600*, 124912. [CrossRef]
44. Chen, F.F.; Chen, J.B.; Li, L.Y.; Peng, F.; Yu, Y. g-C_3N_4 microtubes@$CoNiO_2$ nanosheets p–n heterojunction with a hierarchical hollow structure for efficient photocatalytic CO_2 reduction. *Appl. Surf. Sci.* **2022**, *579*, 151997. [CrossRef]
45. Li, Q.; Sun, Z.X.; Wang, H.Q.; Wu, Z.B. Insight into the enhanced CO_2 photocatalytic reduction performance over hollow-structured Bi-decorated g-C_3N_4 nanohybrid under visible-light irradiation. *J. CO2 Util.* **2018**, *28*, 126–136. [CrossRef]
46. He, W.J.; Wei, Y.C.; Xiong, J.; Tang, Z.L.; Wang, Y.L.; Wang, X.; Deng, J.G.; Yu, X.L.; Zhang, X.; Zhao, Z. Boosting selective photocatalytic CO_2 reduction to CO over Dual-core@shell structured Bi_2O_3/Bi_2WO_6@g-C_3N_4 catalysts with strong interaction interface. *Sep. Purif. Technol.* **2022**, *300*, 121850. [CrossRef]
47. Chen, L.Y.; Xie, X.L.; Su, T.M.; Ji, H.B.; Qin, Z.Z. Co_3O_4/CdS p-n heterojunction for enhancing photocatalytic hydrogen production: Co-S bond as a bridge for electron transfer. *Appl. Surf. Sci.* **2021**, *567*, 150849. [CrossRef]
48. Wan, S.P.; Ou, M.; Zhong, Q.; Cai, W. Haloid Acid Induced Carbon Nitride Semiconductors for Enhanced Photocatalytic H_2 Evolution and Reduction of CO_2 under Visible Light. *Carbon* **2018**, *138*, 465–474. [CrossRef]
49. Liu, B.; Ye, L.Q.; Wang, R.; Yang, J.F.; Zhang, Y.X.; Guan, R.; Tian, L.H.; Chen, X.B. Phosphorus-Doped Graphitic Carbon Nitride Nanotubes with Amino-Rich Surface for Efficient CO_2 Capture, Enhanced Photocatalytic Activity, and Product Selectivity. *ACS Appl. Mater. Interfaces* **2018**, *10*, 4001–4009. [CrossRef]
50. Liu, W.Z.; Sun, M.X.; Ding, Z.P.; Gao, B.W.; Ding, W. Ti_3c_2 MXene Embellished g-C_3N_4 Nanosheets for Improving Photocatalytic Redox Capacity. *J. Alloys Compd.* **2021**, *877*, 160223. [CrossRef]
51. Liu, X.L.; Wang, P.; Zhai, H.S.; Zhang, Q.Q.; Huang, B.B.; Wang, Z.Y.; Liu, Y.Y.; Dai, Y.; Qin, X.Y.; Zhang, X.Y. Synthesis of Synergetic Phosphorus and Cyano Groups (Cn) Modified g-C_3N_4 for Enhanced Photocatalytic H_2 Production and CO_2 Reduction under Visible Light Irradiation. *Appl. Catal. B Environ.* **2018**, *232*, 521–530. [CrossRef]
52. Wang, W.F.; Qiu, L.Q.; Chen, K.H.; Li, H.R.; Feng, L.F.; He, L.N. Morphology and Element Doping Effects: Phosphorus-Doped Hollow Polygonal g-C_3N_4 Rods for Visible Light-Driven CO_2 Reduction. *New J. Chem.* **2022**, *46*, 3017–3025. [CrossRef]

Disclaimer/Publisher's Note: The statements, opinions and data contained in all publications are solely those of the individual author(s) and contributor(s) and not of MDPI and/or the editor(s). MDPI and/or the editor(s) disclaim responsibility for any injury to people or property resulting from any ideas, methods, instructions or products referred to in the content.

Article

Treatment of Mixture Pollutants with Combined Plasma Photocatalysis in Continuous Tubular Reactors with Atmospheric-Pressure Environment: Understanding Synergetic Effect Sources

Lotfi Khezami [1,*] and Aymen Amin Assadi [2,3,*]

1. Department of Chemistry, College of Sciences, Imam Mohammad Ibn Saud Islamic University (IMSIU), P.O. Box 5701, Riyadh 11432, Saudi Arabia
2. College of Engineering, Imam Mohammad Ibn Saud Islamic University (IMSIU), P.O. Box 5701, Riyadh 11432, Saudi Arabia
3. Univ. Rennes, École Nationale Supérieure de Chimie de Rennes, CNRS, ISCR (Institut. des Sciences Chimiques de Rennes)—UMR 6226, Campus de Beaulieu, Av. du Général Leclerc, 35700 Rennes, France
* Correspondence: lhmkhezami@imamu.edu.sa (L.K.); aaassadi@imamu.edu.sa or aymen.assadi@ensc-rennes.fr (A.A.A.); Tel.: +33-2-23238152 (L.K.)

Abstract: This study investigates the pilot-scale combination of nonthermal plasma and photocatalysis for removing Toluene and dimethyl sulfur (DMDS), examining the influence of plasma energy and initial pollutant concentration on the performance and by-product formation in both pure compounds and mixtures. The results indicate a consistent 15% synergy effect, improving Toluene conversion rates compared to single systems. Ozone reduction and enhanced CO_2 selectivity were observed when combining plasma and photocatalysis. This process effectively treats pollutant mixtures, even those containing sulfur compounds. Furthermore, tests confirm nonthermal plasma's in-situ regeneration of the photocatalytic surface, providing a constant synergy effect.

Keywords: mixture of pollutants; coupling system; plasma; photocatalysis; synergetic effect; mineralization

1. Introduction

Outdoor air pollutants originate from familiar anthropogenic sources, including industry, transportation, heating, and agriculture [1]. These pollutants have two types of effects: (i) local effects on health and the environment, necessitating short- and medium-term actions, and (ii) global effects on the planet and climate, manifesting in the long term. In response to these concerns, the European directive 2016/2284/EC was published, aiming to reduce national emissions of air pollutants, including NH_3, SO_2, NOx, and volatile organic compounds (VOCs) (excluding CH_4), with aromatic compounds posing the highest health risks [2]. Recognizing the urgency, France pledged 2005 to reduce its NMVOC, NOx, and SO_2 emissions by 52 to 77% and ammonia by 13% [1–3]. Meeting this challenge necessitates the development of advanced treatment processes for industrial effluents, specifically at the emission step, to restrict the concentrations and fluxes released into the environment [4]. Current processes employing a liquid phase and oxido-reduction mechanisms, such as Stretford, Ferrifloc, Sulfurex, Burner-Scrubber, Catalyst-Scrubber, and Ozone processes, rely on large-volume equipment prone to corrosion caused by aggressive solutions [4].

Consequently, these processes incur high investments and maintenance costs, substantial chemical reagent consumption, and pose environmental issues during waste effluent disposal [5]. Thus, treatment processes often fail to meet industry requirements due to cost concerns. Therefore, an alternative process that does not necessitate the use of reagents but only relies on a power supply while generating only mineralizable inorganic by-products

(CO_2, H_2O, etc.) would align better with industry needs, especially if space requirements and energy consumption are manageable [6].

In recent years, investigations have demonstrated the effectiveness of dielectric barrier discharge (DBD) reactors for hazardous pollutant removal from gas streams with low VOC concentrations at ambient temperatures [7]. The simultaneous reduction in coexisting pollutants has also been studied [8]. Despite its good attributes, DBD plasma has some drawbacks, such as the formation of toxic by-products like CO, NO, NO_2, and O_3. Achieving the desired total oxidation of CO_2 and H_2O is often challenging [9–12]. To address these challenges, coupling DBD plasma with photocatalysis presents a 'zero waste and zero reagent' technology with the potential for synergy and low energy consumption [13,14]. Several studies have investigated the coupling of DBD plasma and photocatalysis using different reactor types and targeting various odorous compounds, including isovaleraldehyde, isovaleric acid, trimethylamine, ammonia, and dimethyl disulfide (DMDS) [15]. It is known that the presence of a catalyst in the plasma enhances performance [16–19]. To avoid catalytic surface poisoning and maintain energetic efficiency, producing a higher concentration of reactive species through the controlled adjustment of the pulsed discharge conditions is crucial, as these species have very short lifetimes [18–21].

This study aims to investigate VOC removal by coupling plasma and photocatalysis, focusing on new electrode configurations, on the performance in handling pollutant mixtures, and on understanding the catalytic surface poisoning mechanisms through tests on the reuse part of the combined system. In fact, the regeneration effect of plasma on the material surface and the understanding of the reactional mechanism in order to lift the scientific barriers was studied in detail in this paper.

2. Materials and Methods

2.1. Experimental Setup

The oxidation (photocatalysis and plasma) runs were conducted using the experimental setup shown in Figure 1. It consists of a tubular cylindrical reactor formed by two concentric Pyrex tubes (100 cm long), one outer tub of 76 mm, and an inner tube of 58 mm. Their wall thickness was about 4 mm. The reactor can be combined with (i) photocatalysis by using an external UV lamp (Philips TL 40W/05 (Philips, Canton of Flayosc, France)) and/or (ii) DBD plasma (Dielectric Barrier Discharge system) by applying high voltage power. Experiments were realized at ambient temperature (25 °C) and atmospheric pressure. A TESTO sensor is used to measure the temperature and relative humidity. Before photocatalytic experiments, the UV lamp (100 cm long, in the inner concentric cylinder) is activated for homogeneous irradiation. The flow rate (a maximum of 10 $m^3 \cdot h^{-1}$) at the system's inlet is controlled using a mass flow meter (Bronkhorst In-Flow). For humidity experiments (from 5 to 90 ± 5%), a variable part of the airflow is derived through a packed column where water flows in the counter current (Figure 1). Two syringe/syringe driver systems (KD Scientific Model 100) were used continuously for liquid Toluene and DMDS injecting into the gas stream. A heating band was used in the injection zone to achieve a good evaporation of pollutants (Figure 1).

In the case of DBD plasma equipment, the different experimental parts used to create the plasma are illustrated in Figure 1. Plasma discharge was generated by applying high voltage using a signal generator (BFi OPTILAS (SRS) reference DS 335/1)-USA. The applied tension with sinusoidal waveform was amplified 3000 v/v using a TREK 30A/40 amplifier-(Denver, CO, USA) [1–5]. The outer and inner electrodes were connected to the amplifier (Figure 2). The voltages applied in the plasma reactor are measured using high-voltage probes and recorded with a digital oscilloscope (Lecroy wave surfer 24Xs, 200 MHz).

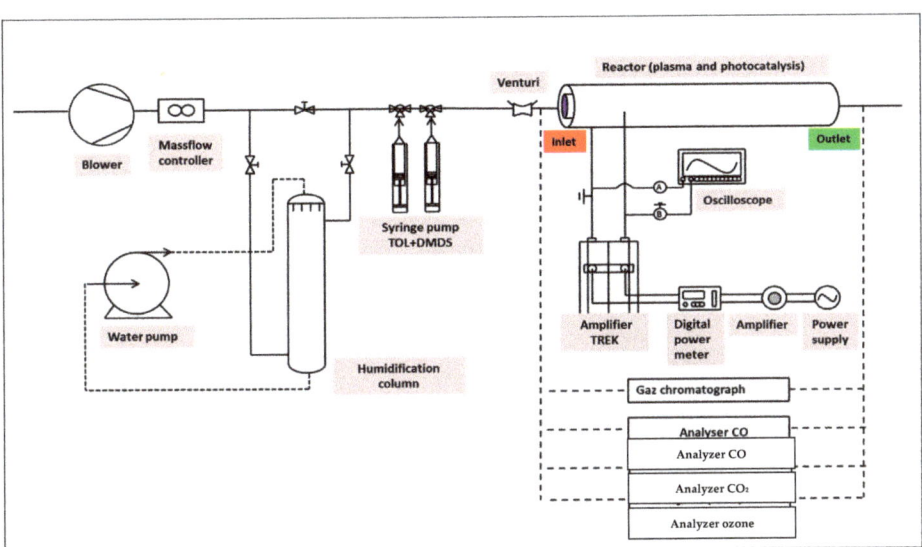

Figure 1. General schematic view of the photocatalysis and non-thermal plasma pilot flowsheeting.

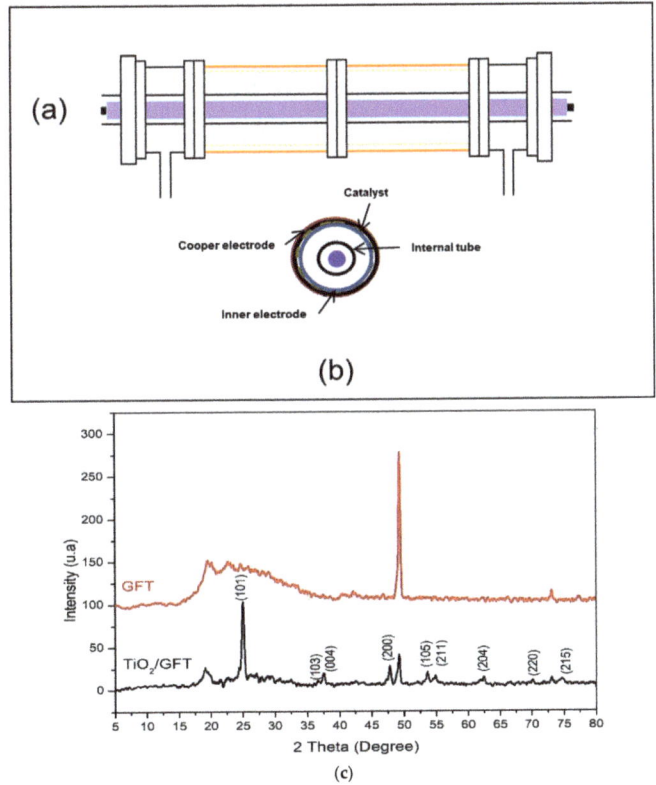

Figure 2. Scheme (**a**) and sectional drawing (**b**) of the cylindrical reactor. (**c**) XRD diffractogram of pristine GFT and GFT coated with TiO_2.

The operating parameter and their ranges are summarized in Table 1.

Table 1. Parameters of combined reactor.

Parameter	Value/Domain
Gas temperature	Ambient (293 K)
Gas pressure	Atmospheric pressure (1 atm)
Relative humidity	$(5, 60, 90) \pm 5\%$
Specific Energy (SE)	4.5–9 $J.L^{-1}$
Target compound concentartion	10–60 $mg.m^{-3}$
Residence time	1.36 s

A commercial Glass Fiber Tissue (BET surface of 300 $m^2.g^{-1}$, 5–10 nm of diameter and 100% Anatase), produced by Ahlstrom Research and Services, was used as a photocatalyst which contains (i) 13 $g.m^{-2}$ of colloidal silica, (ii) 13 $g.m^{-2}$ of titanium dioxide nanoparticles, and (iii) inorganic fibers. To achieve this catalyst, Ahlstrom Research and Services starts with impregnating glass fibers using SiO_2 and TiO_2 nanoparticles suspension in pure water using an industrial-sized press (PC500 Millennium). The second step is a drying step of impregnated fibers [6]. The crystalline phase of the coated photocatalyst on the GFT support was examined using the X-ray diffractometer. The optical band gap of TiO_2 nanoparticles has a value of 3.2 eV.

Figure 2c shows the XRD diffraction pattern of the GFT coated with TiO_2 photocatalyst and the pristine GFT.

The average size of the TiO_2 nanoparticles was calculated following the Scherrer equation considering the intense (101) plane peak, and the obtained value is 16.33 nm.

2.2. Analytical Methods

- The same gas volume (500 µL) was continuously sampled to monitor oxidation phenomena under the photocatalytic plasma reactor. The concentration of Toluene was determined using Gas Chromatography (GC) with a Clarus GC-500 chromatograph equipped with a flame ionization detector (FID) (Salt Lake City, UT, USA) and a 60 m × 0.25 mm polar DB-MS capillary column (film thickness, 0.25 µm). The FID detector was powered by an air and hydrogen mixture (H_2). Helium (He) was used as the carrier gas at a flow rate of 1 $mL.min^{-1}$. The analysis conditions included injection and detection temperatures of 250 °C for both, and the oven temperature was programmed to maintain 90 °C throughout each analysis (analysis time: 4 min). The analysis method for DMDS sulfur pollutant has been described in detail in our previous work [5].
- CO_2 measurements were determined using a Fourier Transform Infrared Spectrophotometer (FTIR) from Environment SA (MIR 9000H). The mineralization step was continuously monitored during the oxidation (plasma/photocatalysis) process using a pump system to control the outlet gas stream. For CO analysis, samples were taken using a gas analyzer (NO/CO ZRE marketed by Fuji Electric France S.A.S.). The SO_2 outlet concentration was measured using a MEDOR gas analyzer (THT MEDOR®-Houston, TX, USA).
- The amount of Ozone generated during the oxidation step with plasma was determined using sodium thiosulfate titration. A membrane pump (KNF lab N86k18) delivered part of the flow exits and then bubbled into a potassium iodide solution (KI, at 10^{-2} M). The chemical reaction between KI and Ozone (Equation (1)) resulted in the appearance of a yellow color, which was then neutralized through titration with a sodium thiosulfate solution ($Na_2S_2O_3$, at 10^{-3} M) until a colorless solution was obtained (Equation (2)) [7]. The titration was carried out in an acid medium by adding concentrated hydrochloric acid (HCl) to the final solution.

$$O_3 + 2I^- \rightarrow I_2 + O_2 + O^- + e^- \qquad (1)$$

$$I_2 + 2\,S_2O_3{}^{2-} \rightarrow 2I^- + S_4O_6{}^{2-} \qquad (2)$$

3. Results and Discussion

The experimental parameters are defined as follows:

- (C_{inlet}) and (C_{outlet}) represent the inlet and outlet concentration of pollutant (mg.m^{-3}), respectively.
- The degradation rate of the pollutant with each process (%) = $(1 - C_{outlet}/C_{inlet}) \times 100$.
- The value of the Synergetic Effect (SE) is calculated using the following expression: SE = $RE_{combined\ process}/[RE_{plasma} + RE_{photocatalysis}]$.
- The Specific Energy SE (J/L) = $[P(W)/Q(m^3.s^{-1})]/1000$, where P is the input power in the function of the applied voltage and Q is the flowrate.
- The sulfur dioxide selectivity (SSO$_2$) = $([SO_2]_{outlet} - [SO_2]_{inlet}) \times 10^4/(n_{s,cov} \times RE \times [C]_{inlet})$.
- The carbon dioxide and monoxide selectivities (SCO$_x$) = $([CO_x]_{outlet} - [CO_x]_{inlet}) \times 10^4/(n_{c,cov} \times RE \times [C]_{inlet})$, where [C] is the concentration of DMDS/Toluene and $n_{c,VOC}$ represents the number of carbons in the molecules (two for DMDS and seven for Toluene).

The degradation studies of (i) Toluene alone (100% C_7H_8), (ii) DMDS alone (100% $C_2H_6S_2$) on a continuous annular reactor, and (iii) their binary mixture (Toluene 50%-DMDS 50%) were investigated. The experiments with the photocatalysis process were carried out under different operating conditions. The plasma process performance was also monitored separately from the photocatalysis (without external UV-lamp), and then the association of the photocatalysis/plasma was studied.

3.1. Photocatalysis Treatment: (i) Effects of Initial Pollutant Concentration and Air Flow on Degradation and (ii) Effect of Water Vapor

This study systematically investigated the impact of varying initial toluene concentrations (10 and 20 mg.m^{-3}) and airflow rates (1–4 m^3.h^{-1}) on the efficiency of the photocatalytic removal of toluene. The experiments were conducted with and without a light source (UV-lamp OFF) during an initial adsorption step to ensure stable inlet toluene concentrations. It was observed that, as the gas flow supplying the photocatalytic reactor was increased, the efficiency of toluene degradation was decreased. This trend can be attributed to the shorter contact time between toluene molecules, active sites on the catalyst, and oxidation species at higher flow rates, resulting in reduced degradation efficiency. The experimental data indicated that toluene degradation was approximately 38.5% at 1 m^3.h^{-1} for an initial concentration of 10 mg.m^{-3} but decreased to around 13% at 4 m^3.h^{-1}. Moreover, an increase in the initial toluene concentration also led to reduced oxidation performance, with degradation rates of 38.5% and 21.9% observed at 1 m^3.h^{-1} for initial concentrations of 10 and 20 mg.m^{-3}, respectively. This behavior aligned with previous research on TiO$_2$-based catalysts [1,6–9].

The influence of humidity levels was also investigated by maintaining a constant inlet toluene concentration (10 mg.m^{-3}) and an airflow rate of 2 m^3.h^{-1} while varying humidity levels at approximately 5%, 60% ± 5, and 90% ± 5 using a humidification column. Two distinct behaviors in toluene removal were revealed in Figure 3b. At lower humidity levels (<60% ± 5), an improvement in toluene removal was observed due to active intermediate species generated under these conditions, enhancing the oxidation step and overall photocatalytic toluene removal. Favorable toluene removal efficiency was demonstrated in the experiments at approximately 60% ± 5 humidity levels, increasing efficiency from 19 to 33.6%. However, at higher humidity levels (>60–90%), competitive adsorption between water vapor and toluene molecules on active sites became more pronounced, decreasing toluene removal efficiency. At high humidity levels (90% ± 5), a slight decrease in toluene removal efficiency to 21.5% was observed. The significance of humidity as an experimental parameter in photocatalytic oxidation processes is highlighted by our comprehensive analysis, with lower humidity favoring toluene removal and higher humidity exerting a negative impact. Thus, increasing the relative humidity inside the reactor results in a net presence of water molecules. The water molecules adsorbed on the surface of the photocatalyst result in photogenerated holes following oxidation leading to the formation of OH radicals known as reactive species in the photocatalytic air treatment. On the other hand, the significant presence of water vapor molecules at high relative humidity levels reverses the trend and reduces the conversion of the pollutant due to the phenomenon of competition between the water molecule and the adsorption of ethylbenzene on the active sites of the photocatalyst [5].

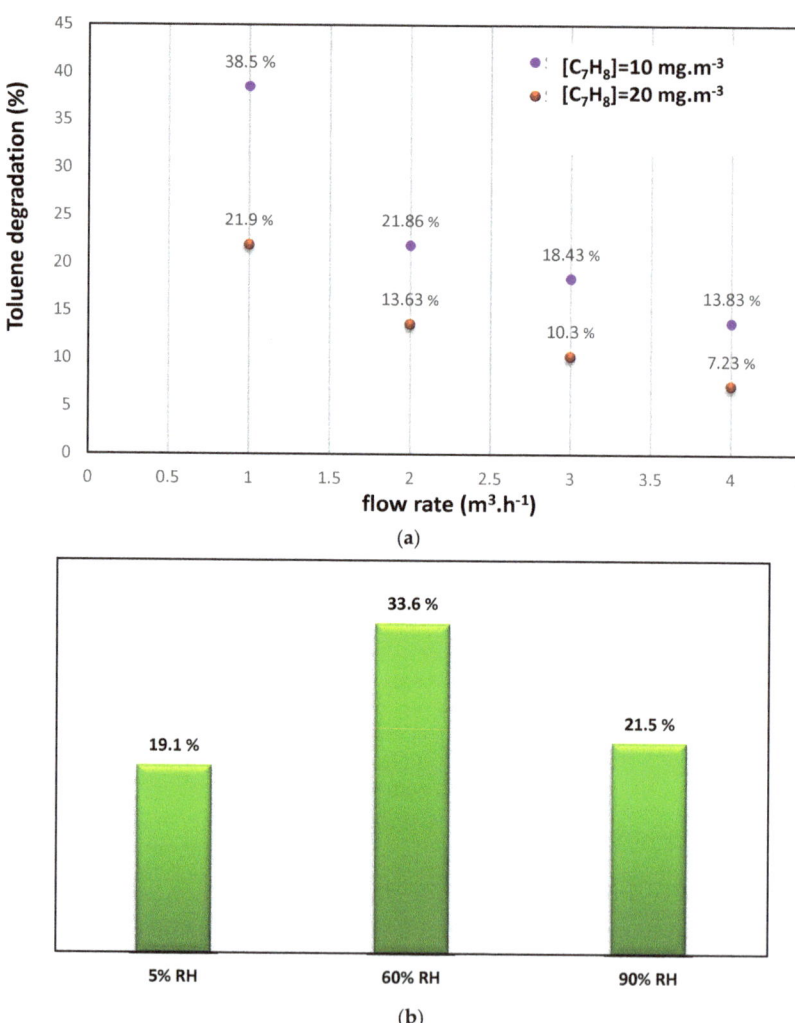

Figure 3. (a) Toluene performance degradation at different [C_7H_8] and flow rates in the photocatalysis process. [T = 20 °C, $UV_{intensity}$ = 20 W m^{-2}]. (b) Toluene performance degradation at different rates of humidity in the photocatalysis process. [Flow rate = 2 m^3 h^{-1}, [C_7H_8] = 10 mg m^{-3}, T = 20 °C, $UV_{intensity}$ = 20 W m^{-2}].

3.2. DBD Direct Plasma Treatment: Effect of Plasma Energy

To study the performance of pollutants' (toluene and DMDS) degradation via a plasma reactor, the degradation study of toluene was performed with (i) humid airflow (2 m^3.h^{-1}, 55% of humidity), (ii) an inlet concentration of 14 ppm, and (iii) a plasma energy of 4.5 J.L^{-1} and 9 J.L^{-1}. The same methodology was applied to DMDS, where the reactor contained a similar pollutant concentration. Figure 4 shows the degradation rate of toluene/DMDS studied separately at different plasma energies. The experimental data via DBD plasma configuration indicate that, with the two pollutants (aromatic and sulfuric compounds), the increase in specific energy (plasma power) leads to an increase in the removal efficiency of contaminants [7–9]. In our previous work, a similar trend of removal efficiency was displayed in fatty acids [10], aldehydes [15], and amines [21], either on a pilot or on an industrial scale. We reported that the removal efficiencies of these molecules strongly depend on the applied voltage. It was observed that increasing energy enhances the level of electrons, which improves the reactive oxygen species formation and consequently leads to greater removal efficiency. In our study, the toluene removal improved from 13% to 25.1% with 4.5 J.L^{-1} and 9 J.L^{-1} of plasma power, respectively. As for DMDS, when the specific energy amount is more and more important (9 J.L^{-1}), the DMDS rate (27.51%) is slightly higher than for toluene (25.1%).

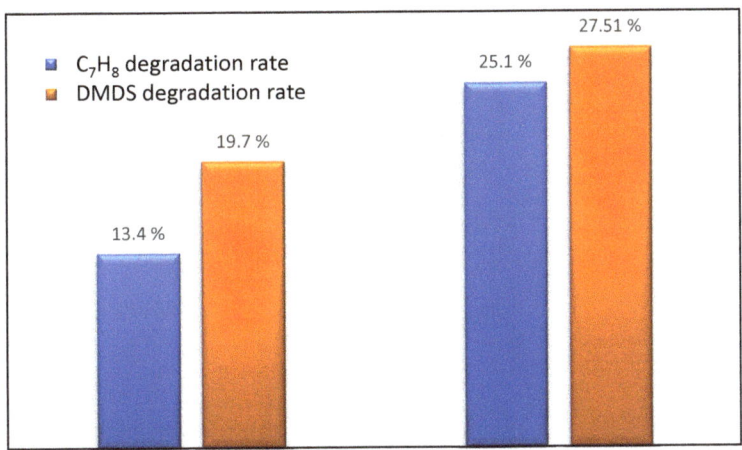

Figure 4. Toluene/DMDS performance degradation at different plasma specific energy SEs in the plasma process. [Q = 2 m^3 h^{-1}, [COV] = 14 ppm, RH = 55%, T = 20 °C].

3.3. Treatment by Coupling Process (Photocatalysis/Plasma): Comparison of Process Performance

The oxidation of toluene was monitored via three processes: (i) photocatalysis, (ii) DBD plasma, and (iii) a combination of photocatalysis and plasma at different plasma energies (4.5 and 9 J.L^{-1}). In this section of the study, these two methods ((i) and (ii)) were used separately and simultaneously (iii) to enhance the degradation performance of the process. Comparing the treatment with the simultaneous application (photocatalysis/plasma) to that with photocatalysis and plasma applied separately, the experimental data (Figure 5a) indicate that the coupling exhibits a higher performance than the sum of photocatalysis alone and plasma alone. At a plasma energy of 9 J.L^{-1}, the combined process achieved a toluene degradation rate of 61%, while the sum of the degradation rate for photocatalysis and plasma separately was 46.2%. Similarly, at low energy (4.5 J.L^{-1}), the toluene degradation reached 40.23% when the combined application was used, compared to 34.5% for the sum of both processes used alone. In this case, an enhancement of 10% ± 3 in the removal efficiency was observed. The same methodology was applied to DMDS (see

Figure 5b). The results indicate a strong synergy between both processes for any pollutant used. Plasma significantly contributes to the desorption of the degraded by-products adsorbed on the TiO_2 surface through the active species ($O_2^{\circ-}$, O°, HO°), leading to the increased catalytic activity of the photocatalyst (in our case, TiO_2 coated on Glass Fiber Tissue) and improved photocatalytic degradation [6,21–27]. For DMDS, the degradation rate during the combined application of photocatalysis and plasma (53%) surpasses the rates achieved with plasma or photocatalysis alone (45.4%). The same behavior has been found by Qi and his coworkers, with toluene removal in a plasma–catalytic hybrid system over $Mn-TiO_2$ and $Fe-TiO_2$ [28]. Moreover, Wang and his collaborators highlighted the synergetic effect on CO_2 reduction in the presence of Dual-plasma enhanced 2D/2D/2D $g-C_3N_4/Pd/MoO_3$ [11].

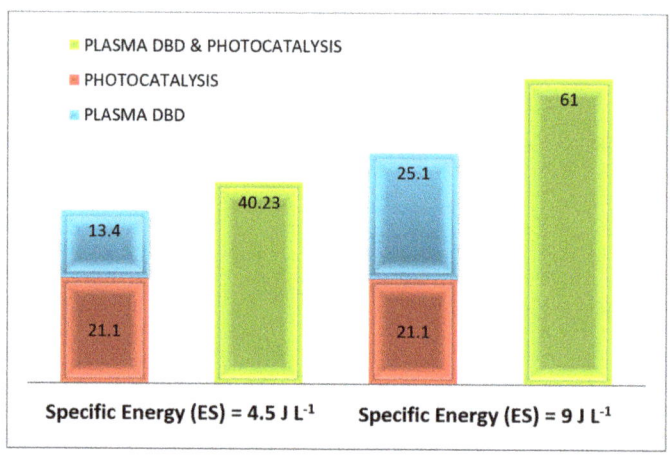

(a)

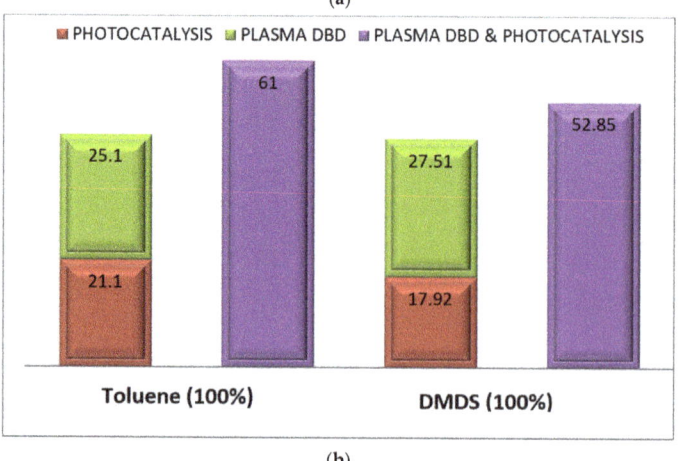

(b)

Figure 5. (a) Toluene performance degradation at different plasma specific energy SE via photocatalysis, plasma, and their combination: Evidence for synergetic effect. [Q = 2 m^3 h^{-1}, [toluene] = 14 ppm, RH = 50%, T = 20 °C, $UV_{intensity}$ = 20 W m^{-2}]. (b) Pollutant performance degradation when applying photocatalysis, plasma, and their combination: Evidence for synergetic effect. [Q = 2 m^3 h^{-1}, [pollutant alone: 100% toluene or DMDS] = 14 ppm, ES= 9 J L^{-1}, RH = 50%, T = 20 °C, SE= 9 J/L, $UV_{intensity}$ = 20 W m^{-2}].

The combination of these processes demonstrates the synergy between DBD plasma and photocatalytic oxidation [13,16–20], which can be attributed to the following:

(i). The action of plasma radicals (N°, O°, H°, OH°), renewing the catalytic site and improving the removal/mineralization step;
(ii). The contribution of the reactive species generated by plasma to photocatalytic mechanisms;
(iii). The Ozone and UV-light reactions generate highly reactive radicals that can activate TiO_2 and enhance performance;
(iv). The enhancement of mass transfer of pollutants by the ionic wind generated by plasma;
(v). The in-situ regeneration of the catalytic surface in the presence of the micro discharge of plasma.

3.4. Treatment by Coupling Process (Photocatalysis/Plasma): Effect of Mixture (Toluene/DMDS) and Plasma Power

The same methodology was applied to the mixture of toluene and DMDS. Figure 6 illustrates the removal efficiency of the toluene/DMDS mixture using three processes: (i) photocatalysis, (ii) DBD plasma, and (iii) photocatalysis/plasma at different plasma energies. For the mixture studied, the reactor was supplied with: a humid air flow (2 $m^3.h^{-1}$, 55% humidity), an inlet concentration of 14 ppm ([Toluene] = [DMDS] = 7 ppm), and plasma energies of 4.5 and 9 $J.L^{-1}$. As depicted in Figure 6, plasma energy is a crucial parameter for monitoring plasma application and it can significantly impact degradation efficiency.

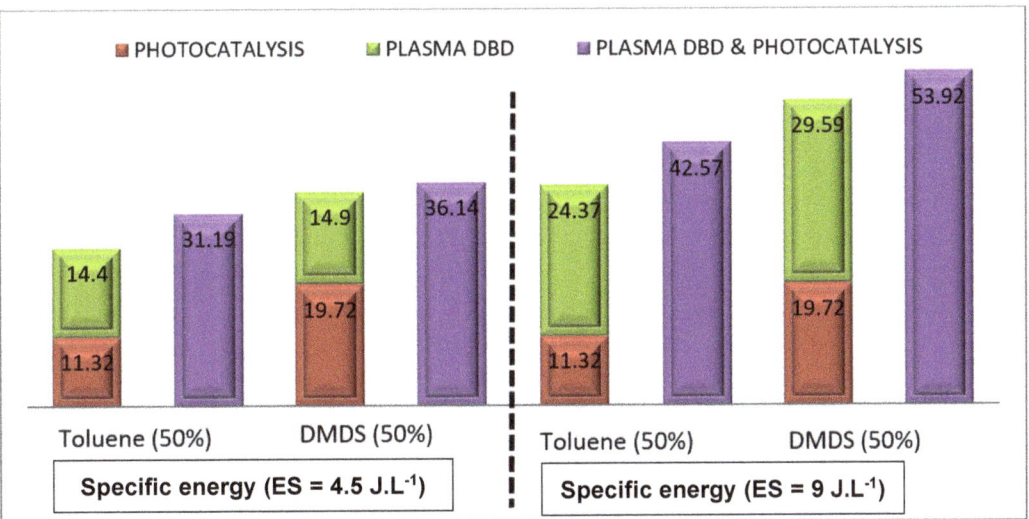

Figure 6. Evolution of toluene/DMDS degradation rate in mixtures at different plasma specific energy SE via photocatalysis, plasma, and their combination: Evidence for synergetic effect. [Q = 2 $m^3 h^{-1}$, [Mixture of pollutant: 50% toluene + 50% DMDS] = 14 ppm, RH = 50%, T = 20 °C, UV intensity = 20 W m^{-2}].

In the case of the mixture with high energy (9 $J.L^{-1}$), the results indicate that (i) toluene degradation reached 42.57% when both applications were combined, compared to 35.69% when the two applications were applied separately, and (ii) DMDS degradation reached 53.92% when both applications were combined, compared to 49.31% for separate applications. Comparing the treatment of the toluene/DMDS mixture to the separate treatment of toluene and DMDS, the experimental data (Figures 6 and 5b, respectively) demonstrate that the coupling exhibits higher removal efficiency, surpassing the sum of photocatalysis and plasma treatments when applied separately.

The toluene/DMDS mixture results indicate a more significant synergetic effect for DMDS degradation than for toluene. When photocatalysis/plasma were used together,

42.57% of the toluene in the mixture and 53.92% of the DMDS in the mixture were decomposed. In contrast, during single treatments (Figure 5b), toluene removal reached 61%, and 52% of the DMDS was decomposed. This different behavior can be attributed to the competitive adsorption/oxidation of the two pollutants in the mixture, which becomes more pronounced [22–25]. The molecular chain structure of DMDS ($C_2H_6S_2$) hampers the adsorption/oxidation and, thus, the photocatalytic oxidation of toluene (C_7H_8). This observation is consistent with the findings reported by Assadi et al. [21], who demonstrated a decrease in oxidation performance (in the case of a VOC mixture) due to the competitive interactions between pollutants, by-products, and active sites.

3.5. Study of By-Products Generation: Ozone Selectivity of $CO/CO_2/SO_2$

3.5.1. Monitoring of the Ozone Formed

Ozone formation, a strong oxidizing by-product, occurs during the operation of the DBD-plasma/photocatalytic reactor. Extensive experiments on oxidation under (i) a DBD-plasma reactor or (ii) coupling plasma/photocatalysis technology have been conducted and detailed in our previous research studies [1,4,6,7,9,23–27]. These studies have shown that increasing the energy of plasma leads to generating a significant amount of ozone in the exhaust. However, experiments with the DBD plasma/photocatalysis combination have been performed to mitigate the excessive ozone production at low plasma energies (4.5 and 9 J.L^{-1}). The humid air flow was fixed at 2 m^3.h^{-1} with 55% humidity, and the concentration of toluene/DMDS was maintained at 14 ppm. UV irradiation tests (with lamp ON/OFF) were conducted to monitor ozone formation at the reactor outlet at various plasma energies. The results, depicted in Figure 7, illustrate the behavior of ozone during the operation of DBD plasma technology and the coupling of plasma/photocatalysis.

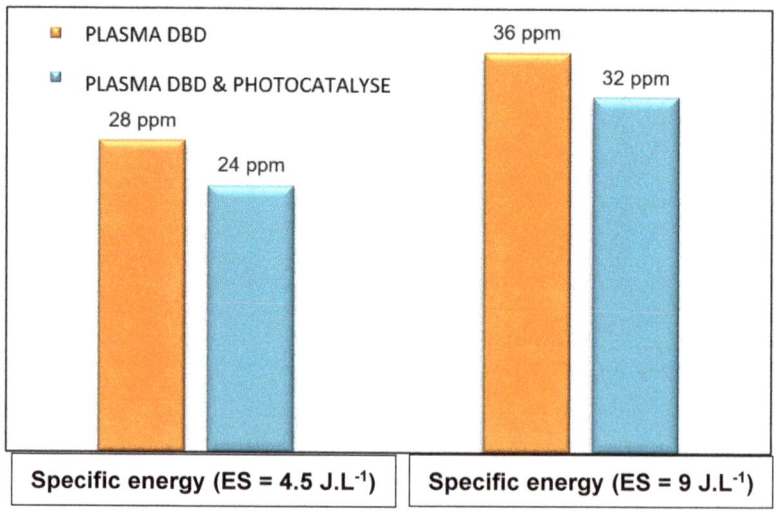

Figure 7. Evolution of ozone concentration (ppm) with plasma specific energy SE. [Q = 2 m^3 h^{-1}, [Mixture of pollutant: 50% toluene + 50% DMDS] = 14 ppm, RH = 50%, T = 20 °C, UV intensity = 20 W m^{-2}].

A slight decrease in ozone concentration was observed (Figure 7) with values of 36 and 32 ppm for (i) plasma alone and (ii) plasma/photocatalysis, respectively. This decrease can be attributed to the decomposition of ozone through reactions (3), (4), and (5), facilitated by the photo-generated radicals (H°, HO°) [26]. It is important to note that adding external UV light to the DBD-plasma system (coupling plasma with UV lamp ON) can play a crucial role in promoting ozone degradation into highly reactive species ($O_2^{\circ-}$ and HO_2°), thereby significantly enhancing the oxidation step.

$$H_2O + e^- \rightarrow H^{\circ} + OH^{\circ} + e^- \quad (3)$$

$$O_3 + OH^{\circ} \rightarrow O_2 + HO_2^{\circ} \quad (4)$$

$$O_3 + H^{\circ} \rightarrow O_2 + OH^{\circ} \quad (5)$$

3.5.2. CO_2, SO_2, CO Selectivity

In this investigation, the selectivity rates of carbon dioxide (CO_2), sulfur dioxide (SO_2), and carbon monoxide (CO) under different oxidation conditions were analyzed. Figure 8 presents the experimental data, showcasing distinct selectivity rates for these pollutants. Carbon dioxide (CO_2) exhibited the highest selectivity rate among the three, achieving 69.81% through photocatalysis alone, 43.75% through plasma alone, and a further increase to 58.98% when plasma and photocatalysis were combined. On the other hand, sulfur dioxide (SO_2) showed a selectivity rate of 17.35% through photocatalysis, while plasma alone achieved 46%. However, the combination of plasma and photocatalysis resulted in a decreased selectivity rate compared to CO_2. For carbon monoxide (CO), selectivity rates were negligible, with only 7% of the CO mineralization rate observed through photocatalysis, plasma, and plasma/photocatalysis.

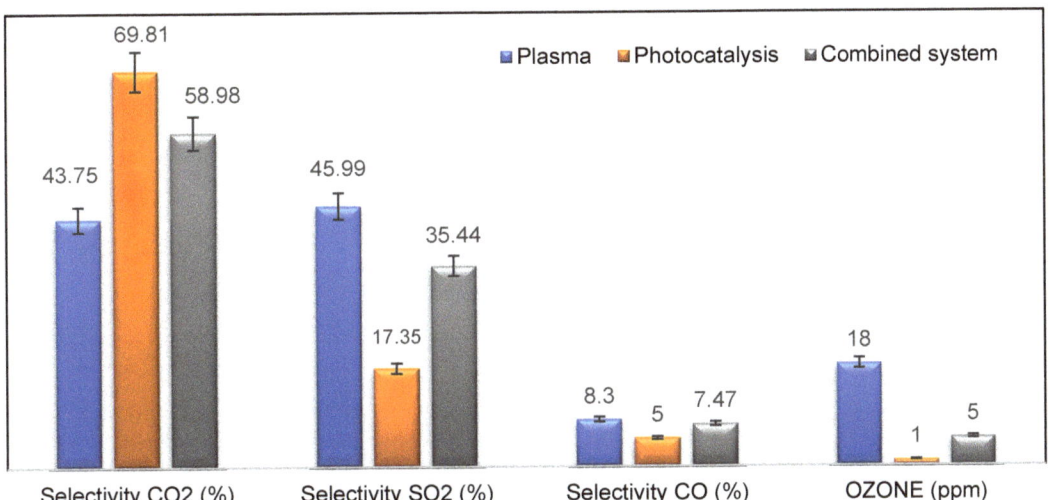

Figure 8. Variation in CO_2, SO_2, and CO (%) and ozone (ppm) during toluene/DMDS removal with photocatalysis/plasma coupling. [Q = 2 m^3 h^{-1}, [Mixture] = 14 ppm, RH = 50%, T = 20 °C, UV$_{intensity}$ = 20 W m^{-2}, SE = 4.5 J L^{-1}].

It is worth noting that ozone (O_3) concentrations decreased compared to plasma alone. Previous research has shown that incorporating external UV light into the DBD–plasma system (via coupling plasma with UV lamp ON) can significantly enhance the mineralization rate and improve the degradation of pollutants and by-products [16–20].

4. Reusability and In Situ Regeneration

A series of experiments consisting of four cycles was conducted to evaluate the photocatalytic stability of the catalyst after multiple cycles. These experiments involved alternating phases: (i) the oxidation step using photocatalysis, plasma, and the coupling of both, and (ii) the catalyst regeneration step. The photoactivity tests were conducted under dry conditions with a continuous flow rate of 2 m^3·h^{-1}, using a toluene/DMDS mixture with a concentration of 14 ppm.

Figure 9 presents the degradation rate of toluene in the mixture and the synergetic effect (SE) value after four cycles (each cycle consisting of experiments with photocatalysis, plasma, and their combination). It can be observed that the degradation rate of toluene slightly decreases after four cycles of continuous oxidation, resulting in a 10% loss of the catalyst's photoactivity. However, a stable degradation rate is observed when the coupling of plasma/photocatalysis is applied. In our case, the regeneration step involved using photocatalysis/plasma.

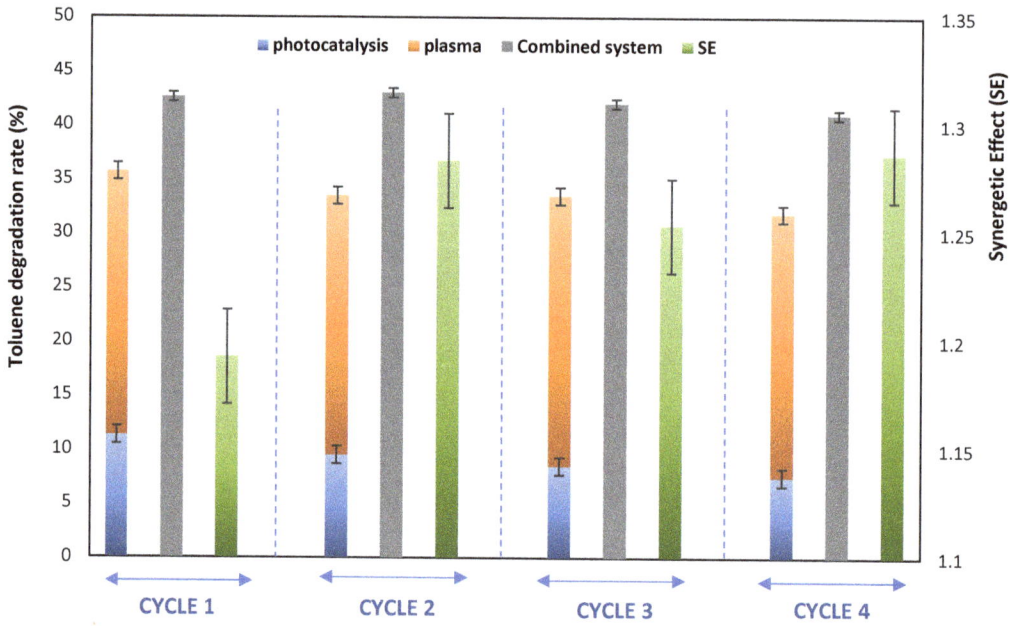

Figure 9. Effect of catalysis reusability on synergetic effect in the case of mixture toluene/DMDS. [Q = 2 m^3 h^{-1}, [Mixture] = 14 ppm, RH = 50%, T = 20 °C, UV$_{intensity}$ = 20 W m^{-2}, SE = 4.5 J L^{-1}].

Previous studies have also observed the deactivation of photocatalysts during the oxidation of sulfur compounds [27]. These studies have shown that plasma can effectively regenerate poisoned catalysts. Therefore, the results indicate that the regeneration process can be enhanced by combining photocatalysis and plasma [28–34]. This finding confirms the in situ regeneration of the photocatalytic support in the presence of plasma [35–41].

5. Conclusions

In this comprehensive study, we conducted a thorough investigation into various parameters, including the synergetic effect (SE), inlet concentrations of toluene (TOL) and DMDS, and relative humidity (RH), to assess their influence on the performance of three distinct processes: DBD plasma, photocatalysis, and the combined DBD plasma/photocatalysis system. Our findings have illuminated the pivotal role of water vapor in VOC removal, revealing optimal RH values that enhance CO_2 selectivity and diminish CO formation. Furthermore, RH was observed to have a mitigating effect on ozone formation.

Across all operational parameters explored, it is evident that coupling DBD plasma with a TiO_2 catalyst under external UV irradiation can yield a synergetic effect, resulting in improved toluene and DMD$_S$ removal. The significant enhancement in CO_2 selectivity during the coupling process is particularly noteworthy, a remarkable 11% improvement compared to DBD plasma alone. The observed reduction in ozone concentration during plasma–photocatalysis coupling can be attributed to the breakdown of ozone into more active oxidizing species facilitated by UV radiation.

These findings hold profound practical implications for pollutant removal and treatment strategies. As demonstrated in this study, the combination of DBD plasma and photocatalysis presents a promising avenue for more efficient and environmentally friendly approaches to addressing VOC pollution. By unraveling the intricate interplay of parameters and processes, we are better positioned to develop cleaner and more sustainable solutions for industrial effluent treatment.

Future research endeavors could delve deeper into optimizing the coupling process, explore additional parameters, and investigate its adaptability in diverse industrial settings. The pursuit of innovative technologies that reduce environmental impact while aligning with industry requirements remains of paramount importance.

Author Contributions: Writing—original draft, Investigation and conceptualization: L.K.; Writing—review and editing and Visualization: A.A.A. All authors have read and agreed to the published version of the manuscript.

Funding: This research received no external funding.

Institutional Review Board Statement: Not applicable.

Informed Consent Statement: Not applicable.

Data Availability Statement: Not applicable.

Acknowledgments: The authors extend their appreciation to the Deputyship for Research & Innovation, Ministry of Education in Saudi Arabia for funding this research through the project number IFP-IMSIU-2023031. The authors also appreciate the Deanship of Scientific Research at Imam Mohammad Ibn Saud Islamic University (IMSIU) for supporting and supervising this project.

Conflicts of Interest: The authors declare no conflict of interest.

References

1. Ren, C.; Yu, C.W.; Cao, S.-J. Development of urban air environmental control policies and measures. *Indoor Built Environ.* **2022**, *32*, 299–304. [CrossRef]
2. Zhang, X.; Fan, M.; Shao, S.; Song, X.; Wang, H. Socioeconomic drivers and mitigating strategies of volatile organic compounds emissions in China's industrial sector. *Environ. Impact Assess. Rev.* **2023**, *101*, 107102. [CrossRef]
3. Hernández-Fernández, J.; Cano, H.; Rodríguez-Couto, S. Quantification and Removal of Volatile Sulfur Compounds (VSCs) in Atmospheric Emissions in Large (Petro) Chemical Complexes in Different Countries of America and Europe. *Sustainability* **2022**, *14*, 11402. [CrossRef]
4. Pan, Q.; Liu, Q.-Y.; Zheng, J.; Li, Y.-H.; Xiang, S.; Sun, X.-J.; He, X.-S. Volatile and semi-volatile organic compounds in landfill gas: Composition characteristics and health risks. *Environ. Int.* **2023**, *174*, 107886. [CrossRef]
5. Saoud, W.A.; Assadi, A.A.; Kane, A.; Jung, A.-V.; Le Cann, P.; Gerard, A.; Bazantay, F.; Bouzaza, A.; Wolbert, D. Integrated process for the removal of indoor VOCs from food industry manufacturing: Elimination of Butane-2,3-dione and Heptan-2-one by cold plasma-photocatalysis combination. *J. Photochem. Photobiol. A Chem.* **2020**, *386*, 112071. [CrossRef]
6. Acayanka, E.; Tarkwa, J.-B.; Nchimi, K.N.; Voufouo, S.A.; Tiya-Djowe, A.; Kamgang, G.Y.; Laminsi, S. Grafting of N-doped titania nanoparticles synthesized by the plasma-assisted method on textile surface for sunlight photocatalytic self-cleaning applications. *Surf. Interfaces* **2019**, *17*, 100361. [CrossRef]
7. Guillard, C.; Baldassare, D.; Duchamp, C.; Ghazzal, M.; Daniele, S. Photocatalytic degradation and mineralization of a malodorous compound (dimethyldisulfide) using a continuous flow reactor. *Catal. Today* **2007**, *122*, 160–167. [CrossRef]
8. Ramírez, M.; Fernández, M.; Granada, C.; Le Borgne, S.; Gómez, J.M.; Cantero, D. Biofiltration of reduced sulphur compounds and community analysis of sulphur-oxidizing bacteria. *Bioresour. Technol.* **2011**, *102*, 4047–4053. [CrossRef]
9. Zhou, W.; Ye, Z.; Nikiforov, A.; Chen, J.; Wang, J.; Zhao, L.; Zhang, X. The influence of relative humidity on double dielectric barrier discharge plasma for chlorobenzene removal. *J. Clean. Prod.* **2021**, *288*, 125502. [CrossRef]
10. Ochiai, T.; Aoki, D.; Saito, H.; Akutsu, Y.; Nagata, M. Analysis of Adsorption and Decomposition of Odour and Tar Components in Tobacco Smoke on Non-Woven Fabric-Supported Photocatalysts. *Catalysts* **2020**, *10*, 304. [CrossRef]
11. Wang, H.; Liu, Q.; Xu, M.; Yan, C.; Song, X.; Liu, X.; Wang, H.; Zhou, W.; Huo, P. Dual-plasma enhanced 2D/2D/2D g-C_3N_4/Pd/MoO_3-x S-scheme heterojunction for high-selectivity photocatalytic CO_2 re-duction. *Appl. Surf. Sci.* **2023**, *640*, 158420. [CrossRef]

12. Młotek, M.; Reda, E.; Jóźwik, P.; Krawczyk, K.; Bojar, Z. Plasma-catalytic decomposition of cyclohexane in gliding discharge reactor. *Appl. Catal. A Gen.* **2015**, *505*, 150–158. [CrossRef]
13. Luo, S.; Lin, H.; Wang, Q.; Ren, X.; Hernández-Pinilla, D.; Nagao, T.; Xie, Y.; Yang, G.; Li, S.; Song, H.; et al. Triggering Water and Methanol Activation for Solar-Driven H_2 Production: Interplay of Dual Active Sites over Plasmonic ZnCu Alloy. *J. Am. Chem. Soc.* **2021**, *143*, 12145–12153. [CrossRef] [PubMed]
14. Zhang, T.; Ren, X.; Ma, F.; Jiang, X.; Wen, Y.; He, W.; Hao, L.; Zeng, C.; Liu, H.; Chen, R.; et al. MOF-derived Co(Ni)Ox species loading on two-dimensional cobalt phosphide: A Janus electrocatalyst toward efficient and stable overall water splitting. *Appl. Mater. Today* **2023**, *34*, 101912. [CrossRef]
15. Ren, X.; Philo, D.; Li, Y.; Shi, L.; Chang, K.; Ye, J. Recent advances of low-dimensional phosphorus-based nanomaterials for solar-driven photocatalytic reactions. *Coord. Chem. Rev.* **2020**, *424*, 213516. [CrossRef]
16. Assadi, A.A.; Loganathan, S.; Tri, P.N.; Gharib-Abou Ghaida, S.; Bouzaza, A.; Tuan, A.N.; Wolbert, D. Pilot scale deg-radation of mono and multi volatile organic compounds by surface discharge plasma/TiO_2 reactor: Investigation of competition and synergism. *J. Hazard. Mater.* **2018**, *357*, 305–313. [CrossRef]
17. Assadi, A.A.; Bouzaza, A.; Wolbert, D. Study of synergetic effect by surface discharge plasma/TiO_2 combination for indoor air treatment: Sequential and continuous configurations at pilot scale. *J. Photochem. Photobiol. A Chem.* **2015**, *310*, 148–154. [CrossRef]
18. Abidi, M.; Hajjaji, A.; Bouzaza, A.; Trablesi, K.; Makhlouf, H.; Rtimi, S.; Assadi, A.; Bessais, B. Simultaneous removal of bacteria and volatile organic compounds on Cu_2O-NPs decorated TiO_2 nanotubes: Competition effect and kinetic studies. *J. Photochem. Photobiol. A Chem.* **2020**, *400*, 112722. [CrossRef]
19. Li, Y.; Wang, W.; Wang, F.; Di, L.; Yang, S.; Zhu, S.; Yao, Y.; Ma, C.; Dai, B.; Yu, F. Enhanced Photocatalytic Degradation of Organic Dyes via Defect-Rich TiO_2 Prepared by Dielectric Barrier Discharge Plasma. *Nanomaterials* **2019**, *9*, 720. [CrossRef] [PubMed]
20. Assadi, A.A.; Bouzaza, A.; Lemasle, M.; Wolbert, D. Removal of trimethylamineand isovaleric acid from gas streams in a con-tinuous flow surface discharge plasma reactor. *Chem. Eng. Res. Des.* **2014**, *93*, 640–651. [CrossRef]
21. Dobslaw, C.; Glocker, B. Plasma Technology and Its Relevance in Waste Air and Waste Gas Treatment. *Sustainability* **2020**, *12*, 8981. [CrossRef]
22. Xu, Z.; Ren, Y.; Deng, X.; Xu, M.; Chai, W.; Qian, X.; Bian, Z. Recent Developments on Gas-Phase Volatile Organic Compounds Abatement Based on Photocatalysis. *Adv. Energy Sustain. Res.* **2022**, *3*, 2200105. [CrossRef]
23. Wood, D.; Shaw, S.; Cawte, T.; Shanen, E.; Van Heyst, B. An overview of photocatalyst immobilization methods for air pollution remediation. *Chem. Eng. J.* **2020**, *391*, 123490. [CrossRef]
24. Li, S.; Li, Y.; Yu, X.; Dang, X.; Liu, X.; Cao, L. A novel double dielectric barrier discharge reactor for toluene abatement: Role of different discharge zones and reactive species. *J. Clean. Prod.* **2022**, *368*, 133073. [CrossRef]
25. Hoshino, M.; Akimoto, H.; Okuda, M. Photochemical Oxidation of Benzene, Toluene, and Ethylbenzene Initiated by OH Radicals in the Gas Phase. *Bull. Chem. Soc. Jpn.* **1978**, *51*, 718–724. [CrossRef]
26. Winayu, B.N.R.; Chen, S.-T.; Chang, W.-C.; Chu, H. rGO doped $S_{0.05}N_{0.1}/TiO_2$ accelerated visible light driven photocatalytic degradation of dimethyl sulfide and dimethyl disulfide. *Appl. Catal. A Gen.* **2023**, *655*, 119113. [CrossRef]
27. Junior, A.G.; Pereira, A.; Gomes, M.; Fraga, M.; Pessoa, R.; Leite, D.; Petraconi, G.; Nogueira, A.; Wender, H.; Miyakawa, W.; et al. Black TiO_2 Thin Films Production Using Hollow Cathode Hydrogen Plasma Treatment: Synthesis, Material Characteristics and Photocatalytic Activity. *Catalysts* **2020**, *10*, 282. [CrossRef]
28. Qi, L.-Q.; Yu, Z.; Chen, Q.-H.; Li, J.-X.; Xue, H.-B.; Liu, F. Toluene degradation using plasma-catalytic hybrid system over Mn-TiO_2 and Fe-TiO_2. *Environ. Sci. Pollut. Res.* **2023**, *30*, 23494–23509. [CrossRef] [PubMed]
29. Maciuca, A.; Batiot-Dupeyrat, C.; Tatibouët, J.M. Synergetic effect by couplingphotocatalysis with plasma for low VOCs con-centration removal from air. *Appl. Catal. B Environ.* **2012**, *43*, 432–438. [CrossRef]
30. Salvadores, F.; Juan Brandi, R.; Alfano, O.M.; de los Milagros Ballari, M. Modelling and experimental validation of reaction chamber simulating indoor air decontamination by photocatalytic paint. *Appl. Catal. A Gen.* **2023**, *663*, 119285. [CrossRef]
31. Zhang, J.; Li, X.; Zheng, J.; Du, M.; Wu, X.; Song, J.; Cheng, C.; Li, T.; Yang, W. Non-thermal plasma-assisted ammonia production: A review. *Energy Convers. Manag.* **2023**, *293*, 117482. [CrossRef]
32. Zhu, B.; Liu, J.-L.; Li, X.-S.; Liu, J.-B.; Zhu, X.; Zhu, A.-M. In Situ Regeneration of Au Nanocatalysts by Atmospheric-Pressure Air Plasma: Regeneration Characteristics of Square-Wave Pulsed Plasma. *Top. Catal.* **2017**, *60*, 914–924. [CrossRef]
33. Malayeri, M.; Haghighat, F.; Lee, C.-S. Modeling of volatile organic compounds degradation by photocatalytic oxidation reactor in indoor air: A review. *Build. Environ.* **2019**, *154*, 309–323. [CrossRef]
34. Ye, H.; Liu, Y.; Chen, S.; Wang, H.; Liu, Z.; Wu, Z. Synergetic effect between non-thermal plasma and photocatalytic oxidation on the degradation of gas-phase toluene: Role of ozone. *Chin. J. Catal.* **2019**, *40*, 631–637. [CrossRef]
35. Adhikari, B.C.; Lamichhane, P.; Lim, J.S.; Nguyen, L.N.; Choi, E.H. Generation of reactive species by naturally sucked air in the Ar plasma jet. *Results Phys.* **2021**, *30*, 104863. [CrossRef]
36. Xu, X.; Zhou, S.; Long, J.; Wu, T.; Fan, Z. The Synthesis of a Core-Shell Photocatalyst Material YF_3:Ho^{3+}@TiO_2 and Investigation of Its Photocatalytic Properties. *Materials* **2017**, *10*, 302. [CrossRef]
37. Parvari, R.; Ghorbani-Shahna, F.; Bahrami, A.; Azizian, S.; Assari, M.J.; Farhadian, M. α-Fe_2O_3/Ag/g-C_3N_4 Core-Discontinuous Shell Nanocomposite as an Indirect Z-Scheme Photocatalyst for Degradation of Ethylbenzene in the Air Under White LEDs Irradiation. *Catal. Lett.* **2020**, *150*, 3455–3469. [CrossRef]

38. Liang, C.; Li, C.; Zhu, Y.; Du, X.; Yao, C.; Ma, Y.; Zhao, J. Recent advances of photocatalytic degradation for BTEX: Materials, operation, and mechanism. *Chem. Eng. J.* **2023**, *455*, 140461. [CrossRef]
39. Liang, W.; Li, J.; Li, J.; Jin, Y. Abatement of toluene from gas streams via ferro-electric packed bed dielectric barrier discharge plasma. *J. Hazard. Mater.* **2009**, *170*, 633–638. [CrossRef] [PubMed]
40. Dong, S.; Wang, Y.; Yang, J.; Cao, J.; Su, L.; Wu, X.; Nengzi, L.-C.; Liu, S. Performance and mechanism analysis of degradation of toluene by DBD plasma-catalytic method with MnOx/Al$_2$O$_3$ catalyst. *Fuel* **2022**, *319*, 123721. [CrossRef]
41. Huang, H.; Ye, D.; Leung, D.Y.; Feng, F.; Guan, X. Byproducts and pathways of toluene destruction via plasma-catalysis. *J. Mol. Catal. A Chem.* **2011**, *336*, 87–93. [CrossRef]

Disclaimer/Publisher's Note: The statements, opinions and data contained in all publications are solely those of the individual author(s) and contributor(s) and not of MDPI and/or the editor(s). MDPI and/or the editor(s) disclaim responsibility for any injury to people or property resulting from any ideas, methods, instructions or products referred to in the content.

Article

Hydrothermal Synthesis of MoS$_2$/SnS$_2$ Photocatalysts with Heterogeneous Structures Enhances Photocatalytic Activity

Guansheng Ma [1], Zhigang Pan [1,2], Yunfei Liu [1,2], Yinong Lu [1,2] and Yaqiu Tao [1,2,*]

1. College of Materials Science and Engineering, Nanjing Tech University, Nanjing 211800, China; 202061203287@njtech.edu.cn (G.M.); panzhigang@njtech.edu.cn (Z.P.); yfliu@njtech.edu.cn (Y.L.); yinonglu@njtech.edu.cn (Y.L.)
2. State Key Laboratory of Materials-Oriented Chemical Engineering, Nanjing 211800, China
* Correspondence: taoyaqiu@njtech.edu.cn; Tel.: +86-137-7078-0496

Abstract: The use of solar photocatalysts to degrade organic pollutants is not only the most promising and efficient strategy to solve pollution problems today but also helps to alleviate the energy crisis. In this work, MoS$_2$/SnS$_2$ heterogeneous structure catalysts were prepared by a facile hydrothermal method, and the microstructures and morphologies of these catalysts were investigated using XRD, SEM, TEM, BET, XPS and EIS. Eventually, the optimal synthesis conditions of the catalysts were obtained as 180 °C for 14 h, with the molar ratio of molybdenum to tin atoms being 2:1 and the acidity and alkalinity of the solution adjusted by hydrochloric acid. TEM images of the composite catalysts synthesized under these conditions clearly show that the lamellar SnS$_2$ grows on the surface of MoS$_2$ at a smaller size; high-resolution TEM images show lattice stripe distances of 0.68 nm and 0.30 nm for the (002) plane of MoS$_2$ and the (100) plane of SnS$_2$, respectively. Thus, in terms of microstructure, it is confirmed that the MoS$_2$ and SnS$_2$ in the composite catalyst form a tight heterogeneous structure. The degradation efficiency of the best composite catalyst for methylene blue (MB) was 83.0%, which was 8.3 times higher than that of pure MoS$_2$ and 16.6 times higher than that of pure SnS$_2$. After four cycles, the degradation efficiency of the catalyst was 74.7%, indicating a relatively stable catalytic performance. The increase in activity could be attributed to the improved visible light absorption, the increase in active sites introduced at the exposed edges of MoS$_2$ nanoparticles and the construction of heterojunctions opening up photogenerated carrier transfer pathways and effective charge separation and transfer. This unique heterostructure photocatalyst not only has excellent photocatalytic performance but also has good cycling stability, which provides a simple, convenient and low-cost method for the photocatalytic degradation of organic pollutants.

Keywords: MoS$_2$; SnS$_2$; photocatalysis; composite catalyst; visible light degradation

Citation: Ma, G.; Pan, Z.; Liu, Y.; Lu, Y.; Tao, Y. Hydrothermal Synthesis of MoS$_2$/SnS$_2$ Photocatalysts with Heterogeneous Structures Enhances Photocatalytic Activity. *Materials* **2023**, *16*, 4436. https://doi.org/10.3390/ma16124436

Academic Editors: Xingwang Zhu and Tongming Su

Received: 29 May 2023
Revised: 12 June 2023
Accepted: 14 June 2023
Published: 16 June 2023

Copyright: © 2023 by the authors. Licensee MDPI, Basel, Switzerland. This article is an open access article distributed under the terms and conditions of the Creative Commons Attribution (CC BY) license (https://creativecommons.org/licenses/by/4.0/).

1. Introduction

Today's industrialized and urbanized world is facing severe energy shortages and environmental pollution problems. The excessive use of organic dye and the indiscriminate release of organic pollutants cause serious damage to ecosystems and have serious effects on future generations [1]. Moreover, the organic ingredients in our living environment are difficult to degrade and toxic in nature. There have been many methods to solve the problem of organic dye contamination in environment, such as adsorption [2,3], membrane separation [4,5], biological decomposition [6], chemical oxidation [7], electrocatalysis [8] and photocatalysis [9,10] decomposition. Among them, the use of solar photocatalysis for the degradation of organic pollutants is considered one of the most promising and efficient strategies [11,12].

Transition metal sulfides have attracted a lot of attention in wastewater treatment because of their high specific surface area, high surface activity and special microstructure. In recent years, MoS$_2$ has been widely used in organic dye decomposition due to its low cost, high abundance and noble-metal-like activities [13–15]. MoS_2 has a graphene-like layered structure with three crystal phases: 1T, 2H and 3R [16]. In a natural state, MoS_2 is usually present in the steady 2H phase, which exhibits semiconducting properties [17,18]. However, the low density of active sites and relatively poor conductivity of $2H-MoS_2$ lead to limited photocatalytic activity [19,20]. Compared with $2H-MoS_2$, the metallic $1T-MoS_2$ phase has the advantages of significant conductivity and a high density of marginal active sites at room temperature and shows better performance in photocatalysis [21–24]. So far, most $1T-MoS_2$ is fabricated as two-dimensional nanosheets to construct hybrid structures with nanoconjunctions [25–27].

Li et al. [28] prepared two-dimensional heterostructured $MoS_2/g-C_3N_4$ (graphite-C_3N_4) photocatalysts using a facile impregnation–calcination method. The experimental results showed that surface MoS_2 nanosheets were successfully loaded horizontally onto $g-C_3N_4$ nanosheets. Meanwhile, the two-dimensional heterojunction formed between $g-C_3N_4$ nanosheets and MoS_2 nanosheets improved the separation efficiency and charge transfer rate of photogenerated electrons. One of the synthesized samples, MCNNs-3 (3 wt% MoS_2 in $MoS_2/g-C_3N_4$ heterojunction), with a catalyst content of 0.8 g/L, reduced the concentration of rhodamine B (RhB) by about 96% after 20 min of irradiation. Chen et al. [29] prepared $MoS_2/TaON$ (tantalum oxynitride) hybrid nanostructures by a hydrothermal method. This work showed that the photocatalytic degradation of rhodamine B (RhB) on Ta1Mo1 (mass ratio of $TaON:MoS_2 = 1:1$) was about 65% after 2 h of visible light irradiation, which was about five times higher than that of pure TaON. In addition, $MoS_2/SiO_2/TaON$ ternary photocatalysts were constructed to further improve the photocatalytic performance. When the mass ratio of Ta8Si1 ($TaON:SiO_2 = 8:1$) to MoS_2 was 1:1, the degradation rate of RhB reached 75% under 2 h of visible light irradiation. Yin et al. [30] synthesized two kinds of MoS_2 and $PbBiO_2Cl$ nanosheets by the solvothermal method and then prepared a novel 2D/2D $MoS_2/PbBiO_2Cl$ photocatalyst by mechanical stirring at room temperature. The resulting experiments showed that 1 wt% of $MoS_2/PbBiO_2Cl$ showed stronger photocatalytic performance and 80% of rhodamine B (RhB) could be completely degraded within 120 min, whereas the photocatalytic activity decreased when the content of MoS_2 was higher.

Composites containing MoS_2 with other similar materials are an effective way to enhance the photocatalytic ability of the material. SnS_2 has a narrow band gap of 2.0 to 2.3 eV and is a low-cost, non-toxic CdI_2-type layered semiconductor [31–33]. According to the literature [34–36], SnS_2 is a relatively stable visible-light-driven photocatalyst in the degradation of organic compounds. However, like most semiconductor photocatalysts, SnS_2 also has the disadvantage of high recombination rates of photogenerated electrons and holes, resulting in low photocatalytic efficiency [37]. Among the modification strategies explored to improve the photocatalytic efficiency of SnS_2, the combination with a suitable semiconductor or other components (e.g., graphene) facilitates the separation of photogenerated electrons and holes through interfacial charge transfer [38–40]. Zhang et al. [41] prepared 2D/2D-type $SnS_2/g-C_3N_4$ (graphite–C_3N_4) heterojunction photocatalysts using an ultrasonic dispersion method. The electron microscopic characterization analysis showed that a large contact zone was induced at the heterojunction interface due to the lamellar structure of both the SnS_2 and $g-C_3N_4$ materials. In the photoluminescence spectra, it can also be shown that the photo-coordination effect of the $SnS_2/g-C_3N_4$ heterojunction effectively enhances the interfacial carrier transfer, leading to enhanced charge separation during the photocatalytic reaction.

Due to the outstanding reactivity of both MoS_2 and SnS_2 in the photocatalytic degradation of organic dyes, the structures of MoS_2 and SnS_2 were coupled to construct a heterogeneous structure to enhance the degradation of organic dyes. Compared with previous work, this experiment is further improved: firstly, by changing the synthesis method of the sample and the synthesis conditions, the high temperature and high energy consumption in the experiment as well as the shortened reaction time are avoided; secondly, the synthesis steps are simpler and less cumbersome; and finally, the reactants are easily available throughout the experiment and the heterogeneous structure is binary, which can efficiently solve the cost problem in the application.

In this work, we used a convenient hydrothermal method to obtain MoS_2/SnS_2 composite catalysts with SnS_2 nanosheets grown on MoS_2 nanoparticles. By adjusting the hydrothermal time of the reaction (12 h, 14 h and 16 h) and changing the molar ratio of the substances (1:1, 2:1, 3:1 and 4:1 atomic molar ratio of molybdenum–tin), a heterogeneous structure was constructed between MoS_2 and SnS_2 after the hydrothermal reaction, resulting in a semiconducting composite photocatalyst with a narrow band gap. Theoretically, the narrowing of the band gap of the material can effectively improve the absorption of visible light and the catalyst material can produce a large number of electrons and holes when light can be irradiated. In addition, the heterostructure can effectively modulate the electronic structure of the complex system while promoting electron transport between the interfaces more effectively, thus improving the photocatalytic ability. This work highlights that the construction of heterojunctions between two substances may be an attractive method for the removal of pollutants from industrial wastewater.

2. Materials and Methods

2.1. Raw Materials

Thiourea (CH_4N_2S) was purchased from Shanghai Ling Feng Chemical Reagent Co. (Shanghai, China). Tin chloride pentahydrate ($SnCl_4 \cdot 5H_2O$, 99%) was supplied by Shanghai Test Four Hervey Chemical Co. (Shanghai, China). Ammonium molybdate tetrahydrate (($NH_4)_6Mo_7O_{24} \cdot 4H_2O$, 99%) was purchased from Sinopharm Chemical Reagent Co. (Shanghai, China). Hydrochloric acid (HCl) was supplied by Yonghua Chemical Co. (Changshu, China). Methylene blue (MB) was supplied by Tianjin Chemical Reagent Research Co. (Tianjin, China). Except for hydrochloric acid, which is superiorly pure, the rest of the chemical reagents are of analytical grade and were used without further purification.

2.2. Synthesis of Photocatalysts

The fabrication steps involved in the synthesis of the MoS_2/SnS_2 composite catalysts are schematically illustrated in Figure 1. First, the synthesis of MoS_2 nanoparticles [42]: MoS_2 nanoparticles were synthesized by the hydrothermal method. Typically, 1.0592 g of $(NH_4)_6Mo_7O_{24} \cdot 4H_2O$ and 1.828 g of CH_4N_2S were dissolved in 60 mL of deionized water at room temperature with continuous stirring until complete dissolution. The mixed solution was then transferred to a 100 mL Teflon-lined autoclave and kept at 180 °C for 16 h. Then, the obtained black precipitate was dried at 80 °C for 2 h.

Figure 1. Schematic diagram of the synthesis procedure of MoS_2/SnS_2 catalysts.

Preparation of MoS_2/SnS_2 composite catalysts [43] with different reaction times (12 h, 14 h and 16 h): the black MoS_2 powder was weighed according to the ratio and dispersed in 60 mL deionized water to form a suspension. Certain proportions of $SnCl_4 \cdot 5H_2O$ and CH_4N_2S were added to the suspension; then, a certain amount of 1 mol/L hydrochloric acid was added to adjust the mixed solution to acidity. Finally, the mixture was transferred to a 100 mL Teflon-lined autoclave and kept at 180 °C for a certain time. After the reaction was completed, the catalyst was collected by centrifugation and dried at 80 °C for 2 h to obtain the catalyst. The catalysts were named MS12-2-H, MS14-2-H and MS16-2-H according to the reaction time and the atomic molar ratio of molybdenum to tin.

Preparation of MoS$_2$/SnS$_2$ composite catalysts with different Mo/Sn atomic molar ratios (1:1, 2:1, 3:1 and 4:1): the reactants were weighed according to the ratios and the synthesis steps were the same as above; the final composite catalysts were named MS14-1-H, MS14-2-H, MS14-3-H and MS14-2-H according to the naming principle.

2.3. Structural Characterization

Powder X-ray diffraction patterns were obtained using a Rigaku Smart Lab diffractometer with Cu Kα (λ = 0.154178 nm) as the radiation source. The morphology of the samples was measured using a JSM-6510 scanning electron microscope. Nitrogen adsorption–desorption isotherms were measured at −196 °C using a specific surface area and pore-size analyzer, the V-Sorb 1800. The samples were all pretreated at 105 °C for 12 h prior to measurement. Electrochemical impedance was tested with a CH1660E electrochemical workstation. X-ray photoelectron spectra were obtained with a KRATOS AXIS SUPRA.

2.4. Photocatalytic Degradation and Photoelectrochemical Test

The catalytic performances of the composite catalysts were evaluated by their ability to degrade the target pollutant, MB, under visible light using a 300 W xenon lamp as a visible light source. In each test, 30 mg of catalyst was dispersed in 80 mL of MB solution (15 mg/L). The mixed solution was stirred in the dark for 30 min prior to the test to achieve an adsorption–desorption equilibrium between the catalyst and the solution. The 7 mL suspension was removed every 10 min under visible light and centrifuged at 3500 r/min for 5 min. The absorbance of the solution at different reaction times was measured by UV–visible spectrophotometer.

The determination of the concentration of organic dyes can be described by the Beer–Lambert law, and the amount of light absorbed by the solution follows the Beer–Lambert law. The specific equation is as follows [44]:

$$A = \varepsilon bc \qquad (1)$$

where A is the absorbance, ε is the light absorption coefficient, b is the solution thickness and c is the dye concentration solution at the time of sampling. According to the Beer–Lambert law, the relationship between dye concentration and absorbed light is linear.

The electrochemical impedance test was carried out in a CH1660E electrochemical workstation with a platinum electrode as the counter electrode and a saturated glycerol electrode as the reference electrode, and the corresponding open circuit voltage and frequency were set. In this experiment [43], catalyst-coated conductive glass was used as the working electrode, namely 20 mg of the catalyst dispersed into a mixture containing 40 uL of 5 wt% Nafion and 0.5 mL of anhydrous ethanol. After mixing well with ultrasound, 200 uL of the suspension was coated onto the surface of the conductive glass with a pipette gun; this was then dried naturally at room temperature. The working electrode for the electrochemical impedance was tested in 0.1 M Na$_2$SO$_4$ solution.

3. Results and Discussion

3.1. Characterization and Properties of Composite Catalysts Synthesized for Different Reaction Times

3.1.1. XRD Characterization

Figure 2 represents the XRD patterns of the MoS$_2$/SnS$_2$ composite catalysts prepared from 1T-MoS$_2$ synthesized by a hydrothermal reaction at 180 °C for different reaction times. It can be seen from the figure that the characteristic peaks of the SnS$_2$ component at 2θ equal to 14.9°, 28.2°, 32.1°, 41.8°, 49.9°, 52.4° and 54.9° are clearly observed, which correspond to the (001), (100), (101), (102), (110), (111) and (103) SnS$_2$ crystalline planes, respectively (JCPDS No. 23-0677) [45]. The SnS$_2$ component is successfully synthesized in the MoS$_2$/SnS$_2$ composite catalysts. The characteristic peaks of MoS$_2$ at 2θ equal to 10.9°, 32.8° and 57.2° are not clearly shown in the figure because of the unique lamellar structure and small grain size of MoS$_2$ [46]. In addition, the layer spacing of MoS$_2$ becomes larger during the reaction process, resulting in a shift of the (002) crystal plane to 10.9°. The presence of both MoS$_2$ and SnS$_2$ components in the synthesized catalysts without spurious peaks in the XRD patterns indicate the successful synthesis of MoS$_2$/SnS$_2$ composite catalysts.

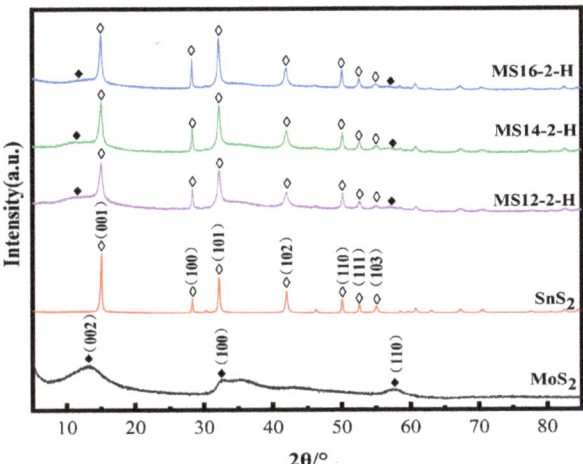

Figure 2. Powder X-ray diffraction patterns of MoS$_2$/SnS$_2$ composite catalysts with different reaction times.

With increasing reaction time, the characteristic peak of MoS$_2$ gradually broadens and the characteristic peaks of SnS$_2$ gradually narrow, indicating the stronger crystallinity of the SnS$_2$ phase, which on the other hand also means that the structure of SnS$_2$ in the reaction process is more complete.

3.1.2. Morphology Analysis

Scanning electron micrographs of the catalysts MS12-2-H, MS14-2-H and MS16-2-H synthesized at different times are shown in Figure 2. In Figure 3a, the hexagonal SnS$_2$ nanosheets are grown on MoS$_2$ nanosphere flowers [1], whereas the hexagonal SnS$_2$ nanosheets are not uniformly distributed and a large portion of the nanoflake particles are not in contact with the nanosheets.

The morphology of the catalysts in Figure 3b changed considerably. The growth of SnS$_2$ nanosheets on the surface of the flower-like morphology of MoS$_2$ was not only more uniformly distributed but also the agglomeration of SnS$_2$ nanosheets was slight, which was obviously different from the other catalysts and could expose more active sites. The SnS$_2$ nanosheets in Figure 3c also have better crystallinity of the grains, although they are more uniformly distributed than in (a), which is consistent with the results for the XRD experiments.

3.1.3. BET Measurements

Table 1 shows the results of the specific surface area test results. The nitrogen adsorption–desorption isotherms were measured at −196 °C after all the samples were pre-treated at 105 °C for 12 h prior to measurement. The specific surface area of the composite catalyst showed a trend of increasing and then decreasing with the increase in the reaction time. The reaction time did not have a great influence on the average pore diameter in the range 2.32–2.34 nm and total pore volume of 0.003 cm^3/g of the composite catalysts, which were almost negligible. Combined with the analysis of the SEM images of the composite catalysts, the specific surface area of the MoS$_2$ and SnS$_2$ materials was greatly enhanced due to the uniform growth of lamellar SnS$_2$ particles on the surface of the flower-like MoS$_2$ particles; thus, effectively improving the catalytic performance [47].

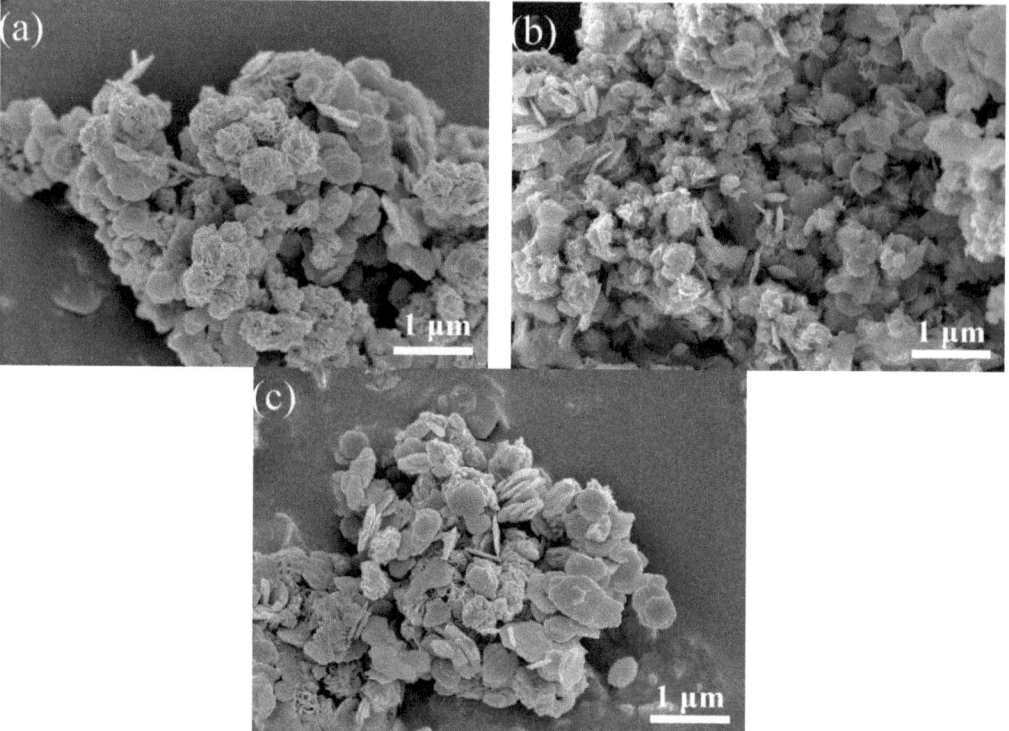

Figure 3. SEM images of MoS_2/SnS_2 composite catalysts: (**a**) MS12-2-H; (**b**) MS14-2-H; (**c**) MS16-2-H.

Table 1. BET analysis of composite catalysts for different reaction times.

Catalysts	BET Surface Area (m^2/g)	Average Pore Width (nm)	Total Pore Volume (cm^3/g)
MoS_2	5.03	2.03	0.001
SnS_2	9.00	2.02	0.003
MS12-2-H	13.07	2.32	0.003
MS14-2-H	17.10	2.34	0.003
MS16-2-H	12.05	2.32	0.003

3.1.4. Electrochemical Impedance Measurement

Figure 4 shows the electrochemical impedance plots for the composite catalysts synthesized for different reaction times. In the electrochemical impedance diagram, the radius of the semicircle in the high-frequency region is positively correlated with the charge transfer resistance, reflecting the transfer characteristics of photogenerated electrons and holes in the catalysts under the light conditions. According to the test results, the impedance radius of the MoS_2/SnS_2 composite catalysts showed a trend of decreasing and then increasing as the reaction time increased, in which the MS14-2-H composite catalyst had the smallest impedance radius and presented a high charge transfer migration efficiency [48]. This also proves that a suitable reaction time has a great impact on the electronic structure of the two components in the catalysts, resulting in the improvement of their charge transfer capacity and thus the photocatalytic degradation performance.

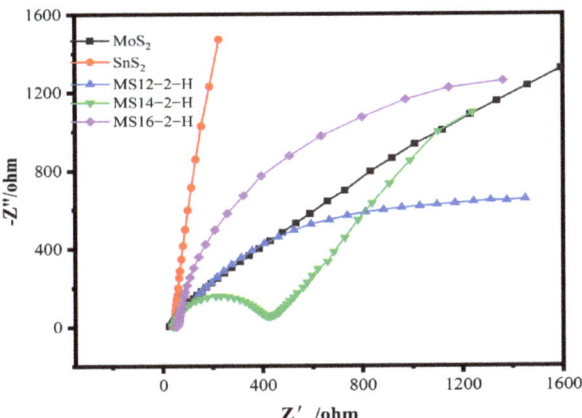

Figure 4. Electrochemical impedance diagram of MoS$_2$/SnS$_2$ composite catalysts at different reaction times.

3.1.5. Photocatalytic Performance

The degradation of methylene blue solution by MoS$_2$/SnS$_2$ composite catalysts formed at different reaction times is shown in Figure 5. From the figures, it can be observed that the composite catalyst degraded MB in visible light. Of the three catalysts, MS14-2-H exhibited the highest MB degradation rates. With the increasing of the synthesis time of the MoS$_2$/SnS$_2$ composite catalysts, a trend of enhancing and then weakening is observed, which is consistent with the test results of the electrochemical impedance of the composite catalysts. The length of the reaction time has an effect on the electronic structure of the synthesized composite catalyst, thus affecting the migration rate of photogenerated charges in visible light and producing an effect on the performance of the degradation of MB. From the degradation data, it was found that the best catalytic performance was achieved by sample MS14-2-H, which had an 83.0% degradation rate after 80 min of visible light irradiation. This was followed by sample MS12-2-H, which had a 55.9% degradation rate. This corroborates the previous results of the XRD, SEM and EIS analyses.

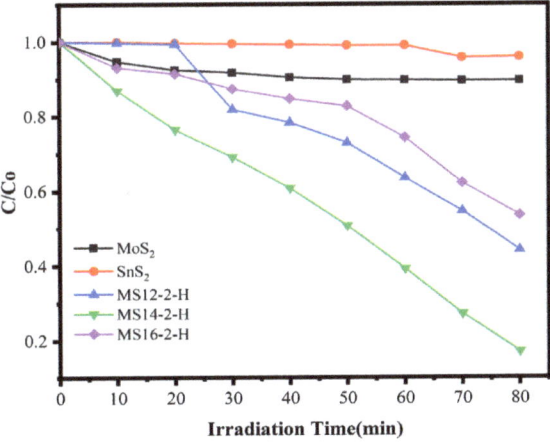

Figure 5. Degradation rate diagram of MoS$_2$/SnS$_2$ composite catalysts at different reaction times.

3.2. Characterization and Properties of Composite Catalysts in Different Mo–Sn Molar Ratios

3.2.1. XRD Characterization

Figure 6 represents the XRD patterns of the composite catalysts synthesized by varying the molybdenum–tin molar ratio in the reactants at a reaction temperature of 180 °C for 14 h. It can be seen from the figures that the characteristic peaks at 2θ equal to 14.9°, 28.2°, 32.1°, 41.8°, 49.9°, 52.4° and 54.9° correspond to the (001), (100), (101), (102), (110), (111) and (103) crystal planes of SnS_2 in the composite catalysts of different molybdenum–tin molar ratios, respectively [45]. In addition, no excess spurious peaks were found in the XRD patterns, indicating that the MoS_2/SnS_2 composite catalysts were successfully synthesized.

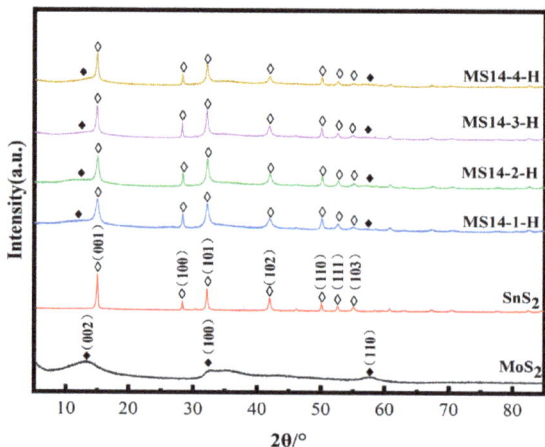

Figure 6. Powder X-ray diffraction patterns of MoS_2/SnS_2 composite catalysts synthesized at different molybdenum–tin molar ratios.

When increasing proportion of Mo in the reactants, the characteristic peaks of both MoS_2 and SnS_2 gradually broadened, especially the (100), (101) and (102) crystal planes in SnS_2.

3.2.2. Morphology Analysis

SEM images of the samples MS14-1-H, MS14-2-H, MS14-3-H and MS14-4-H are shown in Figure 7. From the figures, MoS_2 particles are shown as having flower-like morphology composed of layered nanosheets. Increasing the proportion of Mo leads to more serious MoS_2 agglomeration and the spherical flower particles of MoS_2 become more regular. The growth of SnS_2 nanosheets on the surface of the flower-like MoS_2 morphology exhibits hexagonal morphology and the SnS_2 nanosheets have a size of about 400 nm [1]. With an increase in the MoS_2 fraction, the main change in the morphology is that the SnS_2 nanosheets are not uniformly distributed on the MoS_2 surface, as shown in Figure 7c,d. At lower Mo/Sn ratios, the scanning images of the catalyst changed more, and the MoS_2 morphology was no longer a regular spherical flower shape but a nano-flake shape. In addition, the SnS_2 nanosheets were closely distributed on the MoS_2 surface, as shown in Figure 7a,b. The close distribution of the SnS_2 nanosheets on the MoS_2 surface contributes to the close bonding of the two materials, which can effectively change the electronic structure of the catalyst and increase the photocatalytic active sites.

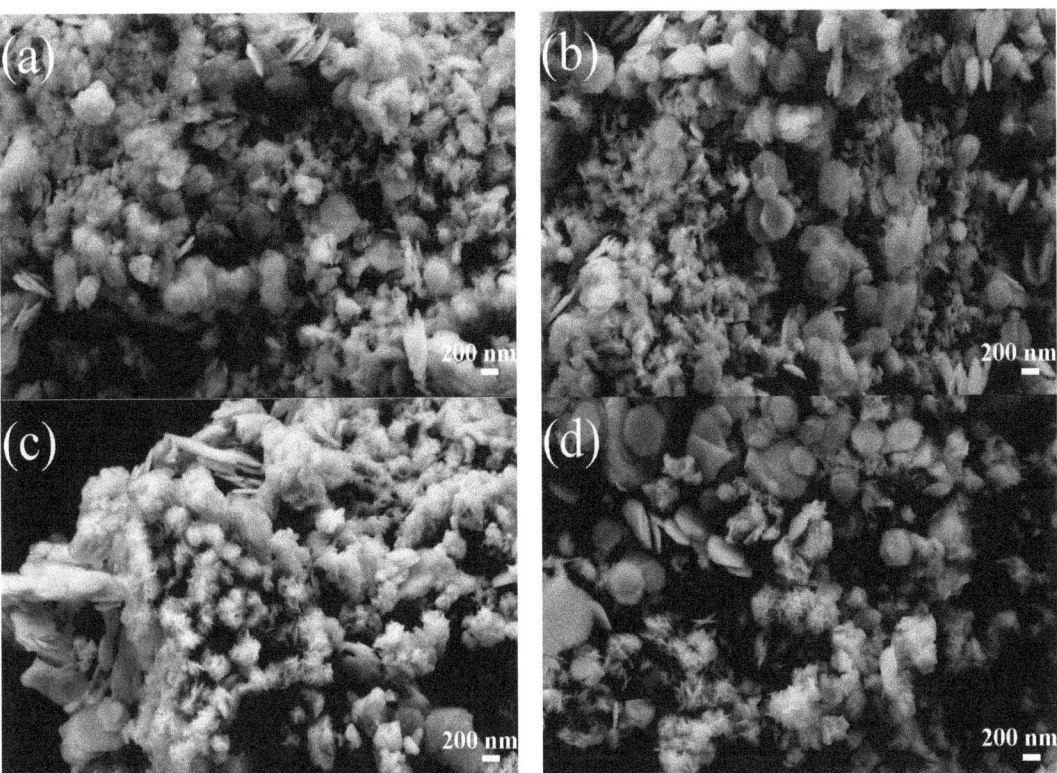

Figure 7. SEM diagrams of composite catalysts. (**a**) MS14-1-H. (**b**) MS14-2-H. (**c**) MS14-3-H. (**d**) MS14-4-H.

Figure 8a shows TEM images of the prepared MS14-2-H catalyst, which further shows that the flake SnS_2 grows on the surface of MoS_2 with a small size. Lattice stripes with distances of 0.68 nm and 0.30 nm are shown in Figure 8b, which could correspond to the (002) plane of MoS_2 and the (100) plane of SnS_2, respectively [49]. These images objectively further explain the coexistence of SnS_2 and MoS_2 in the MS14-2-H composite catalyst. In addition to this, the lattice stripe in the (002) plane of MoS_2 is larger than the standard spacing value (0.62 nm) according to the Bragg equation:

$$2d\sin\theta = n\lambda \tag{2}$$

where d is the lattice spacing, θ is the angle between the incident ray, the reflection line and the reflected crystal plane, λ is the wavelength and n is the number of reflection levels. It is known that under specific conditions, the lattice spacing d is inversely proportional to the angle θ, i.e., at this time, the lattice spacing of the (002) plane of MoS_2 is large and the corresponding diffraction peak angle becomes small, which is similar to the diffraction peak of the (002) corresponding to the MoS_2 phase of the composite catalyst in the XRD analysis shifting to 10.9°.

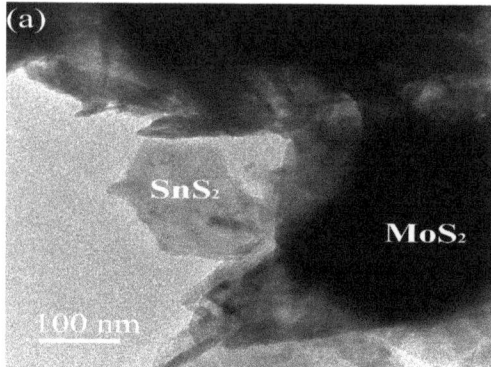 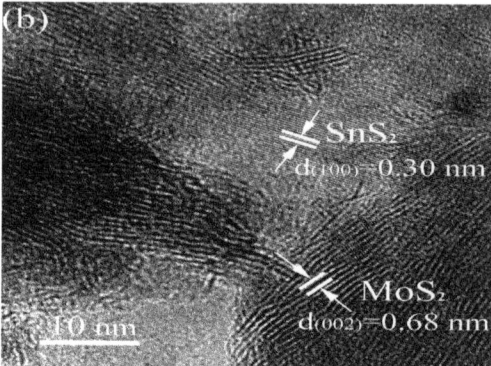

Figure 8. TEM images of MS14-2-H: (**a**) TEM image; (**b**) high-resolution TEM image.

3.2.3. BET Measurements

Table 2 shows the results of the specific surface area tests of the MoS_2/SnS_2 composite catalysts with different molybdenum–tin molar ratios. From the data in the table, it can be observed that the specific surface area of the composite catalysts shows a trend of increasing and then decreasing when increasing the molybdenum–tin molar ratio from 1:1 to 4:1. In addition, the average pore diameter of the catalyst was 2.30~2.34 nm and the total pore volume was 0.003 cm^3/g.

Table 2. BET analysis of composite catalysts with different Mo–Sn molar ratios.

Catalysts	BET Surface Area (m^2/g)	Average Pore Width (nm)	Total Pore Volume (cm^3/g)
MoS_2	5.03	2.03	0.001
SnS_2	9.00	2.02	0.003
MS14-1-H	15.05	2.32	0.003
MS14-2-H	17.10	2.34	0.003
MS14-3-H	12.75	2.30	0.003
MS14-4-H	11.50	2.31	0.003

The effect of molybdenum–tin molar ratio on the average pore diameter and total pore volume of the composite catalysts is almost negligible, and the effect on the catalyst performance is mainly due to the specific surface area, which is 17.10 m^2/g for MS14-2-H, followed by 15.05 m^2/g for MS14-1-H. Combined with the analysis of the SEM images of the composite catalysts, the reason for the large differences in the specific surface areas of the catalysts may be due to the size and shape of the particles of MoS_2 and SnS_2 and the difference in the shape of the particles.

3.2.4. Electrochemical Impedance Measurement

The electrochemical impedance diagrams of the composite catalysts synthesized at different molybdenum–tin molar ratios are shown in Figure 9. According to the test results, the impedance radii of the MoS_2/SnS_2 composite catalysts showed a trend of decreasing and then increasing as the molybdenum–tin molar ratio increased; the MS14-2-H composite catalyst having the smallest impedance radius. Since the radius of the semicircle in the high-frequency region is positively correlated with the charge transfer resistance in the electrochemical impedance diagram, the sample MS14-2-H has the smallest charge transfer resistance [48].

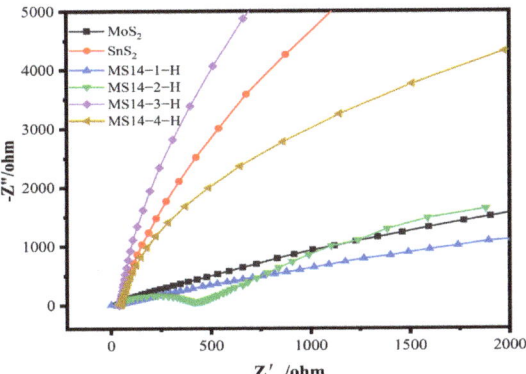

Figure 9. Electrochemical impedance plot of MoS$_2$/SnS$_2$ composite catalysts synthesized with different molybdenum–tin molar ratios.

3.2.5. Structural Composition Analysis of Materials

Figure 10 shows the XPS diagram of the catalyst MS14-2-H. The elemental composition and elemental chemical valence of the sample catalyst can be identified by analyzing the XPS. From Figure 10a, it can be seen that the catalyst MS14-2-H contains C, O, S, Mo and Sn, and the sample contains all the elements of the target substance by the elemental full spectrum.

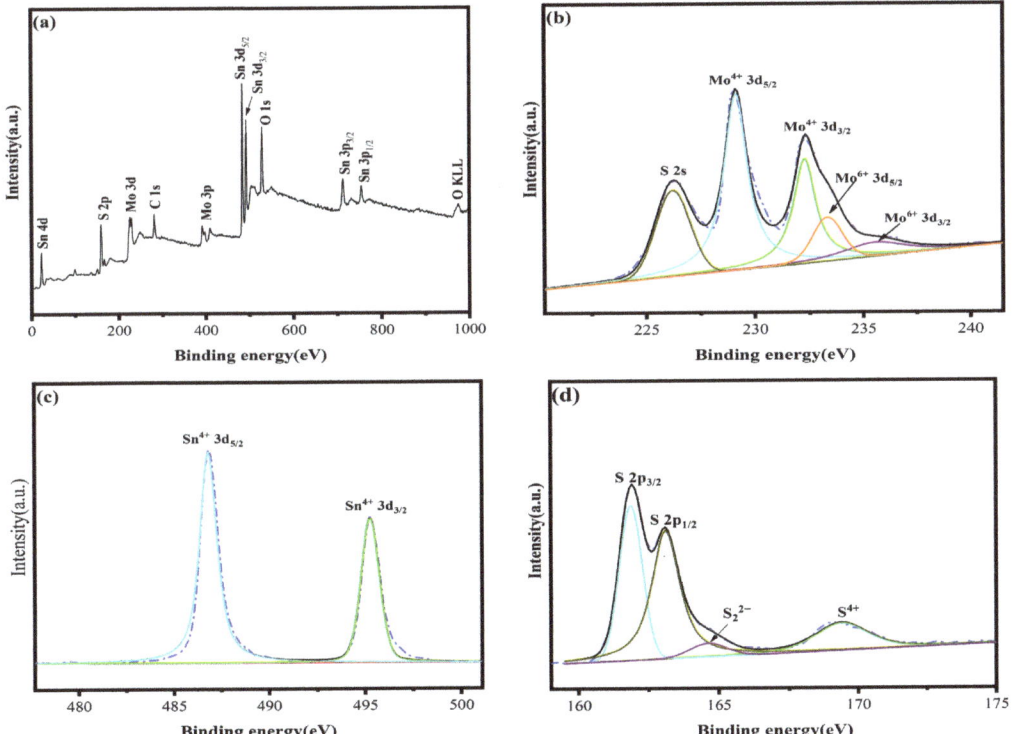

Figure 10. XPS diagram of catalyst MS14-2-H: (**a**) survey spectra; (**b**) Mo 3d; (**c**) Sn 3d; (**d**) S 2p.

Figure 10b shows the XPS spectra of Mo 3d, in which five different characteristic peaks appear. It is known from the split peak fitting and literature review [50] that peaks at 229.1 eV and 232.2 eV correspond to Mo^{4+} $3d_{5/2}$ and Mo^{4+} $3d_{3/2}$, respectively, peaks at 233.3 eV and 235.5 eV correspond to Mo^{6+} $3d_{5/2}$ and Mo^{6+} $3d_{3/2}$, respectively, and the peak at 225.8 eV corresponds to S^{2-} 2s. The Mo^{4+} species belong to MoS_2 and Mo^{6+} signals and may be caused by slight oxidation in air. Only two characteristic peaks appear in Figure 10c, and a review of the literature shows that [51] signals at binding energies of 486.8 eV and 495.3 eV correspond to Sn^{4+} $3d_{5/2}$ and Sn^{4+} $3d_{3/2}$, respectively, whereas the Sn^{4+} species belong to SnS_2. Signals of S^{2-} $2p_{3/2}$ and S^{2-} $2p_{1/2}$ from the composite catalyst present at binding energies equal to 161.9 eV and 163.1 eV [52]. Therefore, the analysis shows that the composition of the synthesized samples is consistent with the target MoS_2/SnS_2 catalyst.

3.2.6. Photocatalytic Performance

Figure 11a shows the performance of the composite catalysts synthesized at different molybdenum–tin molar ratios in the degradation of methylene blue solutions. From the degradation data in the figure, the visible light degradation performances in methylene blue solution of the composite catalysts synthesized at different molybdenum–tin molar ratios are better than that of either pure MoS_2 or SnS_2, indicating that the photocatalytic performance can be effectively improved by using a composite of these two materials. In addition, after visible light irradiation for 80 min, the best catalytic performance was achieved for sample MS14-2-H, which had a degradation rate of 83%, whereas the degradation rate of sample MS14-3-H was 44.6% and that of sample MS14-4-H was 24.7%, indicating that the optimal molybdenum–tin ratio that can effectively improve the photocatalytic performance is 2:1 and that the effect on the photocatalytic performance is limited by only increasing the content of MoS_2 in the composite catalyst. Attention should be paid to the reasonable distribution of the two components in the composite catalysts, which is consistent with the previous results of the XRD, SEM and EIS analyses.

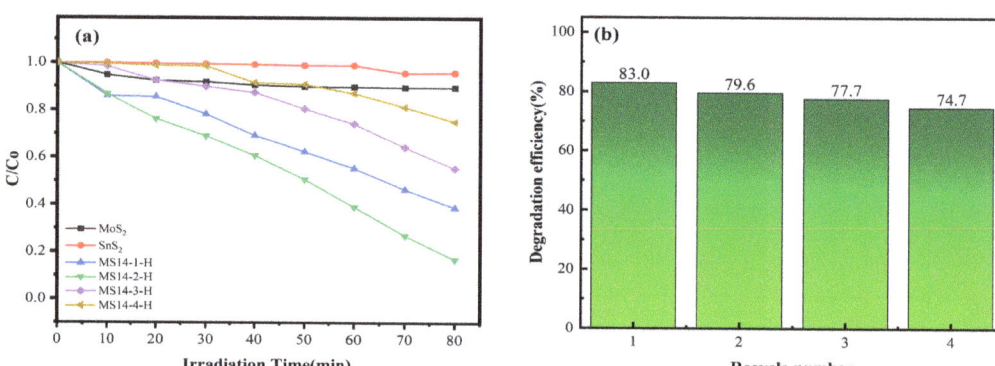

Figure 11. (a) Degradation rate diagram of MoS_2/SnS_2 composite catalysts synthesized with different molybdenum–tin molar ratios. (b) Cycling stability testing of the composite catalyst MS14-2-H.

Figure 11b shows the cycling stability test of the composite catalyst MS14-2-H. It can be seen from the figure that the catalytic degradation efficiency of the composite catalyst MS14-2-H in the MB solution decreased from 83.0% to 74.7% after four visible photocatalytic cycle tests and that the loss of photocatalytic activity was 8.3%. This indicates that the stability and repeatability of the composite catalyst MS14-2-H are good, whereas the loss of photocatalytic activity may be caused by the loss of photocatalysis during the cycle test [1].

3.3. Photocatalytic Mechanism

The photocatalytic mechanism of the composite catalyst is shown in Figure 12. MoS_2 is a p-type semiconductor material with a narrow band structure (e.g., =1.85 eV), whereas

SnS$_2$ is an n-type semiconductor material with a forbidden band width of 2.08 eV [48]. Because the two semiconductors have opposite conductivity types, the electrons and holes of these two semiconductor materials are transferred when they are in close contact to form a heterojunction until the Fermi energy levels of the two semiconductor materials are equal, at which point the p–n heterojunction is in thermal equilibrium and a stable built-in electric field is formed.

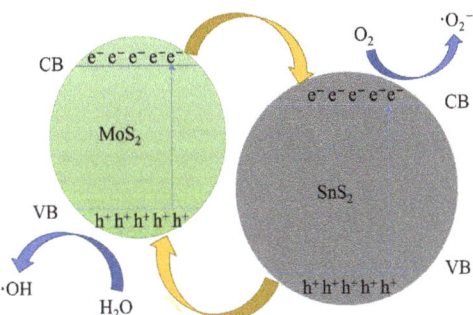

Figure 12. Possible photocatalytic mechanism for the degradation of methylene blue over a composite catalyst.

In the mechanism diagram of the composite catalyst, the CB and VB of MoS$_2$ are higher than that of SnS$_2$, the energy band structures of both are staggered and the heterogeneous structure of the composite catalyst is of type II. When irradiated by visible light, a large number of photogenerated electrons accumulate in the conduction band and a large number of photogenerated holes accumulate in the valence band of both semiconductor materials. Under the effect of potential difference, electrons in the conduction band of MoS$_2$ are transferred to the conduction band of SnS$_2$, whereas holes in the valence band of SnS$_2$ are transferred to the valence band of MoS$_2$. In this way, the electrons and holes can be separated to the maximum extent.

The photocatalytic degradation of MB by composite catalysts under visible light is mainly based on the chemical reaction of the photogenerated electron reduction transferred to the surface of the photocatalyst with dissolved oxygen, which produces strongly oxidizing superoxide radicals (·O^{2-}), and the chemical reaction of the strongly oxidizing holes transferred to the surface of the photocatalyst with hydroxyl radicals (OH$^-$) in water and aqueous solutions, which produces hydroxyl radicals (·OH) [1]. The photocatalytic reaction process is as follows:

$$\text{MoS}_2/\text{SnS}_2 + h\nu \rightarrow e^- + h^+ \quad (3)$$

$$e^- + O_2 \rightarrow \cdot O^{2-} \quad (4)$$

$$h^+ + H_2O/OH^- \rightarrow \cdot OH \quad (5)$$

$$\cdot O^{2-} + MB \rightarrow CO_2 + H_2O \quad (6)$$

$$\cdot OH + MB \rightarrow CO_2 + H_2O \quad (7)$$

4. Conclusions

In summary, a novel MoS$_2$/SnS$_2$ heterostructure was successfully prepared by growing SnS$_2$ nanosheets on MoS$_2$ nanospheres by a facile multi-step hydrothermal method. Based on the measurements of the XRD, SEM, HRTEM and XPS analyses, the present composite sample was found to have high crystalline quality and excellent heterojunction formation. By constructing heterojunctions between the two sulfides, an improved photocatalytic performance was achieved, which greatly solved the problems of low visible light utilization and photogenerated charge recombination. Compared with pure MoS$_2$ or SnS$_2$, this easily accessible and simple composition photocatalyst shows higher photocatalytic activity and good photostability; these effects are attributed to the constructed heterostructure, better light trapping and rapid separation and migration of light-induced electron and hole pairs with the assistance of the MoS_2 metal phase. The optimal MoS_2/SnS_2 photocatalyst (i.e., the one that achieved the best photocatalytic performance) had a degradation efficiency of 83.0% for MB solution, which was 8.3 times higher than the degradation with pure MoS_2 and 16.6 times higher than the degradation with pure SnS_2. The experimental results indicate that this construction of heterojunctions between semiconductors can effectively improve the photocatalytic ability of MoS_2/SnS_2 catalysts in terms of MB degradation.

Author Contributions: Conceptualization, Y.T.; software, Y.L. (Yunfei Liu) and Y.L. (Yinong Lu); data curation, G.M.; writing—original draft preparation, G.M.; writing—review and editing, Y.T. and Z.P. All authors have read and agreed to the published version of the manuscript.

Funding: This work was supported by the priority academic program development of Jiangsu Higher Education Institution (PAPD).

Institutional Review Board Statement: Not applicable.

Informed Consent Statement: Not applicable.

Data Availability Statement: The data presented in this study are available on request from the corresponding author.

Acknowledgments: The authors gratefully acknowledge the assistance from Yaqiu Tao and Zhigang Pan from NJTECH and the staff from State Key Laboratory of Materials-Oriented Chemical Engineering.

Conflicts of Interest: The authors declare no conflict of interest.

References

1. Chen, C.; Shen, M.; Li, Y. One pot synthesis of 1T@2H-MoS_2/SnS_2 heterojunction as a photocatalyst with excellent visible light response due to multiphase synergistic effect. *Chem. Phys.* **2021**, *548*, 11230. [CrossRef]
2. Fang, Y.Y.; Huang, Q.Z.; Liu, P.Y.; Shi, J.F.; Xu, G. A facile dip-coating method for the preparation of separable MoS_2 sponges and their high-efficient adsorption behaviors of Rhodamine B. *Inorg. Chem. Front.* **2018**, *5*, 827–834. [CrossRef]
3. Omar, A.M.; Metwalli, O.I.; Saber, M.R.; Khabiri, G.; Ali, M.E.M.; Hassen, A.; Khalil, M.M.H.; Maarouf, A.A.; Khalil, A.S.G. Revealing the role of the 1T phase on the adsorption of organic dyes on MoS_2 nanosheets. *RSC Adv.* **2019**, *9*, 28345–28356. [CrossRef] [PubMed]
4. Ghadiri, M.; Mohammadi, M.; Asadollahzadeh, M.; Shirazian, S. Molecular separation in liquid phase: Development of mechanistic model in membrane separation of organic compounds. *J. Mol. Liq.* **2018**, *262*, 336–344. [CrossRef]
5. Dong, L.; Li, M.; Zhang, S.; Si, X.; Bai, Y.; Zhang, C. NH_2-Fe_3O_4-regulated graphene oxide membranes with well-defined laminar nanochannels for desalination of dye solutions. *Desalination* **2020**, *476*, 114227. [CrossRef]
6. Maniyam, M.N.; Ibrahim, A.L.; Cass, A.E.G. Decolourization and biodegradation of azo dye methyl red by Rhodococcus strain UCC 0016. *Environ. Technol.* **2020**, *41*, 71–85. [CrossRef]
7. Türgay, O.; Ersöz, G.; Atalay, S.; Forss, J.; Welander, U. The treatment of azo dyes found in textile industry wastewater by anaerobic biological method and chemical oxidation. *Sep. Purif. Technol.* **2011**, *79*, 26–33. [CrossRef]
8. Xie, W.H.; Shi, Y.L.; Wang, Y.X.; Zheng, Y.L.; Liu, H.; Hu, Q.; Wei, S.Y.; Gu, H.B.; Guo, Z.H. Electrospun iron/cobalt alloy nanoparticles on carbon nano fibers towards exhaustive electrocatalytic degradation of tetracycline in wastewater. *Chem. Eng. J.* **2021**, *405*, 126585. [CrossRef]
9. Yin, M.; Li, Z.; Kou, J.; Zou, Z. Mechanism Investigation of Visible Light-Induced Degradation in a Heterogeneous TiO_2/Eosin Y/Rhodamine B System. *Environ. Sci. Technol.* **2009**, *43*, 8361–8366. [CrossRef]
10. Saber, M.R.; Khabiri, G.; Maarouf, A.A.; Ulbricht, M.; Khalil, A.S.G. A comparative study on the photocatalytic degradation of organic dyes using hybridized 1T/2H, 1T/3R and 2H MoS_2 nano-sheets. *RSC Adv.* **2018**, *8*, 26364–26370. [CrossRef]
11. Hu, X.; Cheng, L.; Li, G. One-pot hydrothermal fabrication of basic bismuth nitrate/BiOBr composite with enhanced photocatalytic activity. *Mater. Lett.* **2017**, *203*, 77–80. [CrossRef]
12. Cheng, L.; Hu, X.; Hao, L. Ultrasonic-assisted in-situ fabrication of BiOBr modified $Bi_2O_2CO_3$ microstructure with enhanced photocatalytic performance. *Ultrason. Sonochem.* **2018**, *44*, 137–145. [CrossRef]
13. Lu, D.; Wang, H.; Zhao, X.; Kondamareddy, K.K.; Ding, J.; Li, C.; Fang, P. Highly Efficient Visible-Light-Induced Photoactivity of Z-Scheme g-C_3N_4/Ag/MoS_2 Ternary Photocatalysts for Organic Pollutant Degradation and Production of Hydrogen. *ACS Sustain. Chem. Eng.* **2017**, *5*, 1436–1445. [CrossRef]

14. Cho, S.-Y.; Kim, S.J.; Lee, Y.; Kim, J.-S.; Jung, W.-B.; Yoo, H.-W.; Kim, J.; Jung, H.-T. Highly Enhanced Gas Adsorption Properties in Vertically Aligned MoS$_2$ Layers. *ACS Nano* **2015**, *9*, 9314–9321. [CrossRef]
15. Massey, A.T.; Gusain, R.; Kumari, S.; Khatri, O.P. Hierarchical Microspheres of MoS$_2$ Nanosheets: Efficient and Regenerative Adsorbent for Removal of Water-Soluble Dyes. *Ind. Eng. Chem. Res.* **2016**, *55*, 7124–7131. [CrossRef]
16. Wang, Q.H.; Kalantar-Zadeh, K.; Kis, A.; Coleman, J.N.; Strano, M.S. Electronics and optoelectronics of two-dimensional transition metal dichalcogenides. *Nat. Nanotechnol.* **2012**, *7*, 699–712. [CrossRef]
17. Khabiri, G.; Aboraia, A.M.; Omar, S.; Soliman, M.; Omar, A.M.A.; Kirichkov, M.V.; Soldatov, A.V. The enhanced photocatalytic performance of SnS$_2$@MoS$_2$ QDs with highly-efficient charge transfer and visible light utilization for selective reduction of mythlen-blue. *Nanotechnology* **2020**, *31*, 475602. [CrossRef]
18. Chang, K.; Hai, X.; Pang, H.; Zhang, H.B.; Shi, L.; Liu, G.G.; Liu, H.M.; Zhao, G.X.; Li, M.; Ye, J.H. Targeted Synthesis of 2H-and 1T-Phase MoS$_2$ Monolayers for Catalytic Hydrogen Evolution. *Adv. Mater.* **2016**, *28*, 10033–10041. [CrossRef]
19. Xiao, X.; Wang, Y.; Xu, X.; Yang, T.; Zhang, D. Preparation of the flower-like MoS$_2$/SnS$_2$ heterojunction as an efficient electrocatalyst for hydrogen evolution reaction. *Mol. Catal.* **2020**, *487*, 110890. [CrossRef]
20. Xu, H.; Liu, S.; Ding, Z.; Tan, S.J.R.; Yam, K.M.; Bao, Y.; Nai, C.T.; Ng, M.-F.; Lu, J.; Zhang, C.; et al. Oscillating edge states in one-dimensional MoS$_2$ nanowires. *Nat. Commun.* **2016**, *7*, 12904. [CrossRef]
21. Liang, Z.; Sun, B.; Xu, X.; Cui, H.; Tian, J. Metallic 1T-phase MoS$_2$ quantum dots/g-C$_3$N$_4$ heterojunctions for enhanced photocatalytic hydrogen evolution. *Nanoscale* **2019**, *11*, 12266–12274. [CrossRef] [PubMed]
22. Liu, Q.; Fang, Q.; Chu, W.; Wan, Y.; Li, X.; Xu, W.; Habib, M.; Tao, S.; Zhou, Y.; Liu, D.; et al. Electron-Doped 1T-MoS$_2$ via Interface Engineering for Enhanced Electrocatalytic Hydrogen Evolution. *Chem. Mater.* **2017**, *29*, 4738–4744. [CrossRef]
23. Voiry, D.; Salehi, M.; Silva, R.; Fujita, T.; Chen, M.; Asefa, T.; Shenoy, V.B.; Eda, G.; Chhowalla, M. Conducting MoS$_2$ Nanosheets as Catalysts for Hydrogen Evolution Reaction. *Nano Lett.* **2013**, *13*, 6222–6227. [CrossRef] [PubMed]
24. Wang, H.; Lu, Z.; Kong, D.; Sun, J.; Hymel, T.M.; Cui, Y. Electrochemical Tuning of MoS$_2$ Nanoparticles on Three-Dimensional Substrate for Efficient Hydrogen Evolution. *ACS Nano* **2014**, *8*, 4940–4947. [CrossRef]
25. Acerce, M.; Voiry, D.; Chhowalla, M. Metallic 1T phase MoS$_2$ nanosheets as supercapacitor electrode materials. *Nat. Nanotechnol.* **2015**, *10*, 313–318. [CrossRef] [PubMed]
26. Li, H.; Chen, S.; Jia, X.; Xu, B.; Lin, H.; Yang, H.; Song, L.; Wang, X. Amorphous nickel-cobalt complexes hybridized with 1T-phase molybdenum disulfide via hydrazine-induced phase transformation for water splitting. *Nat. Commun.* **2017**, *8*, 15377. [CrossRef]
27. Liu, Y.; Li, Y.; Peng, F.; Lin, Y.; Yang, S.; Zhang, S.; Wang, H.; Cao, Y.; Yu, H. 2H- and 1T- mixed phase few-layer MoS$_2$ as a superior to Pt co-catalyst coated on TiO$_2$ nanorod arrays for photocatalytic hydrogen evolution. *Appl. Catal. B Environ.* **2019**, *241*, 236–245. [CrossRef]
28. Li, J.; Liu, E.Z.; Ma, Y.N.; Hu, X.Y.; Wan, J.; Sun, L.; Fan, J. Synthesis of MoS$_2$/g-C$_3$N$_4$ nanosheets as 2D heterojunction photocatalysts with enhanced visible light activity. *Appl. Surf. Sci.* **2016**, *364*, 694–702. [CrossRef]
29. Chen, Y.K.; Tan, L.J.; Sun, M.L.; Lu, C.H.; Kou, J.H.; Xu, Z.Z. Enhancement of photocatalytic performance of TaON by combining it with noble-metal-free MoS$_2$ cocatalysts. *J. Mater. Sci.* **2019**, *54*, 5321–5330. [CrossRef]
30. Yin, S.; Ding, Y.; Luo, C.; Hu, Q.S.; Chen, Y.; Di, J.; Wang, B.; Xia, J.X.; Li, H.M. Construction of 2D/2D MoS$_2$/PbBiO$_2$Cl nanosheet photocatalysts with accelerated interfacial charge transfer for boosting visible light photocatalytic activity. *Colloid. Surface A* **2021**, *609*, 125655. [CrossRef]
31. Mondal, C.; Ganguly, M.; Pal, J.; Roy, A.; Jana, J.; Pal, T. Morphology Controlled Synthesis of SnS$_2$ Nanomaterial for Promoting Photocatalytic Reduction of Aqueous Cr(VI) under Visible Light. *Langmuir* **2014**, *30*, 4157–4164. [CrossRef]
32. Tu, J.-R.; Shi, X.-F.; Lu, H.-W.; Yang, N.-X.; Yuan, Y.-J. Facile fabrication of SnS$_2$ quantum dots for photoreduction of aqueous Cr(VI). *Mater. Lett.* **2016**, *185*, 303–306. [CrossRef]
33. Sun, Y.; Li, G.; Xu, J.; Sun, Z. Visible-light photocatalytic reduction of carbon dioxide over SnS$_2$. *Mater. Lett.* **2016**, *174*, 238–241. [CrossRef]
34. Li, X.; Zhu, J.; Li, H. Comparative study on the mechanism in photocatalytic degradation of different-type organic dyes on SnS$_2$ and CdS. *Appl. Catal. B Environ.* **2012**, *123–124*, 174–181. [CrossRef]
35. Lucena, R.; Fresno, F.; Conesa, J.C. Hydrothermally synthesized nanocrystalline tin disulphide as visible light-active photocatalyst: Spectral response and stability. *Appl. Catal. A Gen.* **2012**, *415–416*, 111–117. [CrossRef]
36. Liu, H.; Su, Y.; Chen, P.; Wang, Y. Microwave-assisted solvothermal synthesis of 3D carnation-like SnS$_2$ nanostructures with high visible light photocatalytic activity. *J. Mol. Catal. A Chem.* **2013**, *378*, 285–292. [CrossRef]
37. Zhang, Y.; Zhang, F.; Yang, Z.; Xue, H.; Dionysiou, D.D. Development of a new efficient visible-light-driven photocatalyst from SnS$_2$ and polyvinyl chloride. *J. Catal.* **2016**, *344*, 692–700. [CrossRef]
38. Christoforidis, K.C.; Sengele, A.; Keller, V.; Keller, N. Single-Step Synthesis of SnS$_2$ Nanosheet-Decorated TiO$_2$ Anatase Nanofibers as Efficient Photocatalysts for the Degradation of Gas-Phase Diethylsulfide. *ACS Appl. Mater. Inter.* **2015**, *7*, 19324–19334. [CrossRef]
39. Zhang, X.-Y. Developing bioenergy to tackle climate change: Bioenergy path and practice of Tianguan group. *Adv. Clim. Chang. Res.* **2016**, *7*, 17–25. [CrossRef]
40. Gao, X.; Huang, G.; Gao, H.; Pan, C.; Wang, H.; Yan, J.; Liu, Y.; Qiu, H.; Ma, N.; Gao, J. Facile fabrication of Bi$_2$S$_3$/SnS$_2$ heterojunction photocatalysts with efficient photocatalytic activity under visible light. *J. Alloys Compd.* **2016**, *674*, 98–108. [CrossRef]

41. Zhang, Z.Y.; Huang, J.D.; Zhang, M.Y.; Yuan, L.; Dong, B. Ultrathin hexagonal SnS$_2$ nanosheets coupled with g-C$_3$N$_4$ nanosheets as 2D/2D heterojunction photocatalysts toward high photocatalytic activity. *Appl. Catal. B-Environ.* **2015**, *163*, 298–305. [CrossRef]
42. Liang, Z.Q.; Guo, Y.C.; Xue, Y.J.; Cui, H.Z.; Tian, J. 1T-phase MoS$_2$ quantum dots as a superior co-catalyst to Pt decorated on carbon nitride nanorods for photocatalytic hydrogen evolution from water. *Mater. Chem. Front.* **2019**, *3*, 2032–2040. [CrossRef]
43. Chen, C.Z.; Zang, J.Y.; Wang, Q.; Li, Y.Z. Loading SnS$_2$ nanosheets decorated with MoS$_2$ nanoparticles on a flake-shaped g-C$_3$N$_4$ network for enhanced photocatalytic performance. *Crystengcomm* **2021**, *23*, 4680–4693. [CrossRef]
44. Holler, F.J.; Skoog, D.A.; Crouch, S.R. *Principles of Instrumental Analysis*; Cengage Learning: Boston, MA, USA, 2007.
45. Zhang, Y.C.; Li, J.; Zhang, M.; Dionysiou, D.D. Size-Tunable Hydrothermal Synthesis of SnS$_2$ Nanocrystals with High Performance in Visible Light-Driven Photocatalytic Reduction of Aqueous Cr (VI). *Environ. Sci. Technol.* **2011**, *45*, 9324–9331. [CrossRef]
46. Qiang, T.T.; Chen, L.; Xia, Y.J.; Qin, X.T. Dual modified MoS$_2$/SnS$_2$ photocatalyst with Z-scheme heterojunction and vacancies defects to achieve a superior performance in Cr (VI) reduction and dyes degradation. *J. Clean. Prod.* **2021**, *291*, 125213. [CrossRef]
47. Cui, B.W.; Wang, Y.H.; Zhang, F.; Xiao, X.; Su, Z.Q.; Dai, X.C.; Zhang, H.; Huang, S. Hydrothermal synthesis of SnS$_2$/MoS$_2$ Nanospheres for enhanced adsorption capacity of organic dyes. *Mater. Res. Express* **2020**, *7*, 015016. [CrossRef]
48. Mangiri, R.; Kumar, K.S.; Subramanyam, K.; Ratnakaram, Y.C.; Sudharani, A.; Reddy, D.A.; Vijayalakshmi, R.P. Boosting solar driven hydrogen evolution rate of CdS nanorods adorned with MoS$_2$ and SnS$_2$ nanostructures. *Colloid. Interface Sci. Commun.* **2021**, *43*, 100437. [CrossRef]
49. Yin, S.K.; Li, J.Z.; Sun, L.L.; Li, X.; Shen, D.; Song, X.H.; Huo, P.W.; Wang, H.Q.; Yang, Y.S. Construction of Heterogenous S-C-S MoS$_2$/SnS$_2$/r-GO Heterojunction for Efficient CO$_2$ Photoreduction. *Inorg. Chem.* **2019**, *58*, 15590–15601. [CrossRef]
50. Teng, Y.Q.; Zhao, H.L.; Zhang, Z.J.; Li, Z.L.; Xia, Q.; Zhang, Y.; Zhao, L.N.; Du, X.F.; Du, Z.H.; Lv, P.P.; et al. MoS$_2$ Nanosheets Vertically Grown on Graphene Sheets for Lithium-Ion Battery Anodes. *ACS Nano* **2016**, *10*, 8526–8535. [CrossRef]
51. Man, X.L.; Liang, P.; Shu, H.B.; Zhang, L.; Wang, D.; Chao, D.L.; Liu, Z.G.; Du, X.Q.; Wan, H.Z.; Wang, H. Interface Synergistic Effect from Layered Metal Sulfides of MoS$_2$/SnS$_2$ van der Waals Heterojunction with Enhanced Li-Ion Storage Performance. *J. Phys. Chem. C* **2018**, *122*, 24600–24608. [CrossRef]
52. Tang, T.Y.; Zhang, T.; Zhao, L.N.; Zhang, B.; Li, W.; Xu, J.J.; Zhang, L.; Qiu, H.L.; Hou, Y.L. Multifunctional ultrasmall-MoS$_2$/graphene composites for high sulfur loading Li-S batteries. *Mater. Chem. Front.* **2020**, *4*, 1483–1491. [CrossRef]

Disclaimer/Publisher's Note: The statements, opinions and data contained in all publications are solely those of the individual author(s) and contributor(s) and not of MDPI and/or the editor(s). MDPI and/or the editor(s) disclaim responsibility for any injury to people or property resulting from any ideas, methods, instructions or products referred to in the content.

Article

The Superiority of TiO₂ Supported on Nickel Foam over Ni-Doped TiO₂ in the Photothermal Decomposition of Acetaldehyde

Beata Tryba *, Piotr Miądlicki , Piotr Rychtowski , Maciej Trzeciak and Rafał Jan Wróbel

Department of Catalytic and Sorbent Materials Engineering, Faculty of Chemical Technology and Engineering, West Pomeranian University of Technology in Szczecin, Pułaskiego 10, 70-322 Szczecin, Poland; piotr.miadlicki@zut.edu.pl (P.M.); piotr.rychtowski@zut.edu.pl (P.R.); tm44864@zut.edu.pl (M.T.); rafal.wrobel@zut.edu.pl (R.J.W.)
* Correspondence: beata.tryba@zut.edu.pl

Abstract: Acetaldehyde decomposition was performed under heating at a temperature range of 25–125 °C and UV irradiation on TiO₂ doped by metallic Ni powder and TiO₂ supported on nickel foam. The process was carried out in a high-temperature reaction chamber, "The Praying Mantis^{TM}", with simultaneous in situ FTIR measurements and UV irradiation. Ni powder was added to TiO₂ in the quantity of 0.5 to 5.0 wt%. The photothermal measurements of acetaldehyde decomposition indicated that the highest yield of acetaldehyde conversion on TiO₂ and UV irradiation was obtained at 75 °C. The doping of nickel to TiO₂ did not increase its photocatalytic activity. Contrary to that, the application of nickel foam as a support for TiO₂ appeared to be highly advantageous because it increased the decomposition of acetaldehyde from 31 to 52% at 25 °C, and then to 85% at 100 °C in comparison with TiO₂ itself. At the same time, the mineralization of acetaldehyde to CO₂ doubled in the presence of nickel foam. However, oxidized nickel foam used as support for TiO₂ was detrimental. Most likely, different mechanisms of electron transfer between Ni–TiO₂ and NiO-TiO₂ occurred. The application of nickel foam greatly enhanced the separation of free carriers in TiO₂. As a consequence, high yields from the photocatalytic reactions were obtained.

Keywords: thermo-photocatalysis; nickel foam; Ni-doped TiO₂; acetaldehyde decomposition

1. Introduction

Photothermal catalysis combines photochemical and thermochemical contributions of sunlight and has emerged as a rapidly growing and exciting new field of research [1–3]. Photothermal processes can enhance the yield of photo-Fenton reactions [4,5].

Photothermal catalysis with the application of semiconductors doped by metallic nanoparticles allows for more effective harvesting of the solar spectrum, including low-energy visible and infrared photons.

Certain metallic nanoparticles (NPs), such as Au, Ag, Cu, and others, exhibit unique optical properties related to the localized surface plasmon resonance (LSPR), which can be tuned via varying their size and shape across the entire visible spectrum [6,7].

In the LSPR process, hot charge carriers are generated, which possess higher energies than those induced by direct photoexcitation. These energetic carriers can be transferred to the adsorbate surface or the conductive band of semiconductors or can relax internally and dissipate their energy by local heating of the surroundings, causing a thermal effect on the material. This photothermal effect has been extensively applied in a large number of fields, including cancer therapy, the degradation of pollutants, CO₂ reduction, hydrogen production through water splitting, and other chemical reactions such as ex. catalytic steam reforming or the hydrogenation of olefins [2,8–15]. Photothermal effects can also be obtained in non-plasmonic structures through direct intraband and/or interband electronic

transitions. For instance, Sarina et al. [16] demonstrated that non-plasmonic metal NPs supported on ZrO_2 could catalyze cross-coupling reactions at low temperatures under visible light. According to the authors, upon irradiation with UV light, electrons could shift to high-energy levels through interband transitions, and only those with enough energy could transfer to the LUMO of the adsorbed molecules, just like in the case of plasmonic metal NPs. When excited with low-energy visible-IR light, the electrons were not energetic enough to be injected into adsorbate states, thus contributing to the enhancement of the reaction rate by means of thermal effects. Altogether, photochemical and thermal effects contribute to photothermal performance in non-plasmonic semiconductors.

The size, shape, and quantity of the plasmonic nanoparticles contributing to the Photothermal effect are important. The temperature increase is proportional to the square of the nanoparticle radius. However, it has been theorized that the temperature increment due to the irradiation of a single nanoparticle becomes negligible, but the light-induced heating effect could be strongly enhanced in the presence of a large number of NPs due to collective effects [2]. Theoretical calculations have demonstrated that larger plasmonic NPs provide a larger number of hot carriers, but they display energies close to the Fermi level since most of the absorbed energy is dissipated either by scattering or by heating. In contrast, smaller NPs exhibit higher energies but they show very short lifetimes [2]. Lower-sized plasmonic NPs show blue shift absorption edge due to the quantum size effect [17]. All these parameters are extremely important during the design of nanomaterial for thermophotocatalytic processes.

The most intensively explored plasmonic NPs used in photothermal processes so far are based on Au, Ag, Pd, Cu, and Pt nanoparticles; less information is reported on other metals such as Ni or Co. The thermophotocatalytic decomposition of acetaldehyde was reported for TiO_2 doped by Pt, Ag, Pd, or Au nanoparticles [12,18–21]. Of all these metals doped to TiO_2, Pt had the highest activity due to its strong thermocatalytic effect at elevated temperatures. The activity of metal or metal oxide doped to TiO_2 depended on the oxidation state of the metal and its adsorption abilities towards acetaldehyde species. The other structures, such as metal nanowires coated by TiO_2, have also been prepared and tested for acetaldehyde removal [22,23]. There are some reports on the application of nickel foam with loaded TiO_2 for different purposes, such as anodes for lithium-ion batteries [24], hydrogen evolution from water splitting [25], toluene and acetaldehyde decomposition [26,27], nitrogen oxides removal [28], and others. Nickel foam has also been used as a support and catalyst for different materials in photocatalytic processes [29–31]. The main advantage of the application of nickel foam as a composite with other photocatalyst is the significant improvement in charge separation in the photocatalytic material.

Although some reports on the application of nickel foam with loaded TiO_2 have already been published, in this study, we propose for the first time the mechanism of the thermophotocatalytic decomposition of acetaldehyde under UV light in the presence of TiO_2 supported on nickel foam, and this is compared with others which occur when oxidized nickel foam or nanosized Ni particles added to TiO_2 are utilized.

2. Materials and Methods

The TiO_2 of anatase phase was obtained by a two-step process: hydrothermal treatment of amorphous titania (Police Chemical Factory, Gliwice, Poland) in deionized water at 150 °C under pressure of 7.4 bar for 1 h, followed by annealing in a pipe furnace at 400 °C under Ar flow for 2 h. The obtained material consisted of anatase (97 wt%) with an average crystalline size of 15 nm and small quantity of rutile (3 wt%). The measured BET surface area of this material was 167 m^2/g. Our previous work presented the detailed physicochemical properties of this material [32].

Nickel foam (Jiujiang, China) with purity of 99.8% had the following parameters: thickness, 1.5 mml; porosity, 95–97%; surface density, 300 g/m^2. Nickel powder (Warchem, Poland) had a purity of 99.8%.

The other chemicals used were as follows: p-benzoquinone (HPLC purity, >99.5%, Fluka Analytical, Darmstadt, Germany), terephthalic acid (TA) (purity 98%, Sigma-Aldrich, Saint Louis, MO, USA) ethylenediaminetetraacetic acid (EDTA) (Pure Chemical Standards—Elemental Microanalysis), $AgNO_3$ (Polish Chemical Factory, Gliwice, Poland).

XRD measurements of nickel foams were performed using the diffractometer (PANanalytical, Almelo, The Netherlands) with a Cu X-ray source, λ = 0.154439 nm. Measurements were conducted in the 2θ range of 10–100° with a step size of 0.013. The applied voltage was 35 kV, and the current was 30 mA.

SEM/EDS images were obtained using a field emission scanning electron microscope with high resolution (UHR FE-SEM, Hitachi, Japan).

Oxidation of nickel foams was carried out in a muffle furnace (Czylok SM-2002, Jastrzębie-Zdrój, Poland) at 500 °C for 1 h.

The chemical composition was determined through X-ray photoelectron spectroscopy (XPS) analysis. The measurements were conducted using a commercial multipurpose ultra-high vacuum (UHV) surface analysis system (PREVAC, Rogow, Poland). A nonmonochromatic XPS source and a kinetic electron energy analyzer (SES 2002; Scienta, Taunusstein, Germany) was used. The spectrometer was calibrated using the Ag $3d_{5/2}$ transition. The XPS analysis utilized Mg Kα (h = 1253.6 eV) radiation as the excitation source.

The size of nickel powdered nanoparticles was determined using an atomic force microscope (AFM; NanoScope V Multimode 8, Bruker, Billerica, MA, USA) with a silicon nitride probe in ScanAsyst mode. Prior to measurement, the samples were dispersed in isopropyl alcohol and then drop-casted onto a silicon wafer. Surface topography images were obtained using NanoScope Analysis software (v1.40r1), whereas particles sizes were evaluated via ImageJ software (2023).

Thermophotocatalytic decomposition tests of acetaldehyde were conducted using a high=temperature reaction chamber (Harrick, Pleasantville, NY, USA), as shown in Figure 1. During the test, continuous FTIR measurements were conducted using Thermo Nicolet iS50 FTIR instrument (Thermo, Waltham, MA, USA).

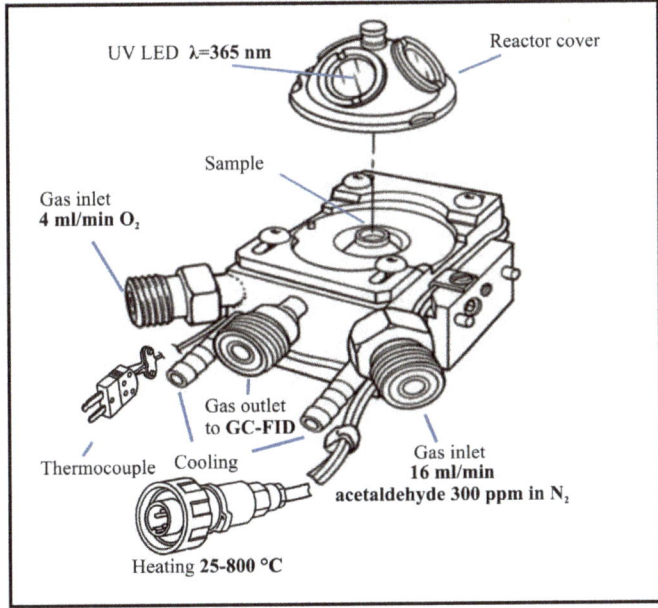

Figure 1. The scheme of the Praying Mantis™ high-temperature reaction chamber.

UV irradiation was conducted through a quartz window, with an illuminator equipped with a fiber optics and a UV-LED 365 nm diode with an optical power of 415 mW (Warszawa, Poland, LABIS). The intensity of incident UV radiation was measured by a photoradiometer, HD2102.1 (TEST-THERM, Kraków, Poland). The obtained value of UV intensity measured at the surface of reactor cover window equaled 20 W/m^2. Gases were supplied from bottles through two inlets using mass flow meters: the first one was acetaldehyde in nitrogen, 300 ppm (Messer, Police, Poland); the second one was a synthetic oxygen with purity 5.0 (Messer). Gases have been mixed in proportions to create a synthetic air composition. After the thermophotocatalytic process, the gas stream was directed to the gas chromatograph (GC) equipped with an automatically dosing sample loop (GC-FID, SRI, Menlo Park, CA, USA). Acetaldehyde concentration was determined from recorded chromatograms in GC. Outlet gas stream from GC was flowing through the CO_2 sensor (APFinder CO_2, ATUT Company, Kraków, Poland) in order to monitor the quantity of CO_2 formed upon acetaldehyde mineralization. The contact time of acetaldehyde species with the photocatalytic material was varied through the application of different flow rates of a gas stream, from 15 to 30 mL/min. Then, the optimal flow rate was selected for all the photocatalytic tests, taking into account the highest quantity of acetaldehyde removal over the irradiated TiO_2.

The catalytic systems that have been used in the tests are as follows: (a) TiO_2; (b) a thin layer (1 mm) of TiO_2 supported on KBr which is inert to acetaldehyde; (c) a thin layer of TiO_2 supported on Ni foam; (d) a thin layer of TiO_2 supported on the oxidized Ni foam; (e) TiO_2 blended with the nickel powder in the various amount, 0.5–5 wt%. In Figure 2, the scheme of the materials compositions used for the photocatalytic tests of acetaldehyde decomposition is illustrated.

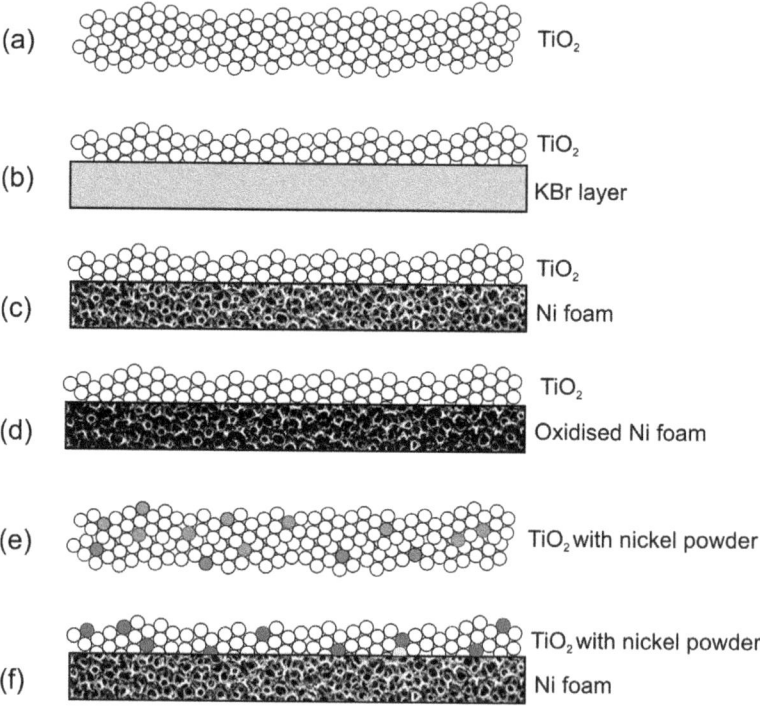

Figure 2. Catalytic systems used during thermophotocatalytic decomposition of acetaldehyde.

To determine the dominant species which take part in the photocatalytic reactions, some measurements were performed in the presence of oxygen radicals and holes scavengers. For that purpose, p-benzoquinone, ethylenediaminetetraacetic acid (EDTA), and terephthalic acids were used as scavengers for $O_2^{-\bullet}$, h^+, and $^{\bullet}OH$ species, respectively. For the quenching reactions, 0.1 g of TiO_2 photocatalyst was mixed with 0.01 g of each scavenger separately, and then such mixture was loaded on the purified nickel foam. Excessive amounts of scavenger reagents were utilized to guarantee the complete capture of related radicals, similar as in the other experiments reported by Q. Zeng et al. [20]. Acetaldehyde decomposition in the presence of TiO_2 and scavenger was carried out in the high-temperature chamber at 50 °C under UV irradiation. As a blank test, the mixture of 0.1 g TiO_2 with 0.01 g of KBr was used because KBr was revealed to be chemically inert for acetaldehyde gas.

3. Results

Preliminary acetaldehyde decomposition was carried out at various flow rates of a gas stream through the reactor, from 15 to 30 mL/min. The diameter of the sample holder equaled 0.6 cm^2. For lower flow rate of a gas, the contact time of acetaldehyde with TiO_2 surface was longer, and the percentage of removed acetaldehyde was higher. However, by increasing the flow rate of a gas, higher loading of TiO_2 with acetaldehyde species was obtained. There is a certain optimum amount of pollutant which can be decomposed on TiO_2 surface at given time. In Figure 3, the dependence of acetaldehyde decomposition over time via the quantity of acetaldehyde loading on the titania surface is illustrated.

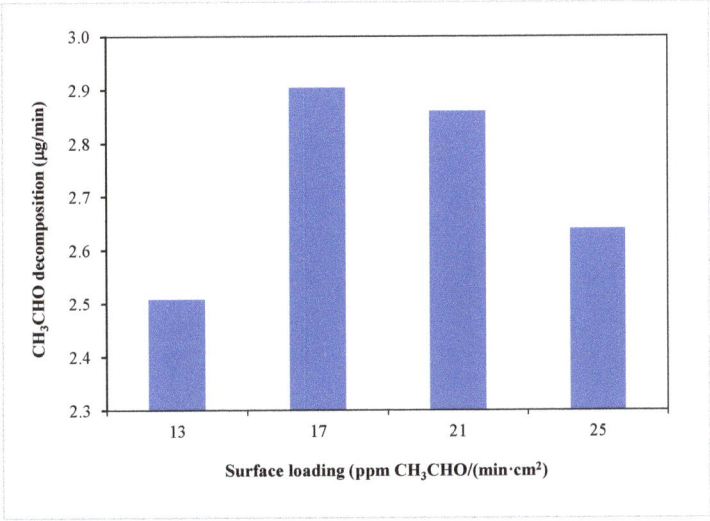

Figure 3. Dependence of acetaldehyde decomposition on its loading on the titania surface at a given time.

These measurements revealed that the maximal quantity of acetaldehyde decomposition could be obtained when the surface loading with these species was 17 ppm per min·cm^2 (for a flow rate of a gas equaled 20 mL/min). For lower flow rate used, such as 15 mL/min, there is higher percentage of acetaldehyde removal from a gas stream; however, the maximum yield of the photocatalytic system has not yet been reached. Therefore, for other photocatalytic tests, the flow rate of a gas stream equal to 20 mL/min was applied.

The results from the thermophotocatalytic decomposition of acetaldehyde over TiO_2 and UV-LED irradiation are illustrated in Figure 4.

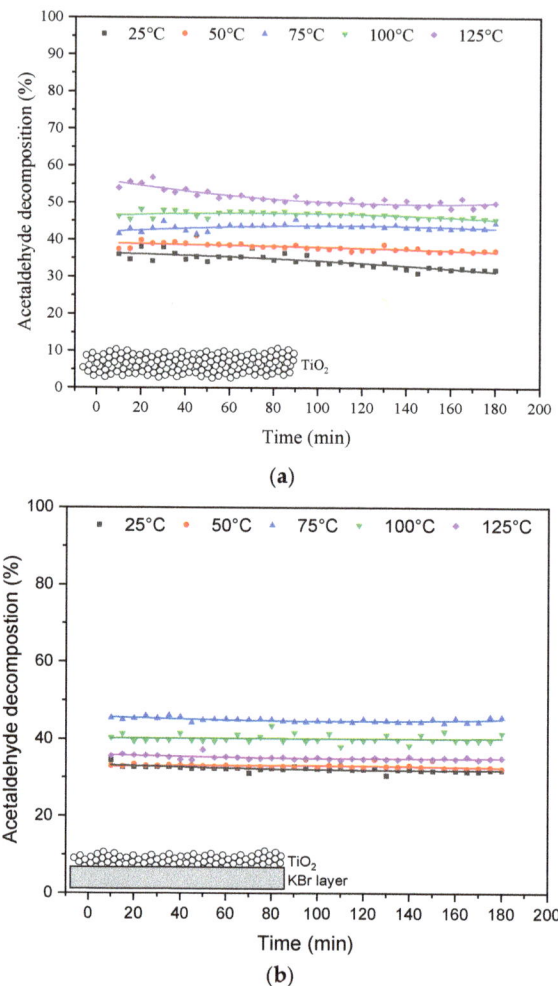

Figure 4. Photocatalytic decomposition of acetaldehyde under UV irradiation at various reaction temperatures in the presence of (**a**) TiO_2; (**b**) TiO_2 supported on KBr.

The performed measurements indicated that the conversion of acetaldehyde on TiO_2 could be enhanced at elevated temperatures. In the case of a thick layer of TiO_2 being used (without KBr) (Figure 4b) the highest acetaldehyde conversion was observed at 125 °C; however, this process was not stable and indicated a gradual falling-off in degradation rate, reaching its maximum after 180 min of UV irradiation with a conversion of 50%. Contrary to that, the application of a thin layer TiO_2 supported on KBr allowed us stabilize the process, but the highest conversion of acetaldehyde was lower than in case of using TiO_2 only and reached 47% at 75 °C. In Table 1, the results from the measurements of CO_2 in the outlet stream of reacted gases are presented.

These results indicated that mineralization of acetaldehyde was higher in the case of the photocatalytic system used with a thin layer of TiO_2. Acetaldehyde can be oxidized over TiO_2 in the dark in the presence of oxygen; however, its complete mineralization occurs in the presence of reactive radicals. Most likely, TiO_2 at the bottom of reactor chamber was not activated by UV light and formed products of acetaldehyde decomposition at the lower part of TiO_2 were not mineralized, but acted as scavengers for reactive radicals formed at the surface of TiO$_2$ on the top of reactor. Therefore, for higher quantities of TiO$_2$ used, acetaldehyde conversion was higher, but its mineralization to CO$_2$ dropped down.

Table 1. Content of CO$_2$ in a gas stream after acetaldehyde decomposition on TiO$_2$.

Reaction Temperature (°C)	CO$_2$ (ppm)	
	TiO$_2$	TiO$_2$ on KBr
25	220	245
50	235	245
75	232	300
100	210	330
125	225	290

Total mineralization of acetaldehyde can be summarized via the following equation:

$$CH_3CHO + 5/2 O_2 \rightarrow 2CO_2 + 2H_2O. \tag{1}$$

From the above equation, it is deduced that from one mole of acetaldehyde, two moles of carbon dioxide are formed. The highest quantity of CO$_2$ (330 ppm) was formed in the conditions of a thin layer of TiO$_2$ and a temperature of 100 °C, where removal of acetaldehyde was determined to be 41%. In the case of total mineralization of acetaldehyde, the quantity of CO$_2$ should be around 197 ppm, but it was 330, which attributes to selectivity much more than 100%. The reason for such an unusual phenomenon is that at the same time acetaldehyde and the adsorbed products of its degradation were mineralized, acetaldehyde was adsorbed onto the TiO$_2$ surface and transformed to other formate and acetate species; hence, other acetaldehyde molecules can be adsorbed on the titania surface after total decomposition of the byproducts. Therefore, the obtained values of CO$_2$ concentration did not refer to the percentage decrease in acetaldehyde concentration at a given time only.

In Figure 5, the results from the thermophotocatalytic decomposition of acetaldehyde over TiO$_2$ supported on the nickel foam are presented.

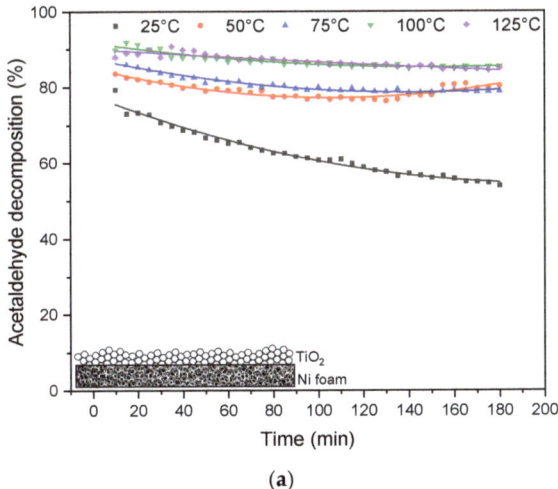

(a)

Figure 5. Cont.

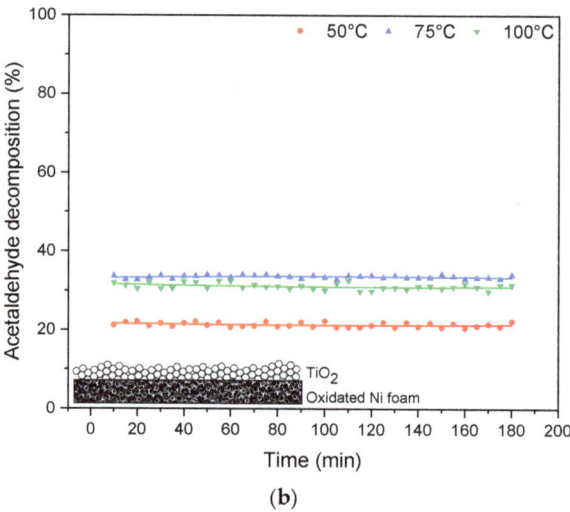

(**b**)

Figure 5. Photocatalytic decomposition of acetaldehyde under UV irradiation at various reaction temperatures in the presence of (**a**) TiO$_2$ supported on Ni foam; (**b**) TiO$_2$ supported on oxidized Ni foam.

High decomposition of acetaldehyde (around 85%) was observed in the presence of TiO$_2$ supported on Ni foam and in reaction temperatures of 100–125 °C. The slight decrease in acetaldehyde decomposition over time is due to by-products blocking the catalyst's active centers. However, the efficiency of this process significantly decreased in the presence of oxidized Ni foam; the decomposition of acetaldehyde dropped down to 33%. It has to be mentioned that Ni foam as received was much less active than that after the photocatalytic process. Therefore, all the results presented in Figure 5a were obtained for reused nickel foam.

In Table 2, the results from measurements of CO$_2$ after the thermophotocatalytic process of acetaldehyde decomposition conducted for TiO$_2$ supported on reused Ni foam and oxidized at 500 °C are presented.

Table 2. Content of CO$_2$ in a gas stream after acetaldehyde decomposition on TiO$_2$.

Reaction Temperature (°C)	CO$_2$ (ppm)	
	TiO$_2$ on Reused Ni Foam	TiO$_2$ on Ni Foam Oxidized at 500 °C
25	364	-
50	409	<100
75	438	<100
100	472	<100
125	423	-

Application of Ni foam with TiO$_2$ not only enhanced the decomposition rate of acetaldehyde but also doubled its mineralization degree. The highest formation of CO$_2$ was noted at 100 °C (472 ppm) and this, again, was higher than total mineralization of acetaldehyde, which was calculated to be 408 ppm for 85% of its decomposition at given time.

In order to explain the mechanism of acetaldehyde decomposition over TiO$_2$ supported on nickel foam, some photocatalytic experiments were performed with the addition of scavengers for some active species. In Figure 6, the results of the measurements are presented.

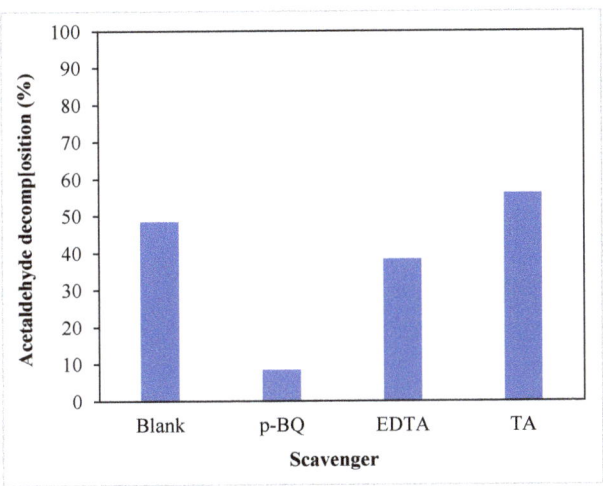

Figure 6. Photocatalytic decomposition of acetaldehyde in the presence of various scavengers.

Addition of p-benzoquinone (p-BQ) to TiO_2, which acted as a scavenger of superoxide anion radicals, greatly suppressed acetaldehyde decomposition. Contrary to that, the addition of terephthalic acid (TA) to TiO_2 (hydroxyl radicals scavenger) enhanced acetaldehyde decomposition in comparison with a blank test performed for TiO_2 without the addition of any scavengers. In the presence of EDTA (hole scavenger), acetaldehyde decomposition decreased but to a lesser extent than in case of p-BQ addition. These experiments showed that superoxide anion radicals played the dominant role in the removal of acetaldehyde. Similar results were obtained by other researchers [20].

In Figure 7 the results from the photocatalytic decomposition of acetaldehyde over TiO_2 blended with nanosized Ni powder are presented.

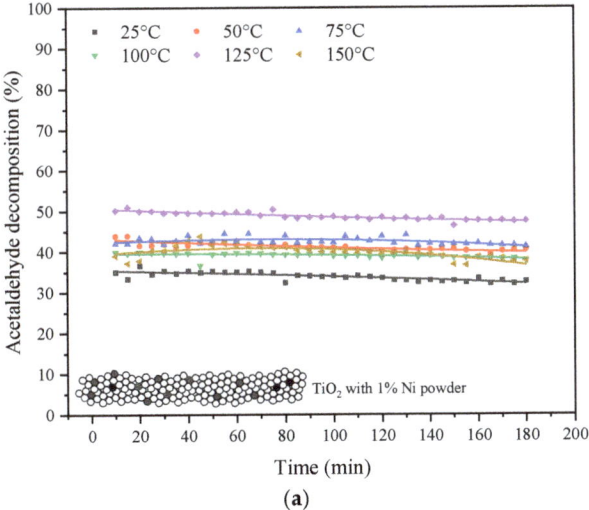

Figure 7. *Cont.*

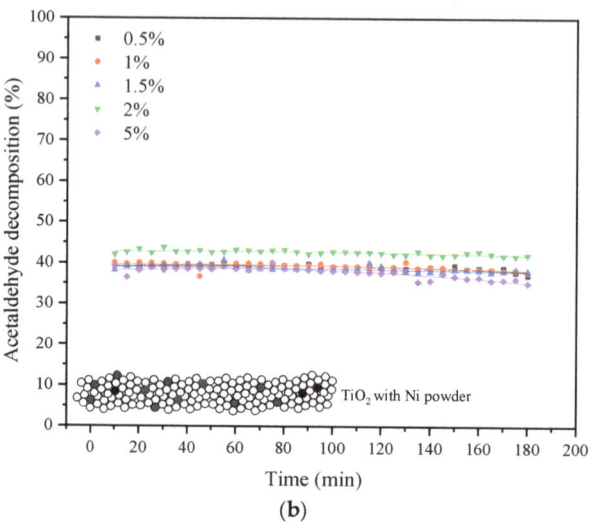

(b)

Figure 7. Photocatalytic decomposition of acetaldehyde under UV irradiation in the presence of TiO_2 doped with Ni powder: (**a**) in various reaction temperatures with 1% of doped Ni; (**b**) for different quantities of doped Ni with a reaction temperature of 100 °C.

The quantity of doped Ni slightly affected the efficiency of acetaldehyde decomposition, which ranged from 38 to 43%. At a temperature of 125 °C, the process yield was the highest and reached around 50% for 1% of doped Ni to TiO_2. Although the acetaldehyde decomposition rate on TiO_2 doped with Ni was quite comparable with that for TiO_2, the mineralization degree was low; the maximum value of formed CO_2 was 150 ppm, which was reached for 1% doped Ni at a reaction temperature of 75 °C; and for the other tests, the quantities of formed CO_2 were lower.

AFM analyses of surface topography with loaded Ni particles indicated that their mean size was c.a. 30 nm, as shown in Figure 8.

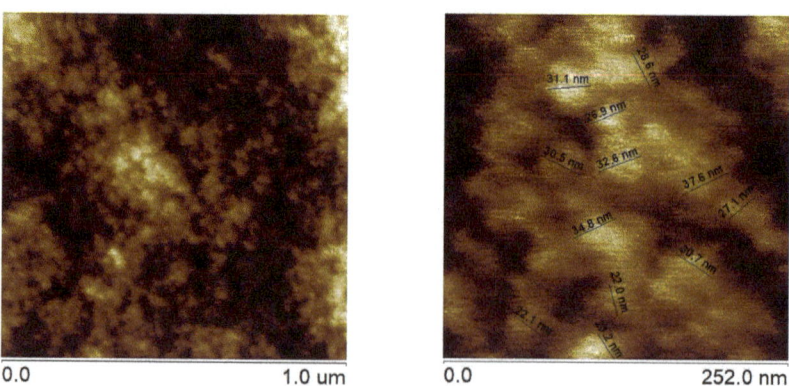

Figure 8. Surface topography of nickel nanoparticles.

In Figure 9, XRD patterns of Ni foam as received and after oxidation at 500 °C in air are illustrated.

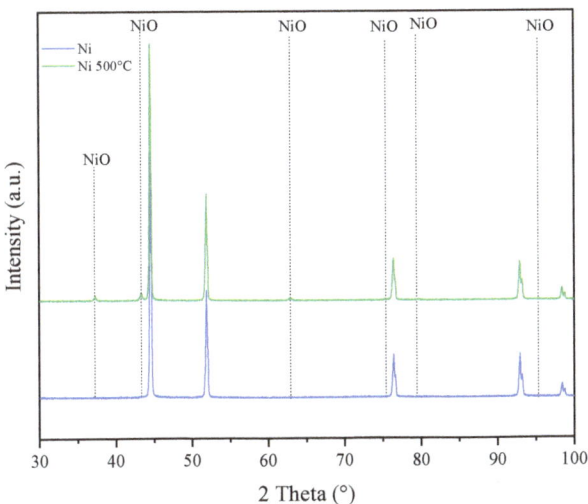

Figure 9. XRD patterns of Ni foam before and after oxidation at 500 °C in air.

After oxidation of Ni foam at a temperature of 500 °C, new reflexes emerged which were assigned to the NiO phase; however, their intensities were much lower than those related to Ni.

In Figure 10 the SEM images of nickel foam are shown at different magnifications.

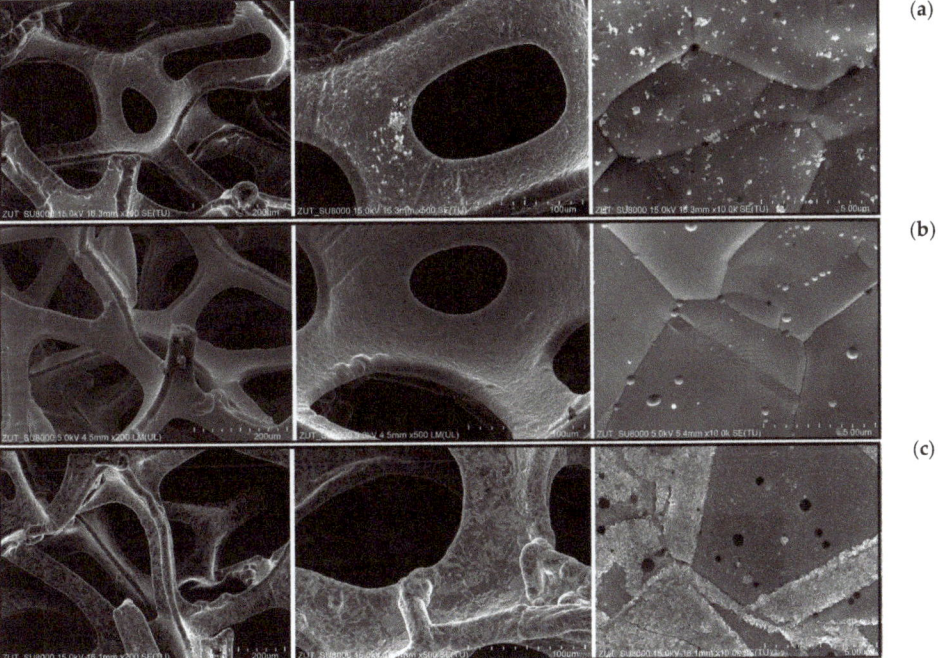

Figure 10. SEM images of a nickel foam: (**a**) as received; (**b**) after photocatalytic process; and (**c**) oxidized at 500 °C in air.

The performed SEM images showed some impurities on the surface of nickel foam as received from the manufacturer (Figure 10a). After the photocatalytic process, these impurities disappeared (Figure 10b). They were removed through the reactive oxygen species formed on TiO$_2$ upon UV irradiation. This was confirmed by the other experiment, in which TiO$_2$ supported on the commercial nickel foam was used with flowing air without acetaldehyde but under UV irradiation. The same effect of nickel foam purification was observed.

The oxidized nickel foam showed a corrugated surface on the grain fringes, indicating the proceeding oxidation process.

In Figure 11, mapping of the oxidized nickel foam is depicted. The green area indicates the oxygen distribution. It is clearly seen that strongly corrugated surface is covered with elemental oxygen.

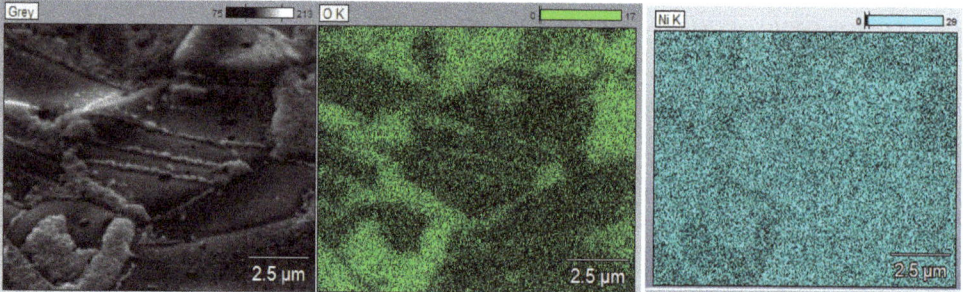

Figure 11. Elemental mapping of nickel foam surface oxidized at 500 °C in air.

In the next step, XPS analyses were performed for nickel foam as received (fresh) after the photocatalytic process and that oxidized at 500 °C in air. Recorded XPS spectra for C1s, Ni2p$_{3/2}$, and O1s signals are presented in Figure 12.

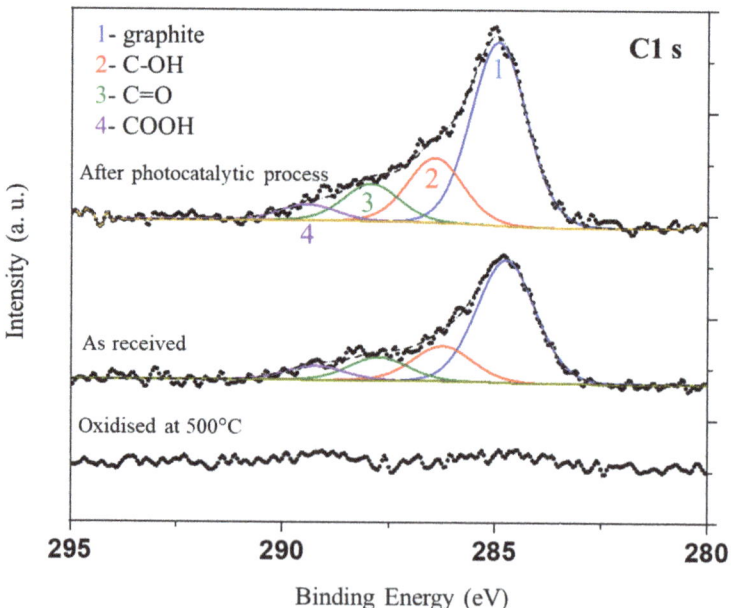

Figure 12. *Cont.*

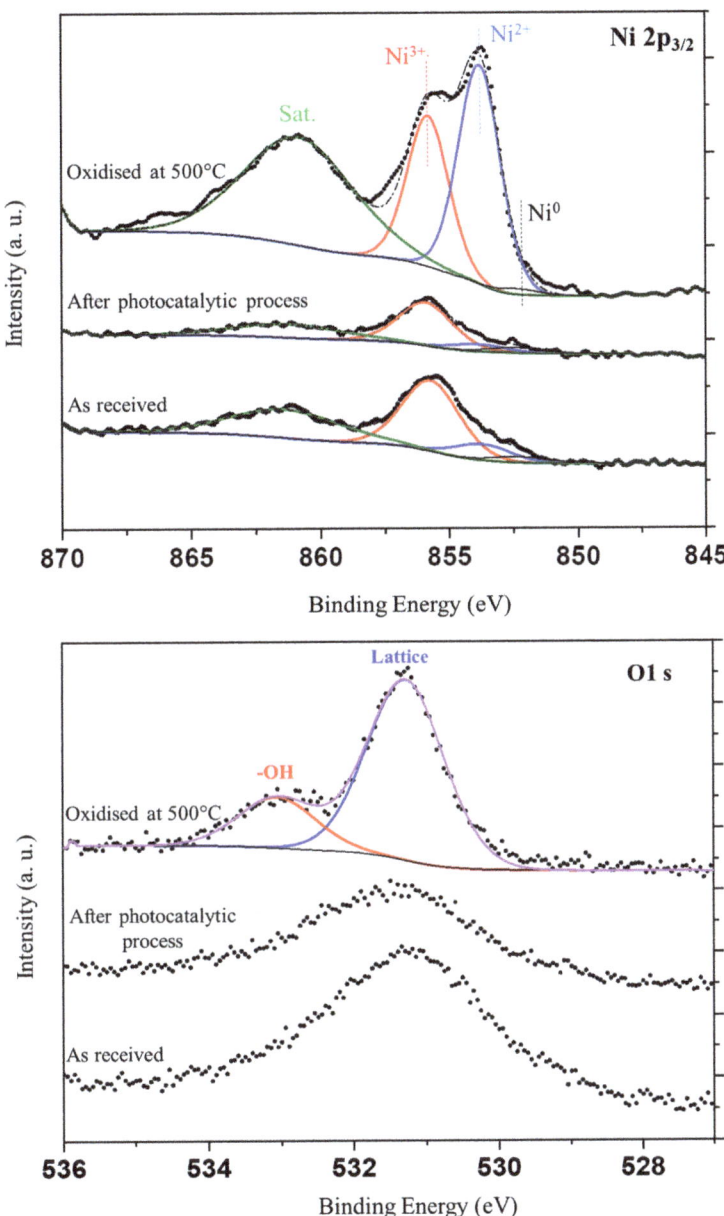

Figure 12. XPS spectra of nickel foams for C1s, Ni2p$_{3/2}$, and O1s signals.

The elemental surface content from XPS analyses is presented in Table 3.

Table 3. Elemental analysis of nickel foam surface.

Nickel Foam	Elemental Surface Content (% at.)		
	O1s	Ni2p	C1s
As received	55	9	36
After photocatalysis	42	6	53
Oxidized at 500 °C	62	38	-

It should be noted that this is an average content of elements from a depth of c.a. 1 nm with an assumption that the elements are homogeneously distributed, which is not valid. Therefore, these values should be taken as an approximation. The nickel foam oxidized at 500 °C exhibits no carbon over the surface. The other nickel foams exhibit a significant amount of carbon, which attenuates the Ni2p signal. This attenuation explains the low values of Ni2p signals.

The deconvolution of C1s signal was performed on the basis of [33]. The carbon species were observed on non-oxidized nickel foam before and after the photocatalytic process. The prominent shoulder on the left-hand side of the signal is related to the carbon–oxygen species present over the surface of carbon. The intensity of C1s signal increased after the process, i.e., the carbon coverage has increased. However, both samples showed similar composition to an oxygen carbon species.

XPS spectra of Ni 2p signals indicated that the highest intensity of Ni species had the sample oxidized at 500 °C. This can be attributed to the absence of carbon over the surface of this sample. In the case of other nickel foams, the intensity of the Ni2p signal was lower due to the carbon coverage, which attenuates the signal of Ni2p. This signal can be deconvoluted to Ni^0, Ni^{2+}, Ni^{3+} (852.3, 853.8, and 855.8 eV, respectively [28]), and satellite signals. The signal of Ni^0 is negligible. The metallic Ni is present in the nickel foam, but the XPS sampling depth is c.a. 1 nm. Therefore, the surface covered with oxides and carbon attenuates the signal from beneath. Nickel foam oxidized at 500 °C clearly showed signals related to Ni^{2+} and Ni^{3+}, whereas the other samples revealed mostly signal of Ni^{3+}.

The O1s can be unambiguously deconvoluted only for the sample oxidized at 500 °C due to the absence of carbon. In the case of other samples, there are carbon–oxygen species which contribute to O1s signal. This renders very difficult the unambiguous deconvolution of the O1s spectrum. In the case of the sample oxidized at 500 °C, two distinct components can be observed: one can be attributed to bulk nickel oxide, and the other to the -OH groups present over the surface of nickel oxide. It is assumed that a thin layer of Ni(oxy)hydroxide was formed on the surface of the oxidized nickel foam.

4. Discussion

Nickel foam used as a support for TiO_2 showed spectacular properties for enhancing the photocatalytic decomposition of acetaldehyde under UV light irradiation. However, the doping of the nanosized Ni particles to TiO_2 did not bring any advantages in either acetaldehyde decomposition or mineralization. Partly oxidized nickel foam also deteriorated the rate of acetaldehyde removal from a gas stream, although some other researchers indicated the superiority of the oxidized nickel foam over that non-oxidized [27]. These different properties of nickel materials could be explained by various reaction mechanisms that occurred in the presence of TiO_2.

The performed photocatalytic tests of acetaldehyde decomposition in the presence of some reactive scavengers species indicated that superoxide anion radicals are the dominant species contributing to acetaldehyde decomposition. These superoxide anion radicals can be formed through the electron capture by the adsorbed oxygen. It is stated that these species are greatly produced at the interfacial border between TiO_2 and nickel foam, where nickel foam is the electron supplier. Therefore, in the presence of nickel foam, acetaldehyde mineralization was greatly enhanced. Mineralized byproducts of acetaldehyde decomposition released space for the adsorption of another acetaldehyde species; therefore, the total efficiency of acetaldehyde removal was increased.

In the case of oxidized Ni foam, there is high probability of electron injection from the conductive band of NiO through the electron transfer zone—Ni (oxy)hydroxide to the TiO_2 due to the p–n heterojunction mechanism between these two semiconductors [34]. XPS measurements showed that there is a thin layer of Ni (oxy)hydroxide formed on the surface of oxidized Ni foam. There are some reports that this layer can facilitate an interfacial electron transfer [35]. On the other hand, holes from the titania valence band can migrate to the lower energy valence band of NiO or can form hydroxyl radicals with hydroxyl groups present on the interfacial border between the oxidized nickel foam and TiO_2. The performed photocatalytic tests with hydroxyl radicals scavenger (TA) indicated that acetaldehyde decomposition increases when there is a quenching of hydroxyl radicals. Therefore, hydroxylated TiO_2 can be disadvantageous for acetaldehyde decomposition. Most likely, the formation of hydroxyl radicals through the reaction of hydroxyl groups with holes is competitive with the reaction of acetaldehyde oxidation:

$$h^+ + {}^-OH \rightarrow {}^\bullet OH \tag{2}$$

$$h^+ + 3CH_3CHO + O_2 \rightarrow 2CH_3COOH + CH_3CO^\bullet + H^+. \tag{3}$$

However, hydroxyl groups take part in the oxidation of byproducts formed upon acetaldehyde decomposition, such as ex. acetic acid:

$$CH_3COOH + {}^\bullet OH \rightarrow CO_2 + H_2O + CH_3^\bullet. \tag{4}$$

Nevertheless, hydroxyl radicals can be also formed in a double-electron oxygen reduction pathway [36]:

$$O_2 + 2H^+ + 2e^- \rightarrow H_2O_2 \tag{5}$$

$$H_2O_2 + e^- \rightarrow {}^\bullet OH + HO^-. \tag{6}$$

The other route of hydroxyl radicals formation can be through single-electron reduction conducted to form superoxide anion radicals, which, in further steps, yield in H_2O_2 production:

$$e^- + O_2 \rightarrow O_2^{-\bullet} \tag{7}$$

$$O_2^{-\bullet} + H^+ \rightarrow HO_2^\bullet \tag{8}$$

$$2\,HO_2^\bullet \rightarrow H_2O_2 + O_2. \tag{9}$$

Therefore, it is stated that the formation of superoxide anion radicals is greatly demanded, together with the formation of hydroxyl radicals through the reaction pathways (5–9). An excess of hydroxyl radicals could recombine with electrons; hence, some reactive species were then extinguished and both acetaldehyde mineralization and decomposition decreased. Some tests of $AgNO_3$ reduction by electrons formed upon excitation of TiO_2 supported on nickel foam (oxidized and not) were performed. The results were placed in the supporting material (Figure S1). These experiments showed a slightly lower reduction of $AgNO_3$ species in the presence of oxidized nickel foam (NiO). It means, that a lower quantity of electrons were formed in the presence of TiO_2/NiO, so non-oxidized nickel foam (Ni) improved the separation of charge carriers in TiO_2. Other researchers also observed improved charge separation in cases of application of nickel foam with other photocatalyst [30,36]. Contrary to that, in the presence of oxidized nickel foam, recombination of charge carriers could take place.

Doping of nanosized Ni powder to TiO_2 did not bring any spectacular effect in boosting the photocatalytic process of acetaldehyde decomposition. Nanosized nickel after excitation can generate electrons, which can be transferred to the conductive band of TiO_2

or to the adsorbed species on its surface. On the other hand, the lifetime of these electrons is very short, and they can undergo back transfer with heat generation.

It was proved that acetaldehyde decomposition could be enhanced at an elevated temperature. It is assumed that in the increased temperature, higher quantity of $O_2^{-\bullet}$ and H_2O_2 species are formed due to the increased mobility of electrons in the Ni-conductive foam. Both H_2O_2 species and the hydroxyl radicals formed upon their reduction can take part in the oxidation of acetaldehyde and its conversion products. Therefore, in the presence of Ni foam, the mineralization degree of acetaldehyde was greatly enhanced.

The other researchers reported increased decomposition of acetaldehyde on TiO_2 doped with 0.5% of Pt due to the spillover of oxygen from Pt to TiO_2 surface, which could oxidize byproducts of acetaldehyde conversion [19].

The other situation was noted in case of Au-doped TiO_2, where Au particles served as the active adsorption centers for acetaldehyde [37]. In that case, Au nanoparticles oxidized acetaldehyde to acetic acid; then, dissociated ions of acetate were transferred to the TiO_2 surface, where they underwent photochemical decomposition.

However, these studies showed that the doping of nanosized Ni species to TiO_2 did not enhance its photocatalytic activity, even at an elevated temperature. The possible direct transfer of electrons from dopant to TiO_2 could be detrimental to this process. Reduction of TiO_2 can increase its hydrophilicity. More advantageous is transfer of electrons from nickel foam to the adsorbed oxygen species.

The mechanism of electron transfer between nickel material and TiO_2 was important in obtaining an increased efficiency of acetaldehyde decomposition. In this case, nickel foam with some adsorbed oxygen species on its surface was suitable because of the enhanced separation of free carriers and increased acetaldehyde mineralization through its contribution to the formation of active radicals.

5. Conclusions

The performed studies on the thermophotocatalytic decomposition of acetaldehyde showed superior properties of nickel foam used as a support for TiO_2. It was evidenced that nickel foam could improve the photocatalytic properties of TiO_2 for acetaldehyde conversion at room temperature form 31 to 52%, and from 40 to 85% at 100 °C, doubling the mineralization degree. Even at lower temperatures, such as 50 °C, the conversion of acetaldehyde on TiO_2 supported on nickel foam was high, reaching 80% (for TiO_2 itself, maximal conversion of acetaldehyde was obtained at 75 °C with value of 46%). These results indicate the synergistic effect between nickel foam and TiO_2, which can also be utilized in other photocatalytic reactions because charge separation in TiO_2 is a key factor responsible for its photocatalytic activity. Using nickel foam as a support for TiO_2 gives much space for the interaction of species between the interface boarder; therefore, high enhancement of the photocatalytic yield was observed. The preparation of photocatalytic composites based on nickel foam creates a new direction in materials development.

Supplementary Materials: The following supporting information can be downloaded at: https://www.mdpi.com/article/10.3390/ma16155241/s1, Figure S1: UV-Vis spectra of TiO_2 doped with $AgNO_3$ and supported on nickel foam (Ni) and oxidised nickel foam (NiO) after irradiation for 10 min under UV LED light.

Author Contributions: B.T.—conceptualization, methodology, supervision and writing—original draft preparation; P.M.—investigation, data curation, visualization; P.R.—investigation, data curation, formal analysis; M.T.—investigations, data curation, R.J.W.—investigation, data curation, formal analysis. All authors have read and agreed to the published version of the manuscript.

Funding: This research was funded by the National Science Centre, Poland, grant number 2020/39/B/ST8/01514.

Institutional Review Board Statement: Not applicable.

Informed Consent Statement: Not applicable.

Data Availability Statement: Data will be available in repozytorium ZUT.

Conflicts of Interest: The authors declare no conflict of interest.

References

1. Czelej, K.; Colmenares, J.C.; Jabłczyńska, K.; Ćwieka, K.; Werner, Ł.; Gradoń, L. Sustainable Hydrogen Production by Plasmonic Thermophotocatalysis. *Catal. Today* **2021**, *380*, 156–186. [CrossRef]
2. Mateo, D.; Cerrillo, J.L.; Durini, S.; Gascon, J. Fundamentals and Applications of Photo-Thermal Catalysis. *Chem. Soc. Rev.* **2021**, *50*, 2173–2210. [CrossRef]
3. Nair, V.; Muñoz-Batista, M.J.; Fernández-García, M.; Luque, R.; Colmenares, J.C. Thermo-Photocatalysis: Environmental and Energy Applications. *ChemSusChem* **2019**, *12*, 2098–2116. [CrossRef] [PubMed]
4. Liu, J.; Chen, H.; Zhu, C.; Han, S.; Li, J.; She, S.; Wu, X. Efficient Simultaneous Removal of Tetracycline Hydrochloride and Cr(VI) through Photothermal-Assisted Photocatalytic-Fenton-like Processes with $CuO_x/\gamma-Al_2O_3$. *J. Colloid Interface Sci.* **2022**, *622*, 526–538. [CrossRef]
5. Yang, L.; Xiang, Y.; Jia, F.; Xia, L.; Gao, C.; Wu, X.; Peng, L.; Liu, J.; Song, S. Photo-Thermal Synergy for Boosting Photo-Fenton Activity with $RGO-ZnFe_2O_4$: Novel Photo-Activation Process and Mechanism toward Environment Remediation. *Appl. Catal. B Environ.* **2021**, *292*, 120198. [CrossRef]
6. Zhang, Y.; He, S.; Guo, W.; Hu, Y.; Huang, J.; Mulcahy, J.R.; Wei, W.D. Surface-Plasmon-Driven Hot Electron Photochemistry. *Chem. Rev.* **2018**, *118*, 2927–2954. [CrossRef]
7. Kumar, A.; Choudhary, P.; Kumar, A.; Camargo, P.H.C.; Krishnan, V. Recent Advances in Plasmonic Photocatalysis Based on TiO_2 and Noble Metal Nanoparticles for Energy Conversion, Environmental Remediation, and Organic Synthesis. *Small* **2022**, *18*, 2101638. [CrossRef]
8. Yao, C.; Zhang, L.; Wang, J.; He, Y.; Xin, J.; Wang, S.; Xu, H.; Zhang, Z. Gold Nanoparticle Mediated Phototherapy for Cancer. *J. Nanomater.* **2016**, *2016*, 5497136. [CrossRef]
9. Adleman, J.R.; Boyd, D.A.; Goodwin, D.G.; Psaltis, D. Heterogenous Catalysis Mediated by Plasmon Heating. *Nano Lett.* **2009**, *9*, 4417–4423. [CrossRef]
10. Yang, Q.; Xu, Q.; Yu, S.-H.; Jiang, H.-L. Pd Nanocubes@ZIF-8: Integration of Plasmon-Driven Photothermal Conversion with a Metal-Organic Framework for Efficient and Selective Catalysis. *Angew. Chem. Int. Ed.* **2016**, *55*, 3685–3689. [CrossRef]
11. Wang, F.; Huang, Y.; Chai, Z.; Zeng, M.; Li, Q.; Wang, Y.; Xu, D. Photothermal-Enhanced Catalysis in Core-Shell Plasmonic Hierarchical Cu_7S_4 Microsphere@zeolitic Imidazole Framework-8. *Chem. Sci.* **2016**, *7*, 6887–6893. [CrossRef]
12. Nikawa, T.; Naya, S.; Kimura, T.; Tada, H. Rapid Removal and Subsequent Low-Temperature Mineralization of Gaseous Acetaldehyde by the Dual Thermocatalysis of Gold Nanoparticle-Loaded Titanium(IV) Oxide. *J. Catal.* **2015**, *326*, 9–14. [CrossRef]
13. Morikawa, T.; Ohwaki, T.; Suzuki, K.; Moribe, S.; Tero-Kubota, S. Visible-Light-Induced Photocatalytic Oxidation of Carboxylic Acids and Aldehydes over N-Doped TiO_2 Loaded with Fe, Cu or Pt. *Appl. Catal. B Environ.* **2008**, *83*, 56–62. [CrossRef]
14. Liu, X.; Ye, L.; Ma, Z.; Han, C.; Wang, L.; Jia, Z.; Su, F.; Xie, H. Photothermal Effect of Infrared Light to Enhance Solar Catalytic Hydrogen Generation. *Catal. Commun.* **2017**, *102*, 13–16. [CrossRef]
15. Zhou, L.; Swearer, D.F.; Zhang, C.; Robatjazi, H.; Zhao, H.; Henderson, L.; Dong, L.; Christopher, P.; Carter, E.A.; Nordlander, P.; et al. Quantifying Hot Carrier and Thermal Contributions in Plasmonic Photocatalysis. *Science* **2018**, *362*, 69–72. [CrossRef] [PubMed]
16. Sarina, S.; Zhu, H.-Y.; Xiao, Q.; Jaatinen, E.; Jia, J.; Huang, Y.; Zheng, Z.; Wu, H. Viable Photocatalysts under Solar-Spectrum Irradiation: Nonplasmonic Metal Nanoparticles. *Angew. Chem. Int. Ed.* **2014**, *53*, 2935–2940. [CrossRef] [PubMed]
17. Hong, S.-J.; Mun, H.-J.; Kim, B.-J.; Kim, Y.-S. Characterization of Nickel Oxide Nanoparticles Synthesized under Low Temperature. *Micromachines* **2021**, *12*, 1168. [CrossRef] [PubMed]
18. Kim, W.; Tachikawa, T.; Kim, H.; Lakshminarasimhan, N.; Murugan, P.; Park, H.; Majima, T.; Choi, W. Visible Light Photocatalytic Activities of Nitrogen and Platinum-Doped TiO_2: Synergistic Effects of Co-Dopants. *Appl. Catal. B Environ.* **2014**, *147*, 642–650. [CrossRef]
19. Falconer, J.L.; Magrini-Bair, K.A. Photocatalytic and Thermal Catalytic Oxidation of Acetaldehyde on Pt/TiO_2. *J. Catal.* **1998**, *179*, 171–178. [CrossRef]
20. Honda, M.; Ochiai, T.; Listiani, P.; Yamaguchi, Y.; Ichikawa, Y. Low-Temperature Synthesis of Cu-Doped Anatase TiO_2 Nanostructures via Liquid Phase Deposition Method for Enhanced Photocatalysis. *Materials* **2023**, *16*, 639. [CrossRef]
21. Sano, T.; Negishi, N.; Uchino, K.; Tanaka, J.; Matsuzawa, S.; Takeuchi, K. Photocatalytic Degradation of Gaseous Acetaldehyde on TiO_2 with Photodeposited Metals and Metal Oxides. *J. Photochem. Photobiol. A Chem.* **2003**, *160*, 93–98. [CrossRef]
22. Zeng, Q.; Xie, X.; Wang, X.; Wang, Y.; Lu, G.; Pui, D.Y.H.; Sun, J. Enhanced Photocatalytic Performance of $Ag@TiO_2$ for the Gaseous Acetaldehyde Photodegradation under Fluorescent Lamp. *Chem. Eng. J.* **2018**, *341*, 83–92. [CrossRef]
23. Zhu, S.; Xie, X.; Chen, S.-C.; Tong, S.; Lu, G.; Pui, D.Y.H.; Sun, J. Cu-Ni Nanowire-Based TiO_2 Hybrid for the Dynamic Photodegradation of Acetaldehyde Gas Pollutant under Visible Light. *Appl. Surf. Sci.* **2017**, *408*, 117–124. [CrossRef]
24. Jiang, Y.; Chen, G.; Xu, X.; Chen, X.; Deng, S.; Smirnov, S.; Luo, H.; Zou, G. Direct Growth of Mesoporous Anatase TiO_2 on Nickel Foam by Soft Template Method as Binder-Free Anode for Lithium-Ion Batteries. *RSC Adv.* **2014**, *4*, 48938–48942. [CrossRef]

25. Yan, Y.; Cheng, X.; Zhang, W.; Chen, G.; Li, H.; Konkin, A.; Sun, Z.; Sun, S.; Wang, D.; Schaaf, P. Plasma Hydrogenated TiO_2/Nickel Foam as an Efficient Bifunctional Electrocatalyst for Overall Water Splitting. *ACS Sustain. Chem. Eng.* **2019**, *7*, 885–894. [CrossRef]
26. Zhang, Q.; Li, F.; Chang, X.; He, D. Comparison of Nickel Foam/Ag-Supported ZnO, TiO_2, and WO_3 for Toluene Photodegradation. *Mater. Manuf. Process.* **2014**, *29*, 789–794. [CrossRef]
27. Hu, H.; Xiao, W.; Yuan, J.; Shi, J.; Chen, M.; Shang Guan, W. Preparations of TiO_2 Film Coated on Foam Nickel Substrate by Sol-Gel Processes and Its Photocatalytic Activity for Degradation of Acetaldehyde. *J. Environ. Sci.* **2007**, *19*, 80–85. [CrossRef]
28. Zeng, Q.; Chen, J.; Wan, Y.; Ni, J.; Ni, C.; Chen, H. Immobilizing TiO_2 on Nickel Foam for an Enhanced Photocatalysis in NO Abatement under Visible Light. *J. Mater. Sci.* **2022**, *57*, 15722–15736. [CrossRef]
29. Jia, J.; Li, D.; Cheng, X.; Wan, J.; Yu, X. Construction of Graphite/TiO_2/Nickel Foam Photoelectrode and Its Enhanced Photocatalytic Activity. *Appl. Catal. A Gen.* **2016**, *525*, 128–136. [CrossRef]
30. Ding, X.; Liu, H.; Chen, J.; Wen, M.; Li, G.; An, T.; Zhao, H. In Situ Growth of Well-Aligned Ni-MOF Nanosheets on Nickel Foam for Enhanced Photocatalytic Degradation of Typical Volatile Organic Compounds. *Nanoscale* **2020**, *12*, 9462–9470. [CrossRef]
31. Hu, H.; Xiao, W.; Yuan, J.; Shi, J.; He, D.; Shangguan, W. High Photocatalytic Activity and Stability for Decomposition of Gaseous Acetaldehyde on TiO_2/Al_2O_3 Composite Films Coated on Foam Nickel Substrates by Sol-Gel Processes. *J. Sol-Gel Sci. Technol.* **2008**, *45*, 1–8. [CrossRef]
32. Tryba, B.; Rychtowski, P.; Srenscek-Nazzal, J.; Przepiorski, J. The Inflence of TiO_2 Structure on the Complete Decomposition of Acetaldehyde Gas. *Mater. Res. Bull.* **2020**, *126*, 110816. [CrossRef]
33. Gęsikiewicz-Puchalska, A.; Zgrzebnicki, M.; Michalkiewicz, B.; Narkiewicz, U.; Morawski, A.W.; Wrobel, R.J. Improvement of CO_2 Uptake of Activated Carbons by Treatment with Mineral Acids. *Chem. Eng. J.* **2017**, *309*, 159–171. [CrossRef]
34. Zheng, D.; Zhao, H.; Wang, S.; Hu, J.; Chen, Z. NiO-TiO_2 p-n Heterojunction for Solar Hydrogen Generation. *Catalysts* **2021**, *11*, 1427. [CrossRef]
35. Wan, K.; Luo, J.; Zhang, X.; Subramanian, P.; Fransaer, J. In-Situ Formation of Ni (Oxy)Hydroxide on Ni Foam as an Efficient Electrocatalyst for Oxygen Evolution Reaction. *Int. J. Hydrog. Energy* **2020**, *45*, 8490–8496. [CrossRef]
36. Xue, Y.; Shao, P.; Yuan, Y.; Shi, W.; Guo, Y.; Zhang, B.; Bao, X.; Cui, F. Monolithic Nickel Foam Supported Macro-Catalyst: Manipulation of Charge Transfer for Enhancement of Photo-Activity. *Chem. Eng. J.* **2021**, *418*, 129456. [CrossRef]
37. Nikawa, T.; Naya, S.; Tada, H. Rapid Removal and Decomposition of Gaseous Acetaldehyde by the Thermo- and Photo-Catalysis of Gold Nanoparticle-Loaded Anatase Titanium(IV) Oxide. *J. Colloid Interface Sci.* **2015**, *456*, 161–165. [CrossRef]

Disclaimer/Publisher's Note: The statements, opinions and data contained in all publications are solely those of the individual author(s) and contributor(s) and not of MDPI and/or the editor(s). MDPI and/or the editor(s) disclaim responsibility for any injury to people or property resulting from any ideas, methods, instructions or products referred to in the content.

Article

Carbon Quantum Dots Accelerating Surface Charge Transfer of 3D PbBiO$_2$I Microspheres with Enhanced Broad Spectrum Photocatalytic Activity—Development and Mechanism Insight

Ruyu Yan [†], Xinyi Liu [†], Haijie Zhang, Meng Ye, Zhenxing Wang, Jianjian Yi, Binxian Gu and Qingsong Hu *

College of Environmental Science and Engineering, Yangzhou University, 196 West Huayang Road, Yangzhou 225127, China
* Correspondence: huqs@yzu.edu.cn
† These authors contributed equally to this work.

Abstract: The development of a highly efficient, visible-light responsive catalyst for environment purification has been a long-standing exploit, with obstacles to overcome, including inefficient capture of near-infrared photons, undesirable recombination of photo-generated carriers, and insufficient accessible reaction sites. Hence, novel carbon quantum dots (CQDs) modified PbBiO$_2$I photocatalyst were synthesized for the first time through an in-situ ionic liquid-induced method. The bridging function of 1-butyl-3-methylimidazolium iodide ([Bmim]I) guarantees the even dispersion of CQDs around PbBiO$_2$I surface, for synchronically overcoming the above drawbacks and markedly promoting the degradation efficiency of organic contaminants: (i) CQDs decoration harness solar photons in the near-infrared region; (ii) particular delocalized conjugated construction of CQDs strength via the utilization of photo-induced carriers; (iii) π–π interactions increase the contact between catalyst and organic molecules. Benefiting from these distinguished features, the optimized CQDs/PbBiO$_2$I nanocomposite displays significantly enhanced photocatalytic performance towards the elimination of rhodamine B and ciprofloxacin under visible/near-infrared light irradiation. The spin-trapping ESR analysis demonstrates that CQDs modification can boost the concentration of reactive oxygen species (O$_2^{\bullet-}$). Combined with radicals trapping tests, valence-band spectra, and Mott–Schottky results, a possible photocatalytic mechanism is proposed. This work establishes a significant milestone in constructing CQDs-modified, bismuth-based catalysts for solar energy conversion applications.

Keywords: PbBiO$_2$I microspheres; CQDs; ionic liquid; charge separation; interface

Citation: Yan, R.; Liu, X.; Zhang, H.; Ye, M.; Wang, Z.; Yi, J.; Gu, B.; Hu, Q. Carbon Quantum Dots Accelerating Surface Charge Transfer of 3D PbBiO$_2$I Microspheres with Enhanced Broad Spectrum Photocatalytic Activity—Development and Mechanism Insight. *Materials* 2023, 16, 1111. https://doi.org/10.3390/ma16031111

Academic Editor: Andrea Petrella

Received: 1 December 2022
Revised: 11 January 2023
Accepted: 21 January 2023
Published: 27 January 2023

Copyright: © 2023 by the authors. Licensee MDPI, Basel, Switzerland. This article is an open access article distributed under the terms and conditions of the Creative Commons Attribution (CC BY) license (https://creativecommons.org/licenses/by/4.0/).

1. Introduction

Semiconductor photocatalysis is deemed as a promising technique to purify water which is contaminated by various pollutants, such as dyes, antibiotics, endocrine disruptors, and so on [1–3]. Taking full advantage of solar energy has already been proven as a "green" strategy to settle environmental contamination and energy crunch [4–6]. For the sake of triggering an effective photocatalytic reaction and achieving a satisfying degradation efficiency, it is essential to boost effective interfacial contact between target organic molecules and reactive species, namely to involve more photo-generated carriers in surface catalysis. Additionally, several studies have shown that the photocatalytic efficiency can be improved via increasing the adsorption capability of photocatalysts towards contaminants [7–9].

Among various Bi-based semiconductor photocatalysts, BiOI displays some fascinating advantages, e.g., high chemical inertness, easy preparation, low toxicity, and broad visible-light absorption range [10]. Moreover, it displays some photocatalytic performance in the field of pollutants elimination, CO$_2$ conversion, N$_2$ reduction, and so forth [11–13]. Nevertheless, individual BiOI is subjected to its inherent weakness, such as the poor oxidation capacity caused by the high value of valence band and low charge separation efficiency. Moreover, during the contaminant degradation process, the contaminants' adsorption and activation capacity over BiOI is still unsatisfactory [14,15]. Therefore, a variety of methods have been adopted to improve the photocatalytic performance of BiOI, e.g., elemental doping, defect regulation, exposed facet control, heterojunctions construction and bismuth-rich strategy [16–19]. In addition, a part of Bi in the $[Bi_2O_2]^{2+}$ layer can be replaced by other main group elements (Pb, Ba, Sr, Ca, etc.) to generate $[ABiO_2]^+$ [20–22], and I^- and $[ABiO_2]^+$ can be arranged alternately to form $ABiO_2I$. The construction of bismuth-based bimetallic oxyiodide can maximize the photocatalytic performance of bismuth oxyiodide under broadband light irradiation. Considering that the radii of Pb^{2+} and Bi^{3+} are very close, the crystalline structure of $PbBiO_2I$ displays no obvious change by substituting Bi^{3+} with Pb^{2+} [23]. More importantly, $PbBiO_2I$ displays preferable band structure with a suitable narrow bandgap of 1.9 eV, which is beneficial to the degradation of contaminants. Therefore, $PbBiO_2I$ displays great potential in the field of environmental purification.

As a key member of the carbon-based nanomaterial family, 0D CQDs have drawn tremendous research attention [24,25]. Particular properties, e.g., high solubility, excellent electron conductivity, and up-conversion performance, endow them with broad applications [26–28]. Herein, CQDs have been extensively employed to modify photocatalysts to heighten their optical and electrochemical properties, and eventually promote their photocatalytic performance. In fact, the introduction of CQDs can boost the separation and transportation of photo-generated electron-hole pairs and enable more reactive oxygen species participating in the degradation of contaminants [29]. In spite of this, in previous references, the core character of CQDs during the photocatalytic reactions and pollutant degradation mechanism have not yet been studied at length. More importantly, considering that the diameter of CQDs is less than 10 nm, the uniform distribution of CQDs around catalyst surface requires further study.

In order to deeply analyze the aforementioned impending issues, CQDs modified $PbBiO_2I$ nanocomposite photocatalysts are obtained via an ionic liquid [Bmim]I assisted solvothermal method. In this approach, [Bmim]I can be employed as template and reaction source to control the growth of $PbBiO_2I$ crystals. In fact, the existence of coulomb force and hydrogen bond between [Bmim]I and CQDs is beneficial to in-situ anchoring more CQDs around $PbBiO_2I$ material [30]. Moreover, CQDs decoration can promote organic pollutants adsorption, boost interface charge separation and transportation, and ultimately enhance the photocatalytic degradation efficiency of rhodamine B (RhB) and ciprofloxacin (CIP) under visible/near-infrared light irradiation. Our research extends the knowledge into developing more CQDs-decorated, bismuth-based bimetallic catalysts with widespread applications in the area of wastewater treatment.

2. Experimental Details

2.1. Sample Preparation

CQDs powder was acquired on the basis of the previous reference and then managed by lyophilization [29]. Hence, 1.06 g citric acid monohydrate was dispersed into deionized water (11 mL), and ethylenediamine (340 μL) was injected and stirred for 1 h. This above clear solution was sealed in Teflon-lined autoclave (25 mL) and reacted at 200 °C for 5 h. After cooling down to room temperature, the brownish red solution was subjected to dialysis for 72 h to obtain the CQDs solution. In the end, CQDs powder was obtained after freeze-drying for 72 h.

The synthetic process of pure $PbBiO_2I$ and $CQDs/PbBiO_2I$ nanocomposite was as follows (Scheme 1) via ionothermal method: First of all, CQDs ($\times$ g), $Bi(NO_3)_3 \cdot 5H_2O$ (0.24 g) and $Pb(NO_3)_2$ (0.16 g) are fully dispersible in ethylene glycol (15 mL) and defined as A. Then, ionic liquid [Bmim]I (0.13 g) was dispersed uniformly into ethylene glycol (5 mL) and defined as B. B was added into A bit by bit and stirred for 1 h. After that, the mixture was sealed in a Teflon-lined autoclave (25 mL) and reacted at 180 °C for 24 h. Subsequently, the sediment was collected via high-speed centrifugation, rinsed three times with deionized water and absolute ethanol, and dried at 80 °C for 12 h. The loading amount of CQDs in CQDs/PbBiO$_2$I nanocomposite was 1, 3, 5, and 8 wt.%, respectively. Individual PbBiO$_2$I was also obtained without the introduction of CQDs.

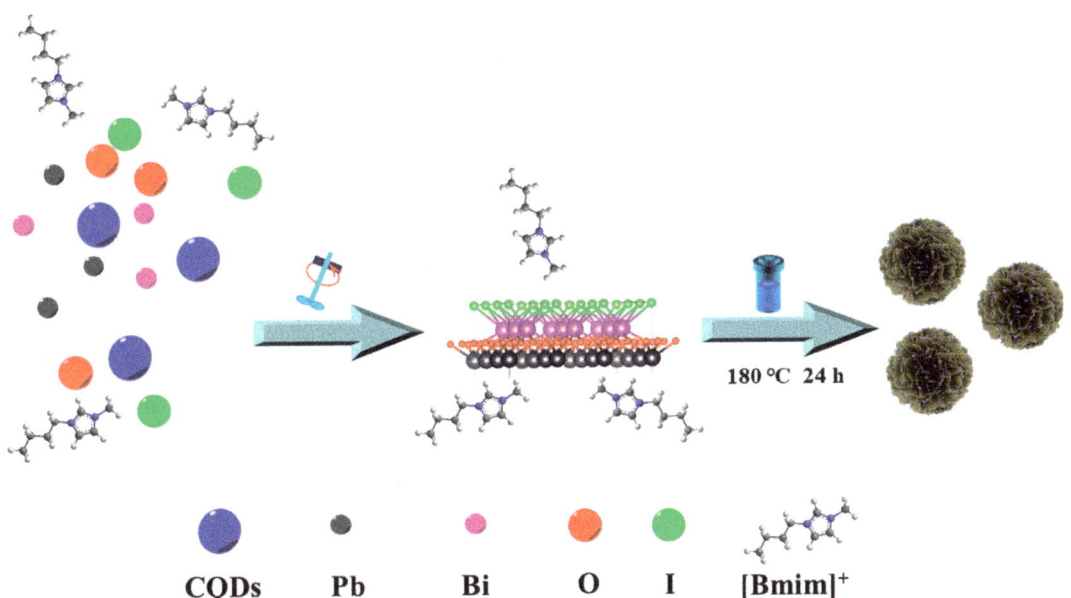

Scheme 1. Schematic diagram for the formation of CQDs/PbBiO$_2$I nanocomposite catalyst.

2.2. Sample Characterization

X-ray diffraction (XRD) was recorded on a D8 Advance diffractometer (Bruker, Germany) using monochromatic Cu Kα radiation. X-ray photoelectron spectroscopy (XPS) spectrum was measured using an ESCALAB 250Xi XPS (Thermo Fisher, Waltham, MA, USA) with monochromatic Mg-Kα radiation as X-ray source for excitation. A laser Raman spectrometer (DXR, Thermo Fisher Scientific, Waltham, MA, USA) was employed to collect Raman spectra with a 532 nm laser as an excitation source. A specific surface and aperture analyzer (Quadrasorb EVO, Anton Paar, Ashland, VA, USA) was used to analyze the specific surface area and pore diameter of the catalysts by N$_2$ adsorption-desorption isotherms analyzed at 77 K using the Brunauer–Emmett–Teller (BET) method and Barret–Joyner–Halenda (BJH) adsorption dV/dW pore volume distribution. The microstructures of the catalysts were investigated by Tecnai G2 F30 S-TWIN transmission electron microscopy (TEM, FEI, USA). UV-Vis diffuse reflection spectra was acquired on a UV-2450 spectrophotometer (Shimadzu, Japan). The photoluminescence (PL) spectra were obtained by a FLS980 fluorescence spectrometer (Edinburgh, UK). An electron spin resonance (ESR) spectrometer (JES-FA200, Bruker, Germany) was used to capture ESR signals of spin-trapped radicals using 5,5-dimethyl-1-pyrroline-*N*-oxide (DMPO, radical trapping reagent) in water and methanol solutions. The electrochemical data were obtained on a CHI 660B electrochemistry workstation (Chenhua, Shanghai) employing a conventional three-electrode cell (ITO slice, Pt wire and saturated Ag/AgCl).

2.3. Photocatalytic Degradation Test

RhB and CIP were selected as the model contaminants. A 250 W xenon plus equipped with optical filter (λ > 400 nm or λ > 610 nm) was employed as the optical source. The reaction temperature was kept at 25 °C via a recirculating cooling water system. Hence, 0.03 g samples were added in RhB solution (100 mL, 20 mg/L) or CIP solution (100 mL, 10 mg/L) by ultrasonic dispersion. The mixed solution was stirred in the dark for 30 min.

Subsequently, the optical source was switched on, and 4.5 mL reaction mixture was sampled at set intervals. The acquired mixture was centrifuged at 15,000 rpm for 4 min to acquire a clear solution. The concentrations of RhB and CIP were studied using a Shimadzu LC-20A HPLC system (Shimadzu, Japan), including an Agilent TC-C (18) column, two Varian Prostar 210 pumps and a ultraviolet detector. A mobile phase consisting of methanol and pure water in the ratio of 30:70 (v:v) was used at the flow rate of 0.8 mL min^{-1}. The reaction solution (10 µL) was injected. To distinguish the major reactive species participating in photocatalytic reactions, various radical scavengers were introduced. Tert-butanol (t-butanol) can capture hydroxyl radical (•OH), nitrogen (N_2) can inhibit the production of superoxide radical ($O_2^{•-}$), while ammonium oxalate (AO) and triethanolamine (TEA) can capture holes (h^+).

3. Result and Discussion

The crystal structures of the acquired materials were analyzed by the X-ray diffraction patterns (XRD). As depicted in Figure 1a, all the peaks can be indexed to tetragonal $PbBiO_2I$ (JCPDS NO. 38-1007) [31]. For single CQDs, a broad peak located at 25° can be observed. After the decoration of CQDs, the nanocomposite retains the same diffraction peaks of tetragonal $PbBiO_2I$. Additionally, it is noteworthy that a small peak centered around 26.9° appears with the increased loading amount of CQDs. This phenomenon validates the successful introduction of CQDs. Raman spectra was conducted to further confirm the successful loading of CQDs. As shown in Figure 1b, the strong peaks (66.4 and 142.2 cm^{-1}) are ascribed to the "lattice vibration" of $PbBiO_2I$. Compared with single $PbBiO_2I$, 5 wt.% CQDs/$PbBiO_2I$ displays two apparent peaks centered at 1389.1 and 1580.2 cm^{-1}, which represent the disordered (D) and graphitic (G) bands of graphene. The peak intensity of G band is much higher than that of D band, which is in agreement with previous report [32]. There is background interference in the nanocomposite because of the strong fluorescence for CQDs. To explore the textural properties of the obtained samples, BET surface area measurement was carried out. In Figure 1c, the N_2 adsorption/desorption isotherms can be categorized as type IV isotherm, indicating the characteristics of mesoporous [33]. This is in accord with the pore diameter distribution analysis (Figure S1). Moreover, the BET value of $PbBiO_2I$ and 5 wt.% CQDs/$PbBiO_2I$ is determined to be 10.20 and 31.77 m^2 g^{-1}. As is well known to all, a higher surface area is beneficial to provide more exposed reactive sites and transport paths for reactants and products, probably resulting in the enhanced photocatalytic performance [34].

The morphology of CQDs and CQDs/$PbBiO_2I$ was studied by TEM analysis. As shown in Figure S2, the CQDs are monodisperse with a spherical shape diameter of 5 nm. Moreover, the lattice fringe spacing of 0.33 nm corresponds to the (002) plane of CQDs [35]. For CQDs/$PbBiO_2I$ nanocomposite, it displays flower-like spheres with a diameter of 2 µm assembled from nanosheets (Figure 1d,e). To further observe this microstructure, high-resolution TEM (HRTEM) measurement was carried out. As shown in Figure 1f, the lattice fringe spacing of 0.33 nm and 0.30 nm corresponds to the (002) plane of CQDs and (103) crystallographic plane of $PbBiO_2I$. The results of TEM analysis demonstrate that CQDs/$PbBiO_2I$ nanocomposite catalyst have been constructed successfully.

The surface chemical states and compositions of the obtained catalysts were investigated by X-ray photoelectron (XPS) spectra (Figure 2). The survey XPS spectrum shows the co-existence of Pb, Bi, O, I, and C elements in the CQDs/$PbBiO_2I$ nanocomposite (Figure 2a). In Figure 2b, the peaks at roughly 142.5 and 137.6 eV are ascribed to the Pb 4f of Pb^{2+} [35]. In Figure 2c, the strong peaks located at 163.9 and 158.6 eV are assigned to the Bi $4f_{5/2}$ and Bi $4f_{7/2}$, manifesting the presence of trivalent bismuth [36]. What is more, after the decoration of CQDs, the binding energies of Pb 4f and Bi 4f shift to higher values when compared to those of $PbBiO_2I$. The binding energy shift unequivocally proves the occurrence of electron re-distribution between $PbBiO_2I$ and CQDs. Similar phenomena have been reported in other references [29,37]. The binding energy of O 1s centered at 529.2 eV (Figure 2d) is assigned to the oxygen of $PbBiO_2I$. The XPS peaks of I 3d centered at 630.1 and 618.7 eV are ascribed to I $3d_{3/2}$ and I $3d_{5/2}$ of I⁻ [31] (Figure 2e). Figure 2f displays the high-resolution XPS spectra of C 1s, in which three deconvoluted peaks can be ascribed to C-C (284.6 eV), C=C (286.5 eV) and C-N (288.1 eV), respectively [26,29]. The XPS results further prove the presence of PbBiO$_2$I and CQDs in the CQDs/PbBiO$_2$I nanocomposite.

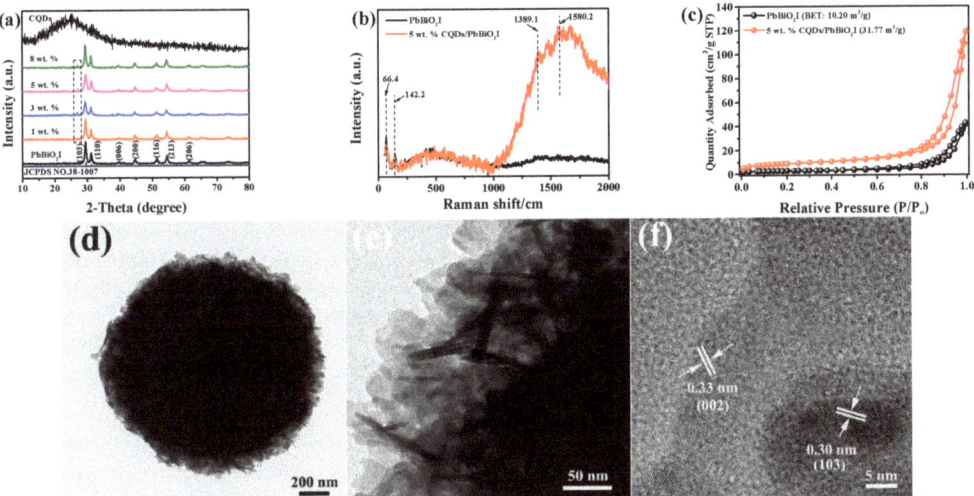

Figure 1. (**a**) XRD patterns of the as-prepared samples; (**b**) Raman spectra and (**c**) Nitrogen adsorption-desorption isotherms of PbBiO$_2$I and 5 wt.% CQDs/PbBiO$_2$I; (**d**,**e**) TEM and (**f**) HRTEM images of 5 wt.% CQDs/PbBiO$_2$I.

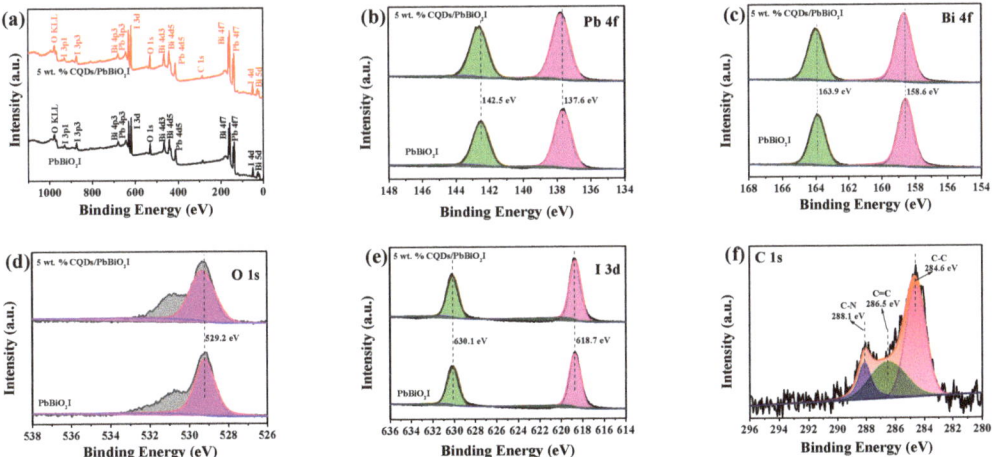

Figure 2. XPS spectra of PbBiO$_2$I and 5 wt.% CQDs/PbBiO$_2$I: (**a**) survey, (**b**) Pb 4f, (**c**) Bi 4f, (**d**) O 1s, (**e**) I 3d, (**f**) C 1s.

The photoreactivity of the as-prepared catalysts was assessed by decomposition of RhB upon irradiation with visible light (λ > 400 nm). As we all know, RhB is broadly used as a coloration in textile and food processing industries and is also a popular water tracer fluorescent. It poses a threat to human beings and animals, and causes irritation of the eyes, skin, and respiratory passage. The carcinogenicity, reproductive and development toxicity and chronic toxicity toward human beings and animals have been proven experimentally [38]. Therefore, it is crucial to reduce RhB concentrations to reach the national standards. Herein, CQDs/PbBiO$_2$I nanocomposite are tentatively employed to remove RhB, and the importance of CQDs can also be confirmed from another angle. The RhB adsorption capacity over different samples is shown in Figure S3. The dark adsorption experiment results demonstrate that the adsorption capacity can be improved thanks to the π-π interactions between CQDs and RhB molecule [36]. In Figure 3a, the photolysis of RhB is almost negligible. For PbBiO$_2$I, the degradation rate of RhB is only 29.7% within 30 min. After the introduction of CQDs, CQDs/PbBiO$_2$I nanocomposites show higher photocatalytic efficiency than that of PbBiO$_2$I, highlighting the key role of CQDs in environmental purification. Notably, 5 wt.% CQDs/PbBiO$_2$I is obviously superior to other CQDs modified PbBiO$_2$I samples. This can be explained as too many CQDs may generate an adverse shielding effect, hindering the PbBiO$_2$I surface from absorbing visible-light photons and overlaying the reactive sites for photocatalysis via CQDs agglomeration [36,39]. Figure 3b shows the photocatalytic kinetics fit of RhB degradation on account of pseudo-first-order model (Langmuir-Hinshelwood model). It can be clearly seen that the degradation efficiency of RhB can be obviously enhanced after adding CQDs. Specifically, the degradation rate of 5 wt.% CQDs/PbBiO$_2$I is approximately 2.27 times larger than that of PbBiO$_2$I. To evaluate the catalyst's reusability, five consecutive cycles were conducted over 5 wt.% CQDs/PbBiO$_2$I photocatalyst. In Figure S4a, the degradation rate remains 90% in the elimination of RhB, manifesting the preferable catalytic stability of the nanocomposite catalyst. Furthermore, XRD patterns further confirm the well-retained structure of the nanocomposite catalyst after photoirradiation reaction (Figure S4b).

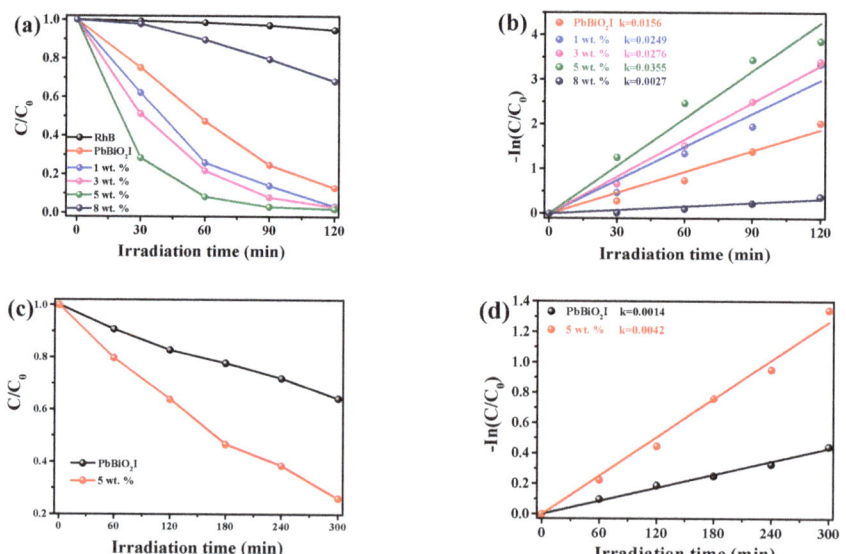

Figure 3. Photocatalytic properties of the catalysts for the degradation of RhB (**a**) and CIP (**c**) under visible light irradiation (λ > 400 nm); reaction kinetics for the degradation of RhB (**b**) and CIP (**d**) under the same condition.

CIP is classified as belonging the second generation of fluoroquinolone antibiotics, being proven to damage the environment and display toxic effects in the surface water and groundwater [40]. This is the first report employing CQDs/PbBiO$_2$I nanocomposite for CIP degradation, and the degradation plots are shown in Figure 3c. The degradation rate of CIP achieves 35.8% within 300 min while employing PbBiO$_2$I as catalyst. Surprisingly, more than 73.9% of CIP can be eliminated over PbBiO$_2$I loading with 5 wt.% CQDs. Further, the calculating rate constant is 0.0042 min^{-1}, which is much higher than that of PbBiO$_2$I (0.0014 min^{-1}) (Figure 3d). Additionally, a total organic carbon (TOC) experiment was carried out to study the mineralization of RhB and CIP over 5 wt.% CQDs/PbBiO$_2$I (Figure S5). Under visible-light irradiation for 120 min, almost 76.2% of RhB is mineralized. For CIP, approximately 40.1% of CIP can be mineralized under illumination for 300 min. This implies that both RhB and CIP can be mineralized effectively over 5 wt.% CQDs/PbBiO$_2$I under visible light irradiation.

The photocatalytic performance of single PbBiO$_2$I and 5 wt.% CQDs/PbBiO$_2$I was further studied under near-infrared photoirradiation (λ > 610 nm). As shown in Figure 4a, only 13.5% RhB can be degraded by PbBiO$_2$I after 120 min near-infrared photoirradiation. Accompanied with the decoration of CQDs, the photocatalytic efficiency is significantly improved and 53.3% RhB is degraded over 5 wt.% CQDs/PbBiO$_2$I under the same condition. The corresponding rate constant for 5 wt.% CQDs/PbBiO$_2$I is 5.36 times higher than pure PbBiO$_2$I (Figure 4b). The ratio of rate constant for 5 wt.% CQDs/PbBiO$_2$I to individual PbBiO$_2$I under near-infrared photoirradiation is analogous to the ratio under visible light irradiation, which indicate the analogical activation pattern. In comparison to visible light condition, the ratio of rate constant for 5 wt.% CQDs/PbBiO$_2$I to individual PbBiO$_2$I under near-infrared light condition is higher, indicating that CQDs can transport photo-generated electrons more efficiently under near-infrared photoirradiation [29]. Given that the dye-sensitization effect involved in the degradation of RhB, the degradation of colorless CIP was carried out under the same condition, pure PbBiO$_2$I only degrades 12.9% CIP after 300 min irradiation. After the modification of CQDs, 5 wt.% CQDs/PbBiO$_2$I can degrade 25.7% CIP after 300 min irradiation (Figure 4c). The reaction rate constant of 5 wt.% CQDs/PbBiO$_2$I is 2.0 times that of PbBiO$_2$I (Figure 4d). The above experimental data demonstrate the key roles of CQDs during a photocatalytic reaction process [29,41].

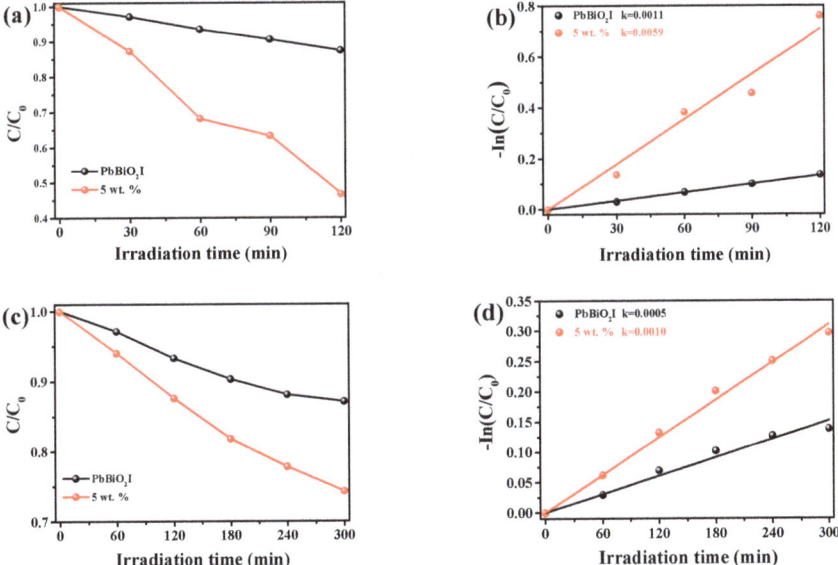

Figure 4. Photocatalytic properties of the catalysts for the degradation of RhB (**a**) and CIP (**c**) under near-infrared light irradiation (λ > 610 nm); reaction kinetics for the degradation of RhB (**b**) and CIP (**d**) under the same condition.

The optical characteristics of the obtained samples across the UV-Vis region are recorded by diffuse reflectance spectra (DRS). In Figure 5a, pristine PbBiO$_2$I displays intrinsic bandgap absorption from 200 nm to 570 nm. This can be attributed to their intrinsic band-to-band transition [36]. After the modification of CQDs, the optical absorption is substantially extended to the near-infrared region. As a result, it may boost the generation of photo-induced carriers thanks to the enhanced light-harvesting capability. Consequently, more reactive species can be involved in the photocatalytic reaction. The E_g values of PbBiO$_2$I and CQDs/PbBiO$_2$I nanocomposite can be obtained employing the Kubelka–Munk function [8]:

$$\alpha h\nu = A\,(h\nu - E_g)^{n/2}$$

where α, h, ν, A, and E_g represent the absorption coefficient, Planck constant, light frequency, a constant, and band gap energy, respectively. As depicted in Figure 5b, the E_g values of PbBiO$_2$I, 1 wt.% CQDs/PbBiO$_2$I, 3 wt.% CQDs/PbBiO$_2$I, 5 wt.% CQDs/PbBiO$_2$I, and 8 wt.% CQDs/PbBiO$_2$I are calculated to be 1.92, 1.83, 1.77, 1.72, and 1.67 eV, respectively.

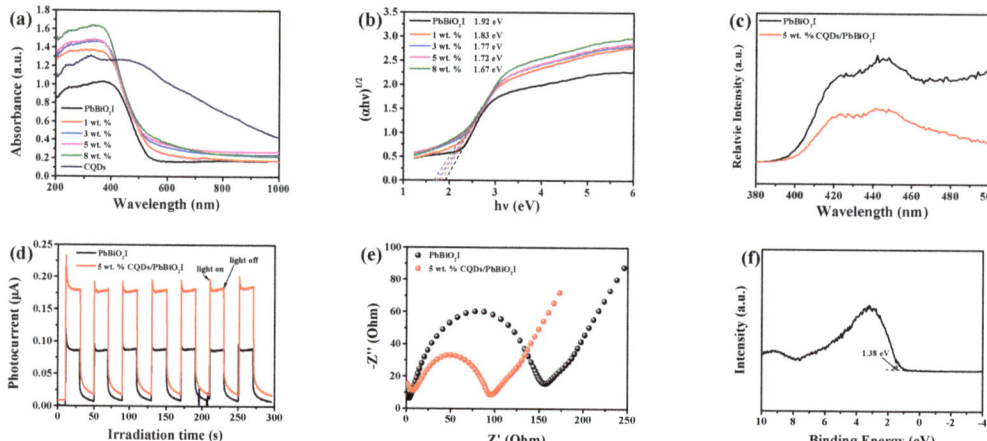

Figure 5. (a) UV-vis DRS of the acquired samples; (b) Bandgap of PbBiO$_2$I and CQDs/PbBiO$_2$I nanocomposite; (c) PL spectra, (d) transient photocurrent spectra, (e) EIS spectra of PbBiO$_2$I and 5 wt.% CQDs/PbBiO$_2$I; (f) XPS Valence-band spectra of PbBiO$_2$I.

To clarify the charge separation and transfer kinetics over the obtained catalysts, photoluminescence (PL) and photoelectrochemical measurements are performed. Figure 5c displays that both PbBiO$_2$I and 5 wt.% CQDs/PbBiO$_2$I possess an obvious emission peak centered at 440 nm. More importantly, 5 wt.% CQDs/PbBiO$_2$I shows an apparent quenching in comparison to PbBiO$_2$I. This characterization result indicates an improving separation probability of photo-generated carriers, which is beneficial for visible/near-infrared light-driven catalytic reactions [42]. The charge separation/transfer process is further monitored by transient photocurrent and electrochemical impedance spectroscopy (EIS) measurements. Figure 5d exhibits the transient photocurrent responses of the two samples under chopped light irradiation. It can be found that the photocurrent density is in the order PbBiO$_2$I < 5 wt.% CQDs/PbBiO$_2$I, which coincide with the trend of photocatalytic performance. The photocurrent results demonstrate that the charge separation efficiency of 5 wt.% CQDs/PbBiO$_2$I is superior to that of PbBiO$_2$I [43]. The results of transient photocurrent are further confirmed by EIS. In Figure 5e, the semicircle arc of 5 wt.% CQDs/PbBiO$_2$I in the Nyquist plot is smaller than that of PbBiO$_2$I, reflecting the lower interfacial charge transfer resistance [44]. Taking the above PL and photoelectrochemical results into account, CQDs modification will boost charge migration and separation, which is favorable for the generation of reactive species.

Apart from the light absorption efficiency and spatial separation efficiency of photoinduced charge carriers, the energy band structure also plays a critical role in determining photocatalytic efficiency. The total density of states of valence band (VB) that can be obtained based on the valence-band XPS spectra with Fermi level (E_f) of semiconductors is 0 eV (Figure 5f). The VB value of PbBiO$_2$I is measured to be 1.38 eV, and the positive slopes of Mott–Schottky curves show that PbBiO$_2$I is defined as a n-type semiconductor (Figure S6). According to the extrapolation of X intercept in the Mott–Schottky plots, the flat band potential of PbBiO$_2$I is measured to be −0.62 V vs. NHE (pH = 7). With regarding to the n-type semiconductors, the Fermi level is close to the flat band potential [45]. As a result, the VB value of PbBiO$_2$I is 0.76 V vs. NHE. On the basis of the E_g, the CB minimum of PbBiO$_2$I occur at approximately −1.16 V vs. NHE.

The ESR (electron spin resonance) technique and free radicals capturing tests are conducted to ascertain the major active species involved in the degradation process [46,47]. The results of ESR analysis are presented in Figure 6a,b. In the darkness, no characteristic signals can be observed for DMPO-O$_2$•− and DMPO-•OH from the two catalysts, and no reactive species can be trapped. Nevertheless, typical characteristic signal peaks of DMPO-O$_2$•− are observed under photoirradiation, and the DMPO-O$_2$•− signals of 5 wt.% CQDs/PbBiO$_2$I are obviously higher than that of PbBiO$_2$I. Furthermore, DMPO-•OH cannot be trapped from the two catalysts upon light illumination. ESR analysis demonstrate that superoxide radical (O$_2$•−) can be generated upon light illumination, and the modification of CQDs is beneficial for the generation of reactive oxygen species. To further verify the presence of these active species, free radical quenching tests are carried out in the presence of 5 wt.% CQDs/PbBiO$_2$I (Figure 6c). After the addition of AO and TEA, notable inhibition of photocatalytic activity can be observed, indicating that direct hole oxidation plays a critical role during visible/near-infrared light-driven catalytic reactions. Moreover, after N$_2$ is pumped into the reaction solution, the degradation efficiency declines greatly [48]. This can be explained as a large amount of O$_2$•− is being generated and acting in a key role during the degradation process. The analysis of radical capture tests are in accordance with ESR results.

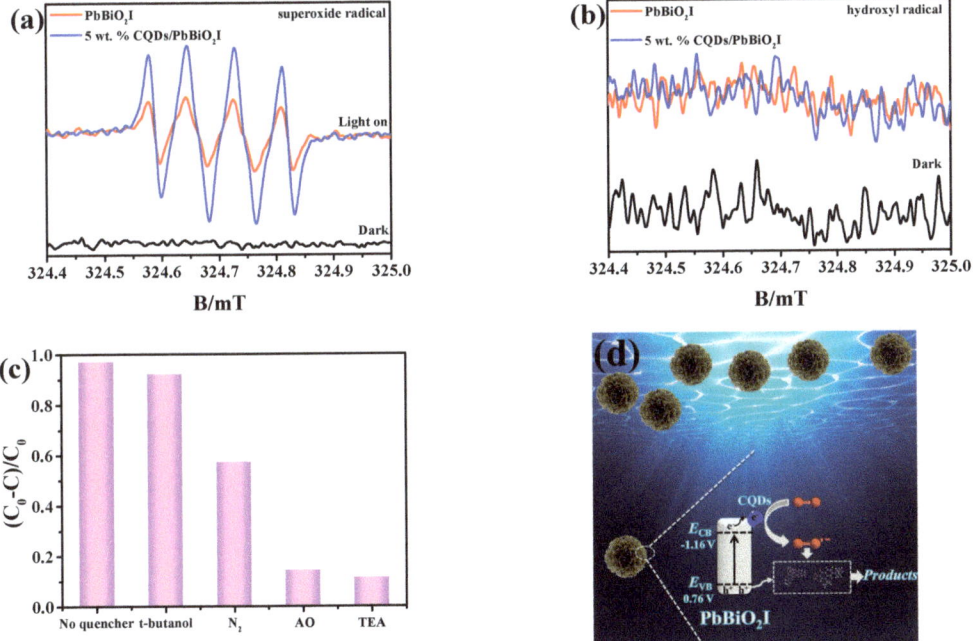

Figure 6. (a) DMPO- O$_2$•− and (b) DMPO-•OH of PbBiO$_2$I and 5 wt.% CQDs/PbBiO$_2$I; (c) trapping experiments of active species during the degradation of RhB; (d) photocatalytic mechanism of CQDs/PbBiO$_2$I nanocomposite in photocatalytic degradation.

Considering the above experimental results, the separation and transformation paths of photo-generated carriers involved in visible/near-infrared light-driven degradation of organic contaminants are presented in Figure 6d. Under broadband light irradiation, electrons in the VB of PbBiO$_2$I can be motivated and then migrated to the CB. Because of the narrow band gap width, the recombination of photo-induced electrons and holes occurs in very little time. After the modification of CQDs, the mobility of photo-generated electrons can be enhanced thanks to the electron acceptor property of CQDs. Consequently, more powerful oxidants participate in photocatalytic reactions. Even so, the holes on the VB cannot thermodynamically oxidize H$_2$O (H$_2$O/•OH 2.34 V vs. NHE) or OH$^-$ (•OH/OH$^-$ 1.99 V vs. NHE) to generate •OH [37,49]. Therefore, the main reactive species, including O$_2$•$^-$ and holes generated under visible/near-infrared light illumination, engage in organic contaminant elimination together, synergistically boosting the enhancement of photocatalytic performance over the CQDs/PbBiO$_2$I nanocomposite.

4. Conclusions

In conclusion, CQDs modified PbBiO$_2$I microspheres have been fabricated via a simple ionothermal method. (i) In this prepared procedure, ionic liquid serves as a high performance template, reactant, and dispersant, which is beneficial for the uniform distribution of CQDs around PbBiO$_2$I microspheres; (ii) the photocatalytic activity for removing organic contaminants is greatly enhanced; and (iii) CQDs modification can successfully reduce charge carrier recombination and accelerate the transformation of photoinduced hole-electron pairs to the catalyst surface. Therefore, more reactive oxygen species can be generated and involved in the elimination of RhB and CIP. This work may offer some insights for constructing a carbon-based material-modified, bismuth-based catalyst, which will be widely welcomed in environmental purification and energy conversion.

Supplementary Materials: The following Supplementary Materials can be downloaded at: https://www.mdpi.com/article/10.3390/ma16031111/s1. Figure S1: The pore size distribution curves of PbBiO$_2$I and 5 wt.% CQDs/PbBiO$_2$I; Figure S2: HR-TEM image of CQDs; Figure S3: The adsorption equilibrium of RhB over various catalysts in the darkness; Figure S4: Cycling runs for RhB degradation over 5 wt.% CQDs/PbBiO$_2$I under visible light irradiation (λ > 400 nm) (a), XRD patterns of 5 wt.% CQDs/PbBiO$_2$I before and after five cycles (b); Figure S5: The decrease of TOC during the photodegradation of RhB (a) and CIP (b) over 5 wt.% CQDs/PbBiO$_2$I; Figure S6: Mott-Schottky plots of PbBiO$_2$I.

Author Contributions: Data curation, investigation, writing—original draft, R.Y.; data curation, investigation, writing—original draft, X.L.; data curation, investigation, H.Z.; data curation, investigation, M.Y.; writing—review & editing, supervision, Z.W.; data curation, investigation, J.Y.; writing—review & editing, resources, B.G.; conceptualization, writing—review & editing, funding acquisition, Q.H. All authors have read and agreed to the published version of the manuscript.

Funding: This work was financially supported by the China Postdoctoral Science Foundation (NO. 2021M691389), Natural Science Foundation of the Jiangsu Higher Education Institutions of China (NO. 22KJB610026), the State Key Laboratory of Pollution Control and Resource Reuse (Project NO. PCRRF20019), and the Science and Technology Innovation Fund Project of Yangzhou University Students (NO. 202211117020Z).

Institutional Review Board Statement: Not applicable.

Informed Consent Statement: Not applicable.

Data Availability Statement: The data presented in this study are available upon request from the corresponding author.

Conflicts of Interest: The authors declare no conflict of interest.

References

1. Long, C.C.; Jiang, Z.X.; Shangguan, J.F.; Qing, T.P.; Zhang, P.; Feng, B. Applications of carbon dots in environmental pollution control: A review. *Chem. Eng. J.* **2021**, *406*, 126848. [CrossRef]
2. Matveev, A.T.; Varlamova, L.A.; Konopatsky, A.S.; Leybo, D.V.; Volkov, I.N.; Sorokin, P.B.; Fang, X.S.; Shtansky, D.V. A new insight into the mechanisms underlying the discoloration, sorption, and photodegradation of methylene blue solutions with and without BNO_x nanocatalysts. *Materials* **2022**, *15*, 8169. [CrossRef]
3. Ashwini, S.; Prashantha, S.C.; Naik, R.; Nagabhushana, H. Enhancement of luminescence intensity and spectroscopic analysis of Eu^{3+} activated and Li^+ charge-compensated Bi_2O_3 nanophosphors for solid-state lighting. *J. Rare Earth* **2019**, *37*, 356–364. [CrossRef]
4. Zhang, J.J.; Zheng, Y.J.; Zheng, H.S.; Jing, T.; Zhao, Y.P.; Tian, J.Z. Porous oxygen-doped g-C_3N_4 with the different precursors for excellent photocatalytic activities under visible light. *Materials* **2022**, *15*, 1391. [CrossRef] [PubMed]
5. Di, J.; Xia, J.X.; Li, H.M.; Guo, S.J.; Dai, S. Bismuth oxyhalide layered materials for energy and environmental applications. *Nano Energy* **2017**, *41*, 172–192. [CrossRef]
6. Manohar, T.; Prashantha, S.C.; Nagaswarupa, H.P.; Naik, R.; Nagabhushana, H.; Anantharaju, K.S.; Vishnu Mahesh, K.R.; Premkumar, H.B. White light emitting lanthanum aluminate nanophosphor: Near ultra violet excited photoluminescence and photometric characteristics. *J. Lumin.* **2017**, *190*, 279–288. [CrossRef]
7. Hu, Q.S.; Dong, J.T.; Chen, Y.; Yi, J.J.; Xia, J.X.; Yin, S.; Li, H.M. In-situ construction of bifunctional MIL-125(Ti)/BiOI reactive adsorbent/photocatalyst with enhanced removal efficiency of organic contaminants. *Appl. Surf. Sci.* **2022**, *583*, 152423. [CrossRef]
8. Wang, H.; Yuan, X.Z.; Wu, Y.; Zeng, G.M.; Chen, X.H.; Leng, L.J.; Li, H. Synthesis and applications of novel graphitic carbon nitride/metal-organic frameworks mesoporous photocatalyst for dyes removal. *Appl. Catal. B Environ.* **2015**, *174–175*, 445–454. [CrossRef]
9. Yu, Y.T.; Huang, H.W. Coupled adsorption and photocatalysis of g-C_3N_4 based composites: Materials synthesis, mechanism, and environmental applications. *Chem. Eng. J.* **2023**, *453*, 139755. [CrossRef]
10. Yang, Y.; Zhang, C.; Lai, C.; Zeng, G.M.; Huang, D.L.; Cheng, M.; Wang, J.J.; Chen, F.; Zhou, C.Y.; Xiong, W.P. BiOX (X=Cl, Br, I) photocatalytic nanomaterials: Applications for fuels and environmental management. *Adv. Colloid Interface Sci.* **2018**, *254*, 76–93. [CrossRef]
11. Ye, L.Q.; Su, Y.R.; Xie, H.Q.; Zhang, C. Recent advances in BiOX (X=Cl, Br and I) photocatalysts: Synthesis, modification, facet effects and mechanisms. *Environ. Sci. Nano* **2014**, *1*, 90–112. [CrossRef]
12. Zeng, L.; Zhe, F.; Wang, Y.; Zhang, Q.L.; Zhao, X.Y.; Hu, X.; Wu, Y.; He, Y.M. Preparation of interstitial carbon doped BiOI for enhanced performance in photocatalytic nitrogen fixation and methyl orange degradation. *J. Colloid Interface Sci.* **2019**, *539*, 563–574. [CrossRef] [PubMed]
13. Ye, L.Q.; Jin, X.L.; Ji, X.X.; Liu, C.; Su, Y.R.; Xie, H.Q.; Liu, C. Facet-dependent photocatalytic reduction of CO_2 on BiOI nanosheets. *Chem. Eng. J.* **2016**, *291*, 39–46. [CrossRef]
14. Guo, J.Y.; Li, X.; Liang, J.; Yuan, X.Z.; Jiang, L.B.; Yu, H.B.; Sun, H.B.; Zhu, Z.Q.; Ye, S.J.; Tang, N.; et al. Fabrication and regulation of vacancy-mediated bismuth oxyhalide towards photocatalytic application: Development status and tendency. *Coord. Chem. Rev.* **2021**, *443*, 214033. [CrossRef]
15. Gao, P.; Yang, Y.N.; Yin, Z.; Kang, F.X.; Fan, W.; Sheng, J.Y.; Feng, L.; Liu, Y.Z.; Du, Z.W.; Zhang, L.Q. A critical review on bismuth oxyhalide based photocatalysis for pharmaceutical active compounds degradation: Modifications, reactive sites, and challenges. *J. Hazard. Mater.* **2021**, *412*, 125186. [CrossRef] [PubMed]
16. Huang, Y.C.; Li, H.B.; Fan, W.J.; Zhao, F.Y.; Qiu, W.T.; Ji, H.B.; Tong, Y.X. Defect engineering of bismuth oxyiodide by IO_3^- doping for increasing charge transport in photocatalysis. *ACS Appl. Mater. Interfaces* **2016**, *8*, 27859–27867. [CrossRef]
17. Xiong, J.; Song, P.; Di, J.; Li, H.M. Bismuth-rich bismuth oxyhalides: A new opportunity to trigger high-efficiency photocatalysis. *J. Mater. Chem. A* **2020**, *8*, 21434–21454. [CrossRef]
18. Xiong, J.; Di, J.; Li, H.M. Interface engineering in low-dimensional bismuth-based materials for photoreduction reactions. *J. Mater. Chem. A* **2021**, *9*, 2662–2677. [CrossRef]
19. Chen, L.P.; Li, C.E.; Zhao, Y.; Wu, J.; Li, X.K.; Qiao, Z.W.; He, P.; Qi, X.M.; Liu, Z.H.; Wei, G.Q. Construction 3D Bi/$Bi_4O_5I_2$ microspheres with rich oxygen vacancies by one-pot solvothermal method for enhancing photocatalytic activity on mercury removal. *Chem. Eng. J.* **2022**, *425*, 131599. [CrossRef]
20. Suzuki, H.; Kunioku, H.; Higashi, M.; Tomita, O.; Kato, D.; Kageyama, H.; Abe, R. Lead bismuth oxyhalides $PbBiO_2X$ (X = Cl, Br) as visible-light-responsive photocatalysts for water oxidation: Role of lone-pair electrons in valence band engineering. *Chem. Mater.* **2018**, *30*, 5862–5869. [CrossRef]
21. Olchowka, J.; Kabbour, H.; Colmont, M.; Adlung, M.; Wickleder, C.; Mentre, O. ABO_2X (A = Cd, Ca, Sr, Ba, Pb; X = halogen) sillen X1 series: Polymorphism versus optical properties. *Inorg. Chem.* **2016**, *55*, 7582–7592. [CrossRef]
22. Lee, A.H.; Wang, Y.C.; Chen, C.C. Composite photocatalyst, tetragonal lead bismuth oxyiodide/bismuth oxyiodide/graphitic carbon nitride: Synthesis, characterization, and photocatalytic activity. *J. Colloid Interface Sci.* **2019**, *533*, 319–332. [CrossRef] [PubMed]
23. Liu, F.Y.; Lin, J.H.; Dai, Y.M.; Chen, L.W.; Huang, S.T.; Yeh, T.W.; Chang, J.L.; Chen, C.C. Preparation of perovskites $PbBiO_2I/PbO$ exhibiting visible-light photocatalytic activity. *Catal. Today* **2018**, *314*, 28–41. [CrossRef]

24. Yu, H.J.; Shi, R.; Zhao, Y.F.; Waterhouse, G.I.N.; Wu, L.Z.; Tung, C.H.; Zhang, T.R. Smart utilization of carbon dots in semiconductor photocatalysis. *Adv. Mater.* **2016**, *28*, 9454–9477. [CrossRef] [PubMed]
25. Lim, S.Y.; Shen, W.; Gao, Z.Q. Carbon quantum dots and their applications. *Chem. Soc. Rev.* **2015**, *44*, 362–381. [CrossRef]
26. Hu, Q.S.; Ji, M.X.; Di, J.; Wang, B.; Xia, J.X.; Zhao, Y.P.; Li, H.M. Ionic liquid-induced double regulation of carbon quantum dots modified bismuth oxychloride/bismuth oxybromide nanosheets with enhanced visible-light photocatalytic activity. *J. Colloid Interface Sci.* **2018**, *519*, 263–272. [CrossRef]
27. Wang, B.Y.; Song, H.Q.; Qu, X.L.; Chang, J.B.; Yang, B.; Lu, S.Y. Carbon dots as a new class of nanomedicines: Opportunities and challenges. *Coord. Chem. Rev.* **2021**, *442*, 214010. [CrossRef]
28. Chung, S.; Revia, R.A.; Zhang, M.Q. Graphene quantum dots and their applications in bioimaging, biosensing, and therapy. *Adv. Mater.* **2021**, *33*, 1904362. [CrossRef]
29. Di, J.; Xia, J.X.; Ji, M.X.; Wang, B.; Yin, S.; Zhang, Q.; Chen, Z.G.; Li, H.M. Carbon quantum dots modified BiOCl ultrathin nanosheets with enhanced molecular oxygen activation ability for broad spectrum photocatalytic properties and mechanism insight. *ACS Appl. Mater. Interfaces* **2015**, *7*, 20111–20123. [CrossRef]
30. Wang, B.; Di, J.; Lu, L.; Yan, S.C.; Liu, G.P.; Ye, Y.Z.; Li, H.T.; Zhu, W.S.; Li, H.M.; Xia, J.X. Sacrificing ionic liquid-assisted anchoring of carbonized polymer dots on perovskite-like $PbBiO_2Br$ for robust CO_2 photoreduction. *Appl. Catal. B Environ.* **2019**, *254*, 551–559. [CrossRef]
31. Li, M.; Yin, S.; Wu, T.; Di, J.; Ji, M.X.; Wang, B.; Chen, Y.; Xia, J.X.; Li, H.M. Controlled preparation of $MoS_2/PbBiO_2I$ hybrid microspheres with enhanced visible-light photocatalytic behavior. *J. Colloid Interface Sci.* **2018**, *517*, 278–287. [CrossRef] [PubMed]
32. Xia, J.X.; Di, J.; Li, H.T.; Xu, H.; Li, H.M.; Guo, S.J. Ionic liquid-induced strategy for carbon quantum dots/BiOX (X=Br, Cl) hybrid nanosheets with superior visible light-driven photocatalysis. *Appl. Catal. B Environ.* **2016**, *181*, 260–269. [CrossRef]
33. Di, J.; Xia, J.X.; Ji, M.X.; Wang, B.; Li, X.W.; Zhang, Q.; Chen, Z.G.; Li, H.M. Nitrogen-doped carbon quantum dots/BiOBr ultrathin nanosheets: In situ strong coupling and improved molecular oxygen activation ability under visible light irradiation. *ACS Sustain. Chem. Eng.* **2016**, *4*, 136–146. [CrossRef]
34. Hu, Q.S.; Chen, Y.; Li, M.; Zhang, Y.; Wang, B.; Zhao, Y.P.; Xia, J.X.; Yin, S.; Li, H.M. Construction of NH_2-UiO-66/BiOBr composites with boosted photocatalytic activity for the removal of contaminants. *Colloids Surf. A PhysicoChem. Eng. Asp.* **2019**, *579*, 123625. [CrossRef]
35. Sun, W.Q.; Hu, Q.S.; Wu, T.; Wang, Z.X.; Yi, J.J.; Yin, S. Construction of 0D/3D carbon quantum dots modified $PbBiO_2Cl$ nicrospheres with accelerated charge carriers for promoted visible-light-driven degradation of organic contaminants. *Colloids Surf. A PhysicoChem. Eng. Asp.* **2022**, *642*, 128591. [CrossRef]
36. Ding, P.H.; Di, J.; Chen, X.L.; Ji, M.X.; Gu, K.Z.; Yin, S.; Liu, G.P.; Zhang, F.; Xia, J.X.; Li, H.M. S,N codoped graphene quantum dots embedded in $(BiO)_2CO_3$: Incorporating enzymatic-like catalysis in photocatalysis. *ACS Sustain. Chem. Eng.* **2018**, *6*, 10229–10240. [CrossRef]
37. Xi, Y.M.; Mo, W.H.; Fan, Z.X.; Hu, L.X.; Chen, W.B.; Zhang, Y.; Wang, P.; Zhong, S.X.; Zhao, Y.L.; Bai, S. A mesh-like BiOBr/Bi_2S_3 nanoarray heterojunction with hierarchical pores and oxygen vacancies for broadband CO_2 photoreduction. *J. Mater. Chem. A* **2022**, *10*, 20934–20945. [CrossRef]
38. Li, K.H.; Xiao, Y.; Zhao, Y.C.; Xia, Y.H.; Ding, J.H.; He, Q.G.; Ling, J.; Li, G.L. A metal-free voltammetric sensor for sensitive determination of Rhodamine B using carboxyl-functionalized carbon nanomaterials. *Inorg. Chem. Commun.* **2022**, *145*, 110025. [CrossRef]
39. Wang, J.J.; Tang, L.; Zeng, G.M.; Deng, Y.C.; Dong, H.R.; Liu, Y.N.; Wang, L.L.; Peng, B.; Zhang, C.; Chen, F. 0D/2D interface engineering of carbon quantum dots modified Bi_2WO_6 ultrathin nanosheets with enhanced photocatalytic for full spectrum light utilization and mechanism insight. *Appl. Catal. B Environ.* **2018**, *222*, 115–123. [CrossRef]
40. Xu, L.; Li, H.N.; Yan, P.C.; Xia, J.X.; Qiu, J.X.; Xu, Q.; Zhang, S.Q.; Li, H.M.; Yuan, S.Q. graphitic carbon nitride/BiOCl composites for sensitive photoelectrochemical detection of ciprofloxacin. *J. Colloid Interface Sci.* **2016**, *483*, 241–248. [CrossRef]
41. Yuan, X.Z.; Zhang, J.; Yang, M.; Si, M.Y.; Jiang, L.B.; Li, Y.F.; Yu, H.B.; Zhang, J.; Zeng, G.M. Nitrogen doped carbon quantum dots promoted the construction of Z-scheme system with enhanced molecular oxygen activation ability. *J. Colloid Interface Sci.* **2019**, *541*, 123–132. [CrossRef] [PubMed]
42. Zhang, X.W.; Wang, P.; Lv, X.Y.; Niu, X.Y.; Lin, X.Y.; Zhong, S.X.; Wang, D.M.; Lin, H.J.; Chen, J.R.; Bai, S. Stacking engineering of semiconductor heterojunctions on hollow carbon spheres for boosting photocatalytic CO_2 reduction. *ACS Catal.* **2022**, *12*, 2569–2580. [CrossRef]
43. Wang, A.; Wu, S.J.; Dong, J.L.; Wang, R.X.; Wang, J.W.; Zhang, J.L.; Zhong, S.X.; Bai, S. Interfacial facet engineering on the Schottky barrier between plasmonic Au and TiO_2 in boosting the photocatalytic CO_2 reduction under ultraviolet and visible light irradiation. *Chem. Eng. J.* **2021**, *404*, 127145. [CrossRef]
44. Chen, Q.; Mo, W.H.; Yang, G.D.; Zhong, S.X.; Lin, H.J.; Chen, J.R.; Bai, S. Significantly enhanced photocatalytic CO_2 reduction by surface amorphization of cocatalysts. *Small* **2021**, *17*, 2102105. [CrossRef]
45. Quan, Y.; Wang, B.; Liu, G.P.; Li, H.M.; Xia, J.X. Carbonized polymer dots modified ultrathin $Bi_{12}O_{17}Cl_2$ nanosheets Z-scheme heterojunction for robust CO_2 photoreduction. *Chem. Eng. Sci.* **2021**, *232*, 116338. [CrossRef]
46. Yan, X.W.; Wang, B.; Zhao, J.Z.; Liu, G.P.; Ji, M.X.; Zhang, X.L.; Chu, P.K.; Li, H.M.; Xia, J.X. Hierarchical columnar $ZnIn_2S_4$/$BiVO_4$ Z-scheme heterojunctions with carrier highway boost photocatalytic mineralization of antibiotics. *Chem. Eng. J.* **2023**, *452*, 139271. [CrossRef]

47. Luo, B.; Wu, C.F.; Zhang, F.Z.; Wang, T.T.; Yao, Y.B. Preparation of porous rllipsoidal bismuth oxyhalide microspheres and their photocatalytic performances. *Materials* **2022**, *15*, 6035. [CrossRef] [PubMed]
48. Hu, Q.S.; Di, J.; Wang, B.; Ji, M.X.; Chen, Y.; Xia, J.X.; Li, H.M.; Zhao, Y.P. In-situ preparation of NH_2-MIL-125(Ti)/BiOCl composite with accelerating charge carriers for boosting visible light photocatalytic activity. *Appl. Surf. Sci.* **2019**, *466*, 525–534. [CrossRef]
49. Chen, X.; Liu, G.P.; Xu, X.Y.; Wang, B.; Sun, S.X.; Xia, J.X.; Li, H.M. Oxygen vacancies mediated $Bi_{12}O_{17}Cl_2$ ultrathin nanobelts: Boosting molecular oxygen activation for efficient organic pollutants degradation. *J. Colloid Interface Sci.* **2022**, *609*, 23–32. [CrossRef]

Disclaimer/Publisher's Note: The statements, opinions and data contained in all publications are solely those of the individual author(s) and contributor(s) and not of MDPI and/or the editor(s). MDPI and/or the editor(s) disclaim responsibility for any injury to people or property resulting from any ideas, methods, instructions or products referred to in the content.

Article

TiO$_2$ Nanotubes Decorated with Mo$_2$C for Enhanced Photoelectrochemical Water-Splitting Properties

Siti Nurul Falaein Moridon [1], Khuzaimah Arifin [1,*], Mohamad Azuwa Mohamed [1,2], Lorna Jeffery Minggu [1], Rozan Mohamad Yunus [1] and Mohammad B. Kassim [1,2]

1. Fuel Cell Institute, Universiti Kebangsaan Malaysia, Bangi 43600, Selangor, Malaysia
2. Department of Chemical Sciences, Faculty of Science and Technology, Universiti Kebangsaan Malaysia, Bangi 43600, Selangor, Malaysia
* Correspondence: khuzaim@edu.ukm.my

Abstract: The presence of Ti^{3+} in the structure of TiO$_2$ nanotube arrays (NTs) has been shown to enhance the photoelectrochemical (PEC) water-splitting performance of these NTs, leading to improved results compared to pristine anatase TiO$_2$ NTs. To further improve the properties related to PEC performance, we successfully produced TiO$_2$ NTs using a two-step electrochemical anodization technique, followed by annealing at a temperature of 450 °C. Subsequently, Mo$_2$C was decorated onto the NTs by dip coating them with precursors at varying concentrations and times. The presence of anatase TiO$_2$ and Ti$_3$O$_5$ phases within the TiO$_2$ NTs was confirmed through X-ray diffraction (XRD) analysis. The TiO$_2$ NTs that were decorated with Mo$_2$C demonstrated a photocurrent density of approximately 1.4 mA cm^{-2}, a value that is approximately five times greater than the photocurrent density exhibited by the bare TiO$_2$ NTs, which was approximately 0.21 mA cm^{-2}. The observed increase in photocurrent density can be ascribed to the incorporation of Mo$_2$C as a cocatalyst, which significantly enhances the photocatalytic characteristics of the TiO$_2$ NTs. The successful deposition of Mo$_2$C onto the TiO$_2$ NTs was further corroborated by the characterization techniques utilized. The utilization of field emission scanning electron microscopy (FESEM) allowed for the observation of Mo$_2$C particles on the surface of TiO$_2$ NTs. To validate the composition and optical characteristics of the decorated NTs, X-ray photoelectron spectroscopy (XPS) and UV absorbance analysis were performed. This study introduces a potentially effective method for developing efficient photoelectrodes based on TiO$_2$ for environmentally sustainable hydrogen production through the use of photoelectrochemical water-splitting devices. The utilization of Mo$_2$C as a cocatalyst on TiO$_2$ NTs presents opportunities for the advancement of effective and environmentally friendly photoelectrochemical (PEC) systems.

Keywords: titanium dioxide; anodization; self-doping; cocatalyst; Mo$_2$C

Citation: Moridon, S.N.F.; Arifin, K.; Mohamed, M.A.; Minggu, L.J.; Mohamad Yunus, R.; Kassim, M.B. TiO$_2$ Nanotubes Decorated with Mo$_2$C for Enhanced Photoelectrochemical Water-Splitting Properties. *Materials* 2023, 16, 6261. https://doi.org/10.3390/ma16186261

Academic Editors: Xingwang Zhu and Tongming Su

Received: 9 June 2023
Revised: 6 July 2023
Accepted: 7 July 2023
Published: 18 September 2023

Copyright: © 2023 by the authors. Licensee MDPI, Basel, Switzerland. This article is an open access article distributed under the terms and conditions of the Creative Commons Attribution (CC BY) license (https://creativecommons.org/licenses/by/4.0/).

1. Introduction

The extreme reliance on fossil fuels for energy generation since the industrial revolution has triggered a global energy crisis and various other environmental problems [1,2]. Therefore, reducing energy dependence on fossil fuels through the provision of clean and renewable energy sources is urgent. One alternative is to use clean and green hydrogen (H$_2$) directly produced from water molecules using solar light energy, known as the photoelectrochemical (PEC) process. H$_2$ can be used as fuel in an electrochemical fuel cell device to produce electricity, with pure water as the only byproduct.

In the PEC process, semiconductor materials are employed as photoelectrodes [2]. To date, numerous semiconductor materials, such as TiO$_2$ [3–5], Co$_3$O$_4$ [6,7], WO$_3$ [8], Cu$_2$O [9], and Fe$_2$O$_3$ [10], have been investigated as photoelectrode materials. Among them, TiO$_2$ has garnered considerable attention due to its photoactivity, low cost, excellent chemical stability, and abundance in nature [11]. However, TiO$_2$ can only be stimulated by

UV light, which accounts for only 3–5% of solar energy radiation, because of its large band gap, further causing quick recombination of photoinduced electron-hole pairs as well as inefficient charge carrier separation [11]. Therefore, various methods have been used to improve the PEC performance and photocatalytic activity of TiO_2, including morphology modification [12–14], synthesis of composite heterojunctions with other materials [15], ion doping [16,17], facet engineering [18], and cocatalyst addition [19].

In terms of morphology, TiO_2 nanotube arrays (NTs) have attracted attention due to their high specific surface area, excellent adsorption capacity, and good structural properties for electron transport. Several methods have been employed to fabricate TiO_2 NT photoelectrodes, and electrochemical anodization is considered one of the most promising methods for fabricating a highly ordered NT structure [20]. Although the electrochemical anodization method is considered a promising fabrication method for producing ordered TiO_2 NTs with defect engineering and doping, the PEC performance of fabricated TiO_2 NTs still does not reach satisfactory levels due to their restricted light harvesting and the high resistance at the interface between the nanotubes and the substrate. Therefore, a synergistic approach combining various strategies, such as the formation of heterojunctions with other semiconductor materials or the addition of cocatalyst materials, could prove excellent for obtaining TiO_2 NTs with efficient light harvesting and charge separation for high PEC water-splitting performance. Currently, 2D MXenes have been investigated as promising catalysts or cocatalysts in many applications. The nomenclature "MXene" has been used to represent a group of compounds that includes transition metal carbides, nitrides, and carbonitrides. The nomenclature "MXene" is derived from its chemical composition, wherein the symbol "M" represents a transition metal, "X" signifies carbon and/or nitrogen, and the suffix "ene" refers to its two-dimensional structural arrangement [21]. Among the reported MXene materials, dimolybdenum carbide (Mo_2C) has been reported to show excellent electrocatalytic performance in the hydrogen evolution reaction (HER) [21]. It has an electronic density of states comparable to that of Pt and excellent electrical conductivity. In addition to being used as a catalyst for the HER, Mo_2C has also been investigated for photocatalytic water splitting for hydrogen generation [22]. Shen et al. reported that the Mo_2C/CdS nanocomposite produced a photocurrent density 7.83 times higher than that of pure CdS [23]. Furthermore, Yue et al. reported that dandelion-like TiO_2 nanoparticles with 1% Mo_2C were able to produce H_2 with a production rate of 39.4 mmol h^{-1} g^{-1}, which is 25 times that obtained with pristine TiO_2 [22].

Although Mo_2C has been shown to be a good cocatalyst for TiO_2, the photocatalytic performance of TiO_2 NTs on Ti foil substrates with Mo_2C has yet to be reported. This study presents the effectiveness of TiO_2 NTs with Mo_2C incorporated as a cocatalyst for PEC water-splitting applications. Here, TiO_2 NTs were prepared by two-step electrochemical anodization, and Mo_2C was inserted by dip coating the NTs into Mo_2C precursors of various concentrations for various dipping times. Our findings indicated that the combined effect of multiple PEC improvement strategies could offer a versatile and systematic way to overcome the intrinsic and extrinsic limitations of TiO_2 NTs for PEC water-splitting applications.

2. Materials and Methods

2.1. Materials

Titanium foil (0.127 nm thickness, obtained from Sigma Aldrich, St. Louis, MI, USA), molybdenum carbide (Mo_2C) (99%, obtained from Sigma Aldrich), Pt mesh (99%, obtained from Sigma Aldrich), ethylene glycol (analytical grade, obtained from Merck, Darmstadt, Germany), ammonium fluoride NH_4F (analytical grade, obtained from Merck, Germany), distilled water, ethanol (analytical grade, obtained from QReC, Kuala Lumpur, Malaysia) and sodium sulfate (Na_2SO_4) were used. All chemicals were used as received from the manufacturer without additional purification.

2.2. Fabrication of TiO$_2$ NTs and TiO$_2$ NTs Decorated with Mo$_2$C

The TiO$_2$ NTs were fabricated using a multiple anodization technique [24]. First, cleaned Ti foil (1.5 cm × 1.5 cm) was used as the anode, and Pt foil was used as the counter electrode connected to a power supply at a voltage of 50 V for one hour. Ethylene glycol containing 0.3 vol. % NH$_4$F and 2 vol. % distilled water was used as the electrolyte. The anodized film was then sonicated in a mixture of ethanol and distilled water (1:1) for 5 min to clean dirt away from the openings of the grown nanotubes. Subsequently, the Ti foil underwent a second anodization process for 30 min in the same electrolyte at the same voltage. Then, ethanol and distilled water were used to flush the samples. The anodized samples were then annealed at 450 °C for 3 h at a ramping rate of 2 °C/min. The best TiO$_2$ NTs that achieved the highest photocurrent were then dip coated in an ethanol/distilled water mixture containing highly dispersed Mo$_2$C at four different concentrations of 5 g/L, 10 g/L, 15 g/L and 20 g/L, and the obtained samples were labeled S-1, S-2, S-3 and S-4, respectively.

2.3. Characterization

X-ray diffraction (XRD) patterns were acquired via a Bruker D-8 Advance (Ettlingen, Germany, Equipment sourced by Bruker Malaysia), and X-ray photoelectron spectroscopy (XPS) was performed using a Kratos/Shimadzu instrument (model: Axis Ultra DLD) (Milton Keynes, UK, Equipment sourced by Shimadzu Malaysia) to determine the chemical phases present in the crystalline substances. The XRD patterns were analyzed using X'Pert HighScore software (Version 2.2b). To investigate the topographic nature of the surface, field emission scanning electron microscopy (FESEM) was carried out using a Zeiss Merlin Compact microscope (Oberkochen, Germany, Equipment sourced by Zeiss Malaysia). The optical properties were analyzed using a Perkin Elmer ultraviolet/visible/near-infrared spectrophotometer (UV–VIS-NIR) (model: Lambda 950) (Waltham, MA, USA, Equipment sourced by Perkin Elmer Malaysia).

2.4. PEC Property Measurements

An Ametek Versastat 4 was used to carry out the PEC analysis. An exposed area of 1 cm^2 was employed for testing the thin films that served as the working electrode in a PEC cell. The counter electrode consisted of a platinum wire; the reference electrode was a Ag/AgCl electrode. The counter electrode measured the potential difference between the two electrodes. In these experiments, 0.5 M Na$_2$SO$_4$ (pH 6.7) was used as the electrolyte. The current density on the thin film surfaces was measured in the dark and under solar AM 1.5 illumination using a xenon lamp (Oriel with an intensity of 100 mW cm^{-2}). Linear sweep voltammetry (LSV) was conducted from 0 to +1.0 V versus Ag/AgCl in 0.5 M Na$_2$SO$_4$ at a scan rate of 5 mV s^{-1}. To obtain a deeper understanding of the charge transport behavior shown by the synthesized photoanodes, Mott-Schottky analysis was performed at 1 kHz. This allowed for calculation of the charge carrier densities, as well as the conduction band (CB). The electrochemical impedance spectra (EIS) Nyquist plots were constructed by utilizing 10 mV sinusoidal perturbations at a frequency of 100 kHz.

3. Results and Discussion

3.1. Physical Characterization of TiO$_2$ NTs

To thoroughly investigate the growth of TiO$_2$ NTs, a detailed analysis was conducted comprising analysis of the morphology, crystal phase, crystallinity, and optical properties. Figure 1 shows the FESEM results that capture the microstructure of the TiO$_2$ NT sample.

Based on the FESEM images, the TiO$_2$ NT sample clearly exhibited non-interconnected single tubes (Figure 1a). The diameter of the TiO$_2$ NTs was ~151–160 nm. Figure 1b displays cross-sectional views of TiO$_2$ NTs. The length of the TiO$_2$ NTs was ~3.4–3.8 μm.

Next, XRD analysis was carried out to identify the phases and determine the chemical composition. Figure 2a shows the XRD patterns of the Ti foil substrate and TiO$_2$ NT sample.

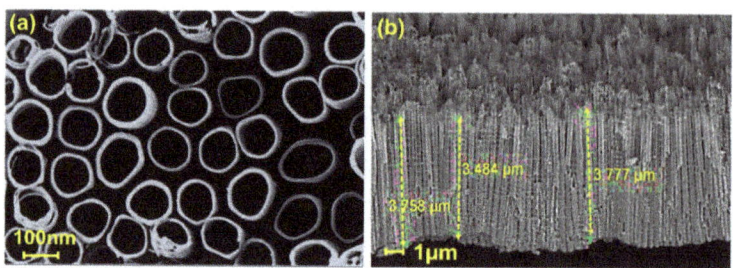

Figure 1. FESEM images of the (**a**) surface morphology and (**b**) cross-section of TiO$_2$ NTs.

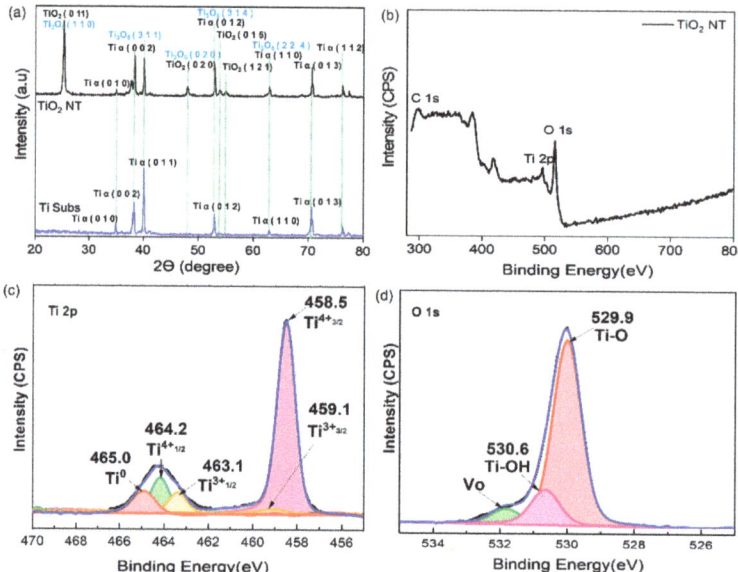

Figure 2. (**a**) XRD patterns and (**b**) XPS survey spectrum of TiO$_2$ NTs; (**c**) Ti 2p and (**d**) O 1s spectra of TiO$_2$ NTs.

The spectra show Ti alpha diffraction peaks representing the Ti foil at 35.01°, 38.27°, 40.10°, 52.99°, 62.89°, 70.66° and 76.18°, corresponding to the (0 1 0), (0 0 2), (0 1 1), (0 1 2), (0 1 3), (1 1 0) and (1 1 2) planes, respectively. However, when the sample was anodized, anatase peaks appeared at 25.28°, 47.83°, 53.09° and 55.02°, corresponding to the (0 1 1), (0 2 0), (0 1 5) and (1 2 1) planes, respectively. The intensity of the Ti alpha peaks was reduced because the surface of the titanium substrate was oxidized during the anodization process, which resulted in the formation of a layer of anatase titanium oxide. The XRD patterns found in this study are similar to those found in Quiroz et al., (2015), which contain a tri-titanium pentoxide (Ti$_3$O$_5$) phase [25]. Based on the XRD library patterns, the Ti alpha peaks overlapped with the Ti$_3$O$_5$ peaks at 38.27°, 52.99° and 70.66° and overlapped with some anatase peaks at 25.28° and 47.83°.

Figure 2b shows the results of the XPS analysis conducted to verify the presence of Ti$_3$O$_5$ in the TiO$_2$ NT samples. The peaks at ~464 eV and 458 eV correspond to Ti 2p, and the Ti 2p peaks of Ti^{3+} were observed at binding energies of 463.1 eV (Ti^{3+} 2p$^{3/2}$) and 459.1 eV (Ti^{4+} 2p$^{1/2}$). The Ti 2p peaks of Ti^{4+} appeared at 458.5 eV (Ti^{4+} 2p$^{3/2}$) and 464.2 eV (Ti^{4+} 2p$^{1/2}$). The XPS spectra of TiO$_2$ NTs revealed a modest shift in position and a change in the size of the original peaks of TiO$_2$ NTs from those in a previous study after self-doping with Ti^{3+}. The observed peak shift indicates that the self-doping of Ti with Ti^{3+} affected its electronic state. As a consequence of this process, some of the Ti^{4+} ions in the lattice are believed to have been replaced by Ti^{3+} ions. Furthermore, the decrease in the Ti^{4+} peaks suggests that there was less TiO_2 present in the sample. The creation of oxygen vacancies in the surface layer during the multistep anodization procedure can be deduced to be responsible for the diminishing area of the Ti^{4+} species peak [26]. Furthermore, the peaks at binding energies of 529.9 eV and 530.6 eV were ascribed to lattice oxygen for TiO_2 NTs. The O 1s spectra provided more evidence demonstrating that more oxygen defects were present in the TiO_2 NTs [27]. The percentage of atomic oxygen vacancies for TiO_2 NTs was 6.25%. The XPS results agreed with the XRD results, suggesting that more Ti^{3+} was produced during the anodization process.

Figure 3a shows the UV–Vis absorption spectra over the wavelength range from 250 to 800 nm, showing that the TiO_2 NTs had higher absorption in the UV range. Next, the band gap of the sample was calculated using Kubelka–Munk theory, and the value for the TiO_2 NTs was 3.15 eV [3].

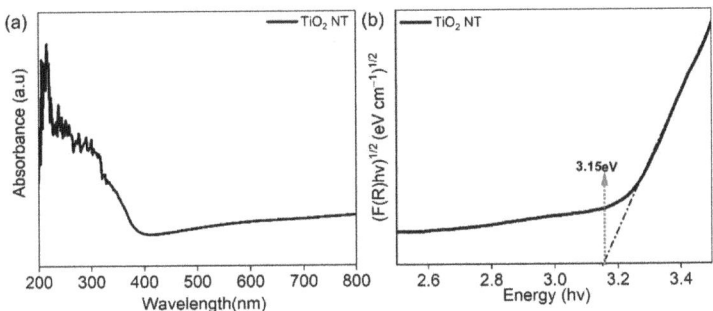

Figure 3. (a) Absorption spectra and (b) band gap determination by Kubelka–Munk plot analysis.

3.2. Physical Characterization of Mo_2C as a Cocatalyst Decorated on TiO_2 NTs

Mo_2C was added to TiO_2 NTs using a dip coating technique. Dip coating is a simple, dependable, and robust process that can be used to cover almost any substrate material by immersing it in a solution and then removing it to drip dry to form a conformal coating.

FESEM analysis was performed to study the effect of different concentrations of the Mo_2C precursor on the morphology of TiO_2 NTs. Figure 4(a1–d3) shows micrographs of TiO_2 NTs for different precursor Mo_2C concentrations.

As shown in Figure 4, increasing the concentration resulted in an increasing deposition amount. Figure 4(a1,b1) illustrate that the distribution of Mo_2C on the surface of the TiO_2 NTs was not uniform. Figure 4(c1) shows that Mo_2C was well distributed inside the TiO_2 NT tubes at a concentration of 15 g/L, which was crucial for the photoelectrode activity. Increasing the concentration to 20 g/L resulted in large nanoclusters of Mo_2C blocking most TiO_2 NTs (Figure 4(d1)). Next, Figure 4(a2–d2) illustrates cross-sectional views of the decorated Mo_2C on the TiO_2 NTs. The length of the TiO_2 NTs increased as the concentration of Mo_2C increased; this finding may support the idea that Mo_2C is distributed on the upper openings of the TiO_2 NTs. In addition, an image of the cross-section of sample S-3 can be seen in the inset of Figure 4(c2); this image demonstrates that Mo_2C decorated the outside wall of the tubes. Energy-dispersive X-ray spectroscopy (EDX) mapping and cross-section analysis were performed to determine the distribution of Mo_2C in sample S-3 with a concentration of 15 g/L. The findings are shown in Figure 4(a3–d3), suggesting that Mo_2C was equally dispersed over the TiO_2 NT surface and in the interstices. This indicates that Mo_2C was efficiently distributed across the sample, resulting in a uniform distribution.

The XRD patterns of TiO_2 NTs after deposition of Mo_2C are presented in Figure 5.

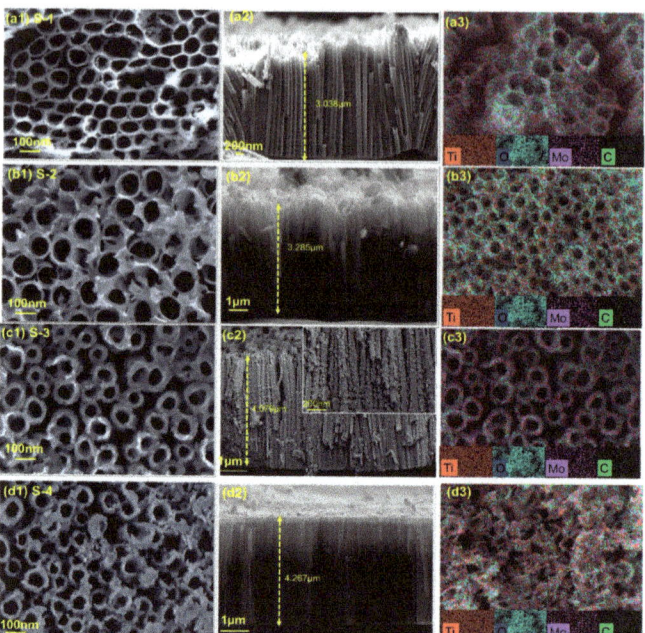

Figure 4. FESEM images of the surface and cross-section, as well as EDX mapping of samples at different concentrations (**a1–a3**) S-1, (**b1–b3**) S-2, (**c1–c3**) S-3 and (**d1–d3**) S-4.

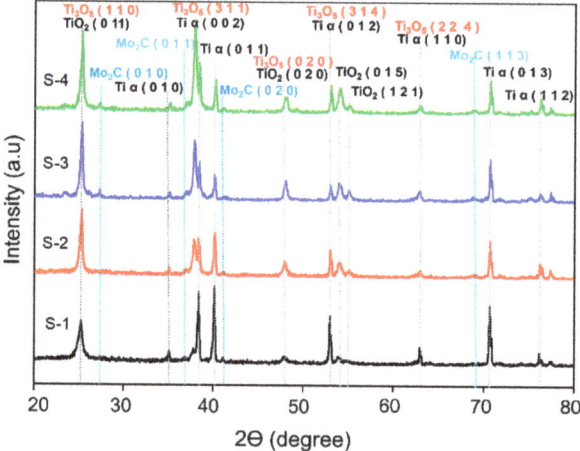

Figure 5. XRD patterns of S-1, S-2, S-3 and S-4.

The diffraction peaks at 25.1°, 37.8°, 48.0°, 52.9° and 62.3° in the pattern of bare TiO_2 NTs were identified as corresponding to the planes of anatase TiO_2 and Ti_3O_5 phases (JCPDS nos. 98-009-4632 and 98-007-1775). Upon deposition of Mo_2C nanoparticles, the patterns displayed additional peaks at 27.2°, 37.2°, 38.3°, 41.1° and 68.8°, which correspond to the standard diffraction peaks of Mo_2C (JCPDS no. 98-006-1705). These findings are consistent with the FESEM results, in which increasing the concentration of Mo_2C leads to increases in the amount of deposited Mo_2C and the intensity of the peaks. The observed peaks indicate the successful deposition of Mo$_2$C nanoparticles onto the TiO$_2$ NT surface, which can potentially enhance the PEC properties of the material.

The UV–Vis absorption spectra of Mo$_2$C/TiO$_2$ NTs with various concentrations are presented in Figure 6a. The TiO$_2$ NTs loaded with Mo$_2$C nanoparticles exhibited a broader absorption in the visible light region (450 nm to 800 nm) compared to pure TiO$_2$ NTs. Among the samples, S-3 showed the highest absorption and thus had the highest PEC activity. The band gap of the samples is displayed in Figure 6b, revealing that the Mo$_2$C-loaded TiO$_2$ NTs had a lower band gap than the pure TiO$_2$ NTs.

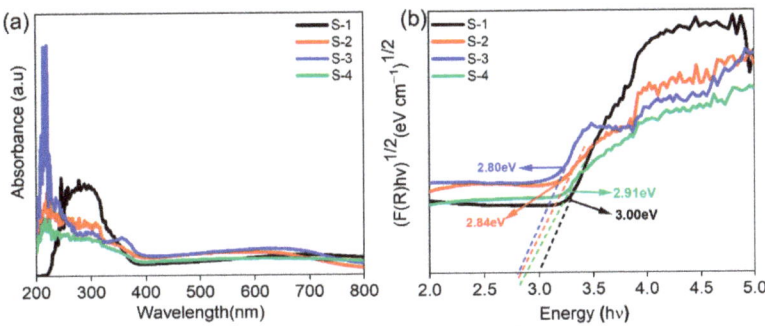

Figure 6. (a) UV–Vis absorption spectra and (b) Kubelka–Munk plots for band gap determination of S-1, S-2, S-3 and S-4.

The band gap of S-3 was determined to be ~2.80 eV, the smallest among the samples. Previous reports suggest that higher absorption in the visible region corresponds to better PEC water-splitting activity. Fine-tuning the band gap and band locations is necessary when creating visible light-responsive photocatalysts for hydrogen production.

3.3. PEC Properties of TiO$_2$ NTs and Mo$_2$C as a Cocatalyst Decorated on TiO$_2$ NTs

Mo$_2$C has garnered interest in the field of PEC applications due to its exceptional stability in challenging environments and remarkable electrical conductivity, making it a promising cocatalyst for such purposes. Mo$_2$C applied onto TiO$_2$ NTs has been observed to function as an electron transfer mediator, thereby facilitating the separation of photo-generated charge carriers and resulting in an improved overall PEC performance of TiO$_2$ NTs. The hybridization of TiO$_2$ NTs with Mo$_2$C has been found to exhibit a synergistic effect, whereby the distinctive characteristics of each material are combined to overcome the constraints of TiO$_2$ NTs. This discussion explores the PEC characteristics of TiO$_2$ NTs and TiO$_2$ NTs that have been decorated with Mo$_2$C as a cocatalyst.

The correlation between the TiO$_2$ NTs with Mo$_2$C as a cocatalyst and the PEC behavior of TiO$_2$ NTs was investigated by chronoamperometric measurements under light chopping, and the test was carried out in 0.5 M Na$_2$SO$_4$ at a bias potential of 0.7 V vs. Ag/AgCl in the presence and absence of illumination (light-off and light-on). The concentration of Mo$_2$C varied, with values of 5 g/L (S-1), 10 g/L (S-2), 15 g/L (S-3) and 20 g/L (S-4), as shown in Figure 7a.

All samples demonstrated a satisfactory photocurrent density as well as a good level of stability after 900 s. The photocurrent density of the TiO$_2$ NT sample was determined to be 0.21 mA cm^{-2}, and this value of photocurrent increased approximately one-fold when compared to the value of pure TiO$_2$ NT due to the presence of oxygen vacancy defects, as reported in previous work [28–30]. Meanwhile, the photocurrent densities produced by samples S-1, S-2 and S-4 were similar to that of bare TiO$_2$ NTs. The significant photocurrent density produced by sample S-3 had a value of ~1.4 mA cm^{-2}, nearly five times higher than that of bare TiO$_2$ NTs.

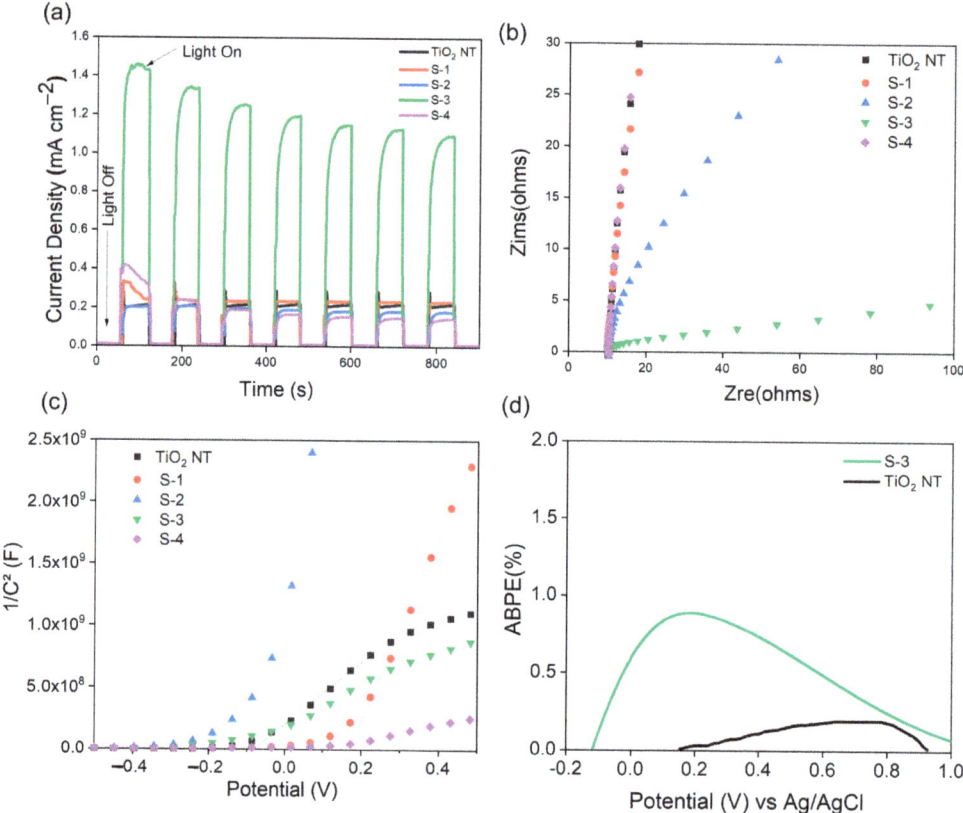

Figure 7. (a) Photocurrent density and stability under light chopping of TiO$_2$ NTs and S1–S4; (b) EIS spectra of the TiO$_2$ NT and S1–S4 samples; (c) N-type Mott–Schottky plots of TiO$_2$ NTs and S1–S4; (d) ABPE % of TiO$_2$ NTs and S-3.

EIS is a trustworthy method for investigating the charge transfer and recombination rate at semiconductor electrolyte interfaces, where "Zre" is the real portion and "Zim" is the imaginary part. Due to the relationship between the arc of the circle and the charge transfer resistance, the Nyquist plots provide sufficient information on the charge transfer. In the Nyquist plot, a smaller arc indicates greater charge carrier separation and higher charge transfer efficiency (conductivity) [30]. Figure 7b shows that sample S-3 has the smallest semicircle radius. This indicates that photogenerated electron-hole pairs are more effectively separated and that electrons may more easily cross the valence band in response to a relatively low-energy excitation. As a consequence, the charge transfer in S-3 is enhanced, proving the presence of a large separation between the holes and electrons.

The Mott–Schottky (M–S) curves of all photoelectrode samples are shown in Figure 7c. The flat band potential (E_{fb}) was estimated by projecting the linear section of the plots onto the potential axis. In addition, the donor density (N_D) was calculated using the slope of the M–S curves and Equation (1) obtained from previous work [25,30]. The N_D and E_{fb} values determined are reported in Table 1, and the E_{fb} values of sample S-3 are less negative, which indicates an upward shift of the Fermi level [30,31].

$$N_D = \left(\frac{2}{e\varepsilon\varepsilon_o A^2}\right)\left[\frac{d\left(\frac{1}{C^2}\right)}{dE}\right]^{-1} \qquad (1)$$

where $\left[\frac{d\left(\frac{1}{C^2}\right)}{dE}\right]$ is the slope of the tangent line in the Mott–Schottky plot, e is the electron charge, ε is the dielectric constant of the TiO_2 film, ε^0 is the vacuum permittivity, and A is the surface area of the TiO_2 NT thin film electrode.

Table 1. Flat band potential (E_{fb}) and donor density (N_D) of TiO_2 NTs and S-3.

Sample	N_D ($\times 10^9$ cm^{-3})	E_{fb} (V)	Band Gap (eV)
TiO_2 NT	3.7918	−0.08	3.17
S-3	3.2629	−0.12	2.80

The efficiency with which a PEC cell converts light into electricity is quantified by the applied bias photon-to-current efficiency (ABPE). To assess how well a PEC cell converts solar energy into a usable form, the ABPE test is crucial. This method is useful for comparing the efficiency of various materials in converting light into electricity and for determining the efficacy of individual materials. To further improve the PEC cell design, the ABPE may be utilized to investigate how an applied bias affects the cell output. In conclusion, the ABPE is a useful metric for assessing PEC cell performance, yielding crucial data for improving future solar energy conversion technologies. Figure 7d shows the ABPE measurements of TiO_2 NTs and S-3. The ABPE value was calculated using Equation (2) [32,33]:

$$ABPE \% = [J_p (E_0 \text{ rev} - E_{app})/I_0] \times (100) \quad (2)$$

where J_p is the photocurrent density (mA/cm^2), I_0 is the illumination intensity (mW/cm^2), E_0 rev is the standard reversible potential for water splitting (1.23 V), and E_{app} is the applied potential. The highest ABPE value for TiO_2 NTs is 0.19% at 0.8 V, while that of S-3 is 0.89% at 0.2 V. Next, the solar-to-hydrogen efficiency (STH) was also calculated for PEC water splitting with a visible light source of irradiance 100 mW cm^{-2} using Equation (3) [32,33]:

$$STH(\%) = J_P \frac{1.23 - V_{App}}{P} \times 100 \quad (3)$$

where J_p is the photocurrent density (mA/cm^2), V_{app} is the applied potential, and P is the intensity of the light source. The maximum STH % for TiO_2 NTs is 0.05%, while that for Mo_2C/TiO_2 NTs (S-3) is 0.32%, as shown in Table 2.

Table 2. Measured parameters of the PEC cell.

Sample Details	Photocurrent Density (mA cm^{-2}) at 1 V vs. Ag/AgCl	Solar to Hydrogen Conversion Efficiency, (η %)
TiO_2 NT	0.23	0.05
S-3	1.4	0.32

The increased PEC properties of TiO_2 NTs decorated with Mo_2C are further explained by the proposed mechanism shown in Figure 8.

In this work, oxygen vacancies produce localized electronic states inside the energy gap, which correspond to Ti^{3+} species, which are in the mid-band gap. These restricted states may serve as traps or recombination sites for electrons and holes created by photons. The Mo_2C deposited onto the TiO_2 photoanode may serve as a cocatalyst to accelerate the HER and oxygen evolution reaction (OER), leading to a higher efficiency in PEC water splitting. Moreover, the Mo_2C cocatalyst may minimize the energy barrier for charge transfer and avoid charge recombination, resulting in an increased photocurrent and better stability.

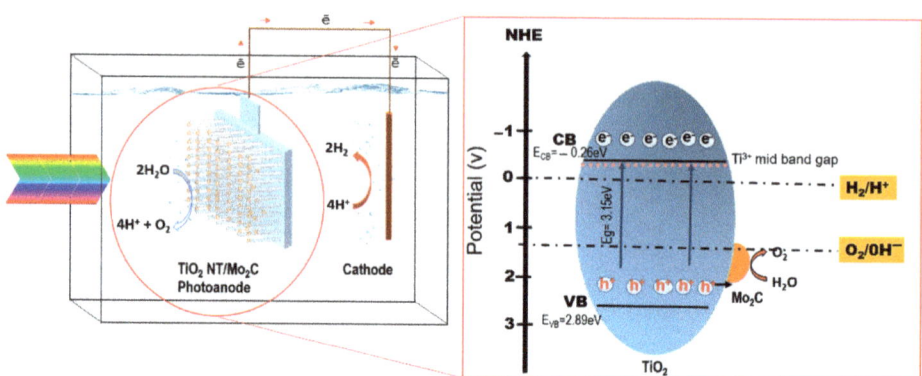

Figure 8. Proposed mechanism of Mo$_2$C as a cocatalyst in TiO$_2$ NTs self-doped by Ti^{3+}.

Upon irradiation with solar light, the TiO$_2$ electrons may be excited into the CB during the process, and electrons move from the TiO$_2$ photoanode to the cathode, resulting in the reduction of protons to hydrogen at the cathode. Mo$_2$C works as a cocatalyst, boosting the oxidation of water molecules by facilitating the flow of holes from the TiO$_2$ photoanode surface to the water molecules. The Mo$_2$C catalyst helps reduce the energy barrier for the OER, which leads to a decrease in the overpotential needed to drive the reaction. This may lead to an increase in the rate of the reaction and consequently a greater efficiency of the entire water-splitting process.

The method for achieving very uniform Mo$_2$C nanoparticles distributed on both the inside and outside vertically aligned TiO$_2$ NTs via the dip coating deposition process has great promise. These novel interactions of Mo$_2$C/TiO$_2$ NTs dramatically enhance both the light absorption and the PEC activity under visible light illumination by 5-fold compared to those of pure TiO$_2$ NTs. These characterization results are consistent with an improved Mo$_2$C/TiO$_2$ NT performance, although the result is not as incredible as that reported by previous researchers on the photocatalytic effect of Mo$_2$C for pristine powder TiO$_2$ [34]. There is still much room to improve the performance, such as by optimizing the length and tube diameter of TiO$_2$ NTs so that they are suitable for Mo$_2$C diffusion, as well as the Mo$_2$C deposition methods.

4. Conclusions

The implementation of Mo$_2$C as a cocatalyst led to a significant enhancement in the photocurrent density of TiO$_2$ NTs. The photocurrent density of the modified TiO$_2$ NTs was observed to be significantly enhanced by a factor of five when compared to the unmodified TiO$_2$ NTs. The implications of these findings are of great importance for the further development of environmentally sustainable PEC water-splitting technologies, specifically in the domain of hydrogen production. The successful decoration of TiO$_2$ NTs with Mo$_2$C was confirmed through XRD and FESEM analysis. Moreover, the incorporation of Mo$_2$C has been observed to significantly decrease the band gap and enhance the light absorption capabilities of TiO$_2$ NTs. Significantly, it was determined that the most favorable concentration of Mo$_2$C was 15 g/L (S-3), exhibiting the highest photoelectrochemical efficiency. In general, this study provides insights into the possible utilization of Mo$_2$C-decorated TiO$_2$ NTs, specifically the Ti$_3$O$_5$ phase, for enhancing the effectiveness and efficacy of photoelectrochemical systems. The utilization of Mo$_2$C as a cocatalyst in PEC water-splitting applications is highly promising due to several factors. These include the achievement of enhanced photocurrent density, the confirmation of Mo$_2$C's presence through X-ray diffraction (XRD) analysis, the reduction of the band gap, and the determination of an optimal concentration.

Author Contributions: Conceptualization, K.A.; methodology, K.A. and S.N.F.M.; validation, S.N.F.M.; formal analysis, S.N.F.M.; investigation, S.N.F.M.; writing—original draft preparation, S.N.F.M.; writing—review and editing, K.A., L.J.M., R.M.Y., M.A.M. and M.B.K.; supervision, K.A.; project administration, K.A. All authors have read and agreed to the published version of the manuscript.

Funding: This work was supported by the Ministry of Education, Malaysia, through FRGS/1/2019/STG01/UKM/03/2.

Institutional Review Board Statement: Not applicable.

Informed Consent Statement: Informed consent was obtained from all subjects involved in the study.

Data Availability Statement: The authors do not have permission to share data.

Conflicts of Interest: The authors declare no conflict of interest.

References

1. Mahmoud, M.; El-Kalliny, A.S.; Squadrito, G. Stacked titanium dioxide nanotubes photoanode facilitates unbiased hydrogen production in a solar-driven photoelectrochemical cell powered with a microbial fuel cell treating animal manure wastewater. *Energy Convers. Manag.* **2022**, *254*, 115225. [CrossRef]
2. Chen, S.; Huang, D.; Xu, P.; Xue, W.; Lei, L.; Cheng, M.; Wang, R.; Liu, X.; Deng, R. Semiconductor-based photocatalysts for photocatalytic and photoelectrochemical water splitting: Will we stop with photocorrosion? *J. Mater. Chem. A* **2020**, *8*, 2286–2322. [CrossRef]
3. Arifin, K.; Yunus, R.M.; Minggu, L.J.; Kassim, M.B. Improvement of TiO_2 nanotubes for photoelectrochemical water splitting: Review. *Int. J. Hydrogen Energy* **2021**, *46*, 4998–5024. [CrossRef]
4. Lin, S.; Ren, H.; Wu, Z.; Sun, L.; Zhang, X.-G.; Lin, Y.-M.; Zhang, K.H.L.; Lin, C.-J.; Tian, Z.-Q.; Li, J.-F. Direct Z-scheme WO_{3-x} nanowire-bridged TiO_2 nanorod arrays for highly efficient photoelectrochemical overall water splitting. *J. Energy Chem.* **2021**, *59*, 721–729. [CrossRef]
5. Arifin, K.; Daud, W.R.W.; Kassim, M.B. Optical and photoelectrochemical properties of a TiO_2 thin film doped with a ruthenium–tungsten bimetallic complex. *Ceram. Int.* **2013**, *39*, 2699–2707. [CrossRef]
6. Moridon, S.N.F.; Salehmin, M.I.; Mohamed, M.A.; Arifin, K.; Minggu, L.J.; Kassim, M.B. Cobalt oxide as photocatalyst for water splitting: Temperature-dependent phase structures. *Int. J. Hydrogen Energy* **2019**, *44*, 25495–25504. [CrossRef]
7. Wang, C.; Zhang, X.; Yuan, B.; Wang, Y.; Sun, P.; Wang, D.; Wei, Y.; Liu, Y. Multi-heterojunction photocatalysts based on WO_3 nanorods: Structural design and optimization for enhanced photocatalytic activity under visible light. *Chem. Eng. J.* **2013**, *237*, 29–37. [CrossRef]
8. Salehmin, M.N.I.; Minggu, L.J.; Mark-Lee, W.F.; Mohamed, M.A.; Arifin, K.; Jumali, M.H.H.; Kassim, M.B. Highly photoactive Cu_2O nanowire film prepared with modified scalable synthesis method for enhanced photoelectrochemical performance. *Sol. Energy Mater. Sol. Cells* **2018**, *182*, 237–245. [CrossRef]
9. Ng, K.H.; Minggu, L.J.; Mark-Lee, W.F.; Arifin, K.; Jumali, M.H.H.; Kassim, M.B. A new method for the fabrication of a bilayer WO_3/Fe_2O_3 photoelectrode for enhanced photoelectrochemical performance. *Mater. Res. Bull.* **2018**, *98*, 47–52. [CrossRef]
10. Huang, X.; Zhang, R.; Gao, X.; Yu, B.; Gao, Y.; Han, Z.-G. TiO_2-rutile/anatase homojunction with enhanced charge separation for photoelectrochemical water splitting. *Int. J. Hydrogen Energy* **2021**, *46*, 26358–26366. [CrossRef]
11. Liang, S.; He, J.; Sun, Z.; Liu, Q.; Jiang, Y.; Cheng, H.; He, B.; Xie, Z.; Wei, S. Improving Photoelectrochemical Water Splitting Activity of TiO_2 Nanotube Arrays by Tuning Geometrical Parameters. *J. Phys. Chem. C* **2012**, *116*, 9049–9053. [CrossRef]
12. Lee, J.; Tan, L.-L.; Chai, S.-P. Heterojunction photocatalysts for artificial nitrogen fixation: Fundamentals, latest advances and future perspectives. *Nanoscale* **2021**, *13*, 7011–7033. [CrossRef] [PubMed]
13. Lin, Y.; Qian, Q.; Chen, Z.; Feng, D.; Tuan, P.D.; Yin, F. Surface Modification of TiO_2 Nanotubes Prepared by Porous Titanium Anodization via Hydrothermal Reaction: A Method for Synthesis High-Efficiency Adsorbents of Recovering Sr Ions. *Langmuir* **2022**, *38*, 11354–11361. [CrossRef] [PubMed]
14. Lakshmanareddy, N.; Rao, V.N.; Cheralathan, K.K.; Subramaniam, E.P.; Shankar, M.V. Pt/TiO_2 nanotube photocatalyst—Effect of synthesis methods on valance state of Pt and its influence on hydrogen production and dye degradation. *J. Colloid Interface Sci.* **2019**, *538*, 83–98. [CrossRef]
15. Cai, J.; Zhang, A.; Tao, H.; Li, R.; Han, J.; Huang, M. Ni-doped hybrids of TiO_2 and two-dimensional Ti_3C_2 MXene for enhanced photocatalytic performance. *Phys. E Low-Dimens. Syst. Nanostruct.* **2023**, *145*, 115476. [CrossRef]
16. Wang, X.; Dai, M.; Chen, Q.; Cheng, X.; Xu, H.; Zhang, J.; Song, H. Enhanced photoelectrochemical performance of NiO-doped TiO_2 nanotubes prepared by an impregnation–calcination method. *J. Chem. Res.* **2021**, *45*, 1076–1082. [CrossRef]
17. Zhang, D.; Chen, J.; Xiang, Q.; Li, Y.; Liu, M.; Liao, Y. Transition-Metal-Ion (Fe, Co, Cr, Mn, Etc.) Doping of TiO_2 Nanotubes: A General Approach. *Inorg. Chem.* **2019**, *58*, 12511–12515. [CrossRef]
18. Cao, F.; Xiong, J.; Wu, F.; Liu, Q.; Shi, Z.; Yu, Y.; Wang, X.; Li, L. Enhanced Photoelectrochemical Performance from Rationally Designed Anatase/Rutile TiO_2 Heterostructures. *ACS Appl. Mater. Interfaces* **2016**, *8*, 12239–12245. [CrossRef]

19. Anitha, V.C.; Goswami, A.; Sopha, H.; Nandan, D.; Gawande, M.B.; Cepe, K.; Ng, S.; Zboril, R.; Macak, J.M. Pt nanoparticles decorated TiO_2 nanotubes for the reduction of olefins. *Appl. Mater. Today* **2018**, *10*, 86–92. [CrossRef]
20. Cho, S.; Yim, G.; Park, J.T.; Jang, H. Surfactant-free one-pot synthesis of Au-TiO_2 core-shell nanostars by inter-cation redox reaction for photoelectrochemical water splitting. *Energy Convers. Manag.* **2022**, *252*, 115038. [CrossRef]
21. Inoue, K.; Matsuda, A.; Kawamura, G. Tube length optimization of titania nanotube array for efficient photoelectrochemical water splitting. *Sci. Rep.* **2023**, *13*, 103. [CrossRef]
22. Yuan, S.; Xia, M.; Liu, Z.; Wang, K.; Xiang, L.; Huang, G.; Zhang, J.; Li, N. Dual synergistic effects between Co and Mo_2C in Co/Mo_2C heterostructure for electrocatalytic overall water splitting. *Chem. Eng. J.* **2022**, *430*, 132697. [CrossRef]
23. Yue, X.; Yi, S.; Wang, R.; Zhang, Z.; Qiu, S. A novel architecture of dandelion-like Mo_2C/TiO_2 heterojunction photocatalysts towards high-performance photocatalytic hydrogen production from water splitting. *J. Mater. Chem. A* **2017**, *5*, 10591–10598. [CrossRef]
24. Ozkan, S.; Nguyen, N.T.; Mazare, A.; Schmuki, P. Optimized Spacing between TiO_2 Nanotubes for Enhanced Light Harvesting and Charge Transfer. *Chemelectrochem* **2018**, *5*, 3183–3190. [CrossRef]
25. Habibi-Hagh, F.; Foruzin, L.J.; Nasirpouri, F. Remarkable improvement of photoelectrochemical water splitting in pristine and black anodic TiO_2 nanotubes by enhancing microstructural ordering and uniformity. *Int. J. Hydrogen Energy* **2023**, *48*, 11225–11236. [CrossRef]
26. Onoda, M. Phase Transitions of Ti_3O_5. *J. Solid State Chem.* **1998**, *136*, 67–73. [CrossRef]
27. Babu, S.J.; Rao, V.N.; Murthy, D.H.K.; Shastri, M.; Murthy, M.; Shetty, M.; Raju, K.A.; Shivaramu, P.D.; Kumar, C.S.A.; Shankar, M.; et al. Significantly enhanced cocatalyst-free H_2 evolution from defect-engineered Brown TiO_2. *Ceram. Int.* **2021**, *47*, 14821–14828. [CrossRef]
28. Zhao, P.-F.; Li, G.-S.; Li, W.-L.; Cheng, P.; Pang, Z.-Y.; Xiong, X.-L.; Zou, X.-L.; Xu, Q.; Lu, X.-G. Progress in Ti_3O_5: Synthesis, properties and applications. *Trans. Nonferrous Met. Soc. China* **2021**, *31*, 3310–3327. [CrossRef]
29. Divyasri, Y.V.; Reddy, N.L.; Lee, K.; Sakar, M.; Rao, V.N.; Venkatramu, V.; Shankar, M.V.; Reddy, N.C.G. Optimization of N doping in TiO_2 nanotubes for the enhanced solar light mediated photocatalytic H_2 production and dye degradation. *Environ. Pollut.* **2021**, *269*, 116170. [CrossRef] [PubMed]
30. Ning, X.; Huang, J.; Li, L.; Gu, Y.; Jia, S.; Qiu, R.; Li, S.; Kim, B.H. Homostructured rutile TiO_2 nanotree arrays thin film electrodes with nitrogen doping for enhanced photoelectrochemical performance. *J. Mater. Sci. Mater. Electron.* **2019**, *30*, 16030–16040. [CrossRef]
31. Wang, Y.; Gao, L.; Huo, J.; Li, Y.; Kang, W.; Zou, C.; Jia, L. Designing novel 0D/1D/2D NiO@La(OH)$_3$/g-C_3N_4 heterojunction for enhanced photocatalytic hydrogen production. *Chem. Eng. J.* **2023**, *460*, 141667. [CrossRef]
32. Li, Z.; Bian, H.; Xu, Z.; Lu, J.; Li, Y.Y. Solution-Based Comproportionation Reaction for Facile Synthesis of Black TiO_2 Nanotubes and Nanoparticles. *ACS Appl. Energy Mater.* **2020**, *3*, 6087–6092. [CrossRef]
33. Wu, C.; Gao, Z.; Gao, S.; Wang, Q.; Xu, H.; Wang, Z.; Huang, B.; Dai, Y. Ti^{3+} self-doped TiO_2 photoelectrodes for photoelectrochemical water splitting and photoelectrocatalytic pollutant degradation. *J. Energy Chem.* **2016**, *25*, 726–733. [CrossRef]
34. Miao, M.; Pan, J.; He, T.; Yan, Y.; Xia, B.Y.; Wang, X. Molybdenum Carbide-Based Electrocatalysts for Hydrogen Evolution Reaction. *Chem. Eur. J.* **2017**, *23*, 10947–10961. [CrossRef] [PubMed]

Disclaimer/Publisher's Note: The statements, opinions and data contained in all publications are solely those of the individual author(s) and contributor(s) and not of MDPI and/or the editor(s). MDPI and/or the editor(s) disclaim responsibility for any injury to people or property resulting from any ideas, methods, instructions or products referred to in the content.

Article

Efficiency Improvement of Industrial Silicon Solar Cells by the POCl₃ Diffusion Process

Xiaodong Xu [1], Wangping Wu [1,*] and Qinqin Wang [2,*]

[1] Electrochemistry and Corrosion Laboratory, School of Mechanical Engineering, Changzhou University, Changzhou 213164, China
[2] School of Mechanical Engineering, Yangzhou University, Yangzhou 225127, China
* Correspondence: wwp3.14@163.com or wuwping@cczu.edu.cn (W.W.); wangqinqin@yzu.edu.cn (Q.W.)

Abstract: To improve the efficiency of polycrystalline silicon solar cells, process optimization is a key technology in the photovoltaic industry. Despite the efficiency of this technique to be reproducible, economic, and simple, it presents a major inconvenience to have a heavily doped region near the surface which induces a high minority carrier recombination. To limit this effect, an optimization of diffused phosphorous profiles is required. A "low-high-low" temperature step of the POCl₃ diffusion process was developed to improve the efficiency of industrial-type polycrystalline silicon solar cells. The low surface concentration of phosphorus doping of 4.54×10^{20} atoms/cm³ and junction depth of 0.31 μm at a dopant concentration of $N = 10^{17}$ atoms/cm³ were obtained. The open-circuit voltage and fill factor of solar cells increased up to 1 mV and 0.30%, compared with the online low-temperature diffusion process, respectively. The efficiency of solar cells and the power of PV cells were increased by 0.1% and 1 W, respectively. This POCl₃ diffusion process effectively improved the overall efficiency of industrial-type polycrystalline silicon solar cells in this solar field.

Keywords: polycrystalline silicon; solar cells; low-high-low; phosphorus diffusion

1. Introduction

Carbon-neutral development strategies have a significant impact on the Earth's environment, and silicon (Si) solar cells have attracted much attention as a means to use solar energy to convert sunlight into electricity [1,2]. However, a compromise between cost reduction and efficiency improvement must be reached [3]. The PN junction is one of the key technologies in crystalline Si solar cells, which affects photoelectric conversion efficiency. Therefore, the PN junction has excellent performance and stable uniformity. To improve the photoelectric conversion efficiency, the high sheet resistance of over 90 Ω/sq, which has low surface doping concentration and a shallow junction process, was accepted [4]. High open-circuit voltage (V_{oc}) and short-circuit current (J_{sc}) values were obtained by this process. Tube furnace diffusion using phosphorus oxychloride (POCl₃) as a dopant precursor is the dominant emitter formation technology for p-type Si solar cells [5]. The majority of the PV industry currently uses POCl₃ diffusion to remove metal impurities, including iron [6,7]. The solar cell emitters are obtained by P (phosphorus) diffusion in p-type Si inside of a diffusion tube furnace under controlled conditions of temperature, pressure, and gas flow to form an emitter layer [3].

The photovoltaic (PV) industry has used a quartz diffusion tube furnace to form an emitter layer of the POCl₃ source. However, there are three flaws in this process [8]: (1) the high sheet resistance, (2) the difficulty in controlling the uniformity of high sheet resistance, and (3) the shallow junction. Many manufacturers tried to reduce the heavy inactive phosphorus concentration and the thickness of the dead zone through an additional step in the industrial process: i.e., chemical etching of the PSG layer after the phosphorus diffusion [9–11]. This solution increased the duration of the industrial process and it is expensive. A one-step diffusion process was a common method in which a single temperature and continuous flow of dopant gas were used to deposit phosphor silicate glass (PSG) and to drive dopants to the desired depth [12]. This is a fast process but it tends to create an excessively doped emitter that deteriorates electrical performance [13]. POCl₃ diffusion could be performed in a two-step process: a PSG deposition step, followed by a drive-in step at variable temperature. During the process, POCl₃ gas is allowed in the PSG layer, and subsequently, dopants are moved deeply from the PSG layer to the Si substrate in the drive-in step [14]. Wolf et al. [5] presented the status and perspective of emitter formation by the POCl₃-diffusion process and discussed the diluted source and in-situ post-oxidation technological options for advanced tube furnace POCl₃-diffusion processes. Cui et al. [15] studied POCl3-based diffusion optimization for the formation of homogeneous emitters and the correlation with metal contact in p-type polycrystalline Si solar cells and found that the sheet resistance is high and that the P surface concentration and emitter saturation current density (J_{oe}) are low. Cho et al. [16] compared POCl₃ diffusion and P ion-implantation induced gettering in solar cells and found that the increase in P implantation dose improved the gettering efficiency by increasing bulk lifetime and decreasing iron concentration, but the process remained inferior to POCl₃ diffusion. POCl₃-diffused cast quasi-mono cells showed 0.4% higher efficiency due to their higher bulk lifetimes compared to P-implanted emitters. Ghembaza et al. [17] studied the optimization of P emitter formation from POCl₃ diffusion for p-type Si solar cells and showed that the emitter standard sheet resistances of ~60 Ω/sq and wafer uniformity <3% were obtained from the low-pressure tube furnace. Li et al. [18] investigated POCl₃ diffusion for the emitter layer formation of industrial Si solar cells and presented the impact of processing parameters on emitter layer formation and electrical performance.

According to the above review, P diffusion can be performed in a single step by controlling a parameter, such as temperature or time [4], or a two-step process, such as ion implantation. In this work, a "low-high-low" (LHL) diffusion process, low-high-low temperature, and three-step diffusion were used to diffuse P elements with different POCl₃ flows. The ECV profile, open-circuit voltage (V_{oc}), fill factor (FF), and overall efficiency of solar cells of this process were studied and simultaneously compared with the baseline using the online conventional process.

2. Experimental

P diffusion emitters were prepared on 156 × 156 mm 0.5–3 Ω·cm p-type mc-Si wafers with a thickness of ~180 μm in the quartz furnace tube. The distance between the wafers was about 2.35 mm. These wafers were vertically inserted into the quartz boat and then placed in the furnace. There were 1000 pieces per batch. There were two diffusion processes. One was the online process, and another was the LHL diffusion process. Figure 1 shows the schematic of the online and LHL diffusion processes.

Table 1 displays the process parameters of low-temperature online diffusion, namely the BKM (Best Known Method) diffusion process and the LHL diffusion process for solar cells. For the low-temperature online diffusion process, the temperature was kept at 810 °C, and the flow of POCl₃, O₂, and N₂ gas was 1600 mL/min, 800 mL/min, and 30,000 mL/min, respectively. For the LHL diffusion process, the first step is a low temperature and high POCl₃ flow diffusion. The low temperature was controlled at about 810 °C, and the flows of POCl₃, O₂, and N₂ gas were set at 1900 mL/min, 800 mL/min, and 30,000 mL/min, respectively. Next, the high temperature was controlled at 825 °C and the flows of POCl₃, O₂, and N₂ gas were fixed at 2100 mL/min, 800 mL/min, and 30,000 mL/min, respectively. Finally, a low temperature and low POCl₃ flow diffusion process was used. The diffusion temperature was the same as in the first step. The flows of POCl₃, O₂, and N₂ gas were 1600 mL/min, 800 mL/min, and 30,000 mL/min, respectively. The three-step variable-temperature diffusion LHL process is useful in the gettering process [19].

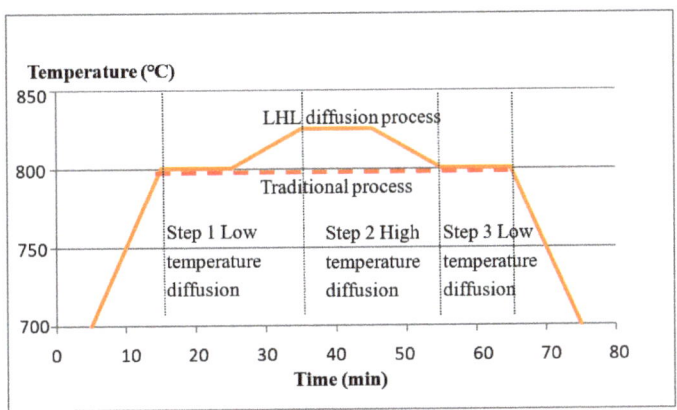

Figure 1. Schematic of BKM and LHL diffusion process.

Table 1. Process parameters of BKM and LHL diffusion processes for p-type mc-Si solar cells.

Condition		Temperature (°C)	POCl$_3$ (sccm)	O$_2$ (sccm)	N$_2$ (sccm)
BKM	1st step	810	1600	800	30,000
LHL	1st step	810	1900	800	30,000
	2nd step	825	2100	800	30,000
	Last step	:0	1600	800	30,000

Three batches of samples for each process were manufactured in order to get the average values. The sheet resistance was measured using four-point probe equipment, and the P diffusion profiles of selected samples were determined using electrochemical capacitance voltage profiling (ECV-profiling, WEP CVP21). The microstructure and morphology of the textured structure and front metalized areas were observed by scanning electronic microscopy (SEM, Quanta FEG 250, FEI). The electrical properties of solar cells were characterized by a Berger cell tester.

3. Results and Discussion

We designed the LHL diffusion process with low-high-low temperature and a three-step diffusion with different POCl$_3$ flows. The specific schematic diagrams are shown in Figure 2. Figure 2a shows the schematic diagram of the conventional primary diffusion process. Firstly, pre-oxidation is carried out with oxygen at a low temperature of 700–800 °C to generate silicon oxides on the surface of Si wafers, which is helpful to the distribution of POCl$_3$ diffusion. Then, POCl$_3$ is deposited at a low temperature of 800 °C, and finally at a high temperature of 850 °C, in order to redistribute the P element. Figure 2b presents the schematic diagram of the LHL diffusion process. LHL diffusion is characterized by three sets of P doping and three sets of redistribution. Variation in temperature is the simplest way to control the phosphorous diffusion profile. As the temperature increases, doping increases, and the formed junctions are deeper. This behavior is explained by the variation of coefficient diffusion and limited solubility with temperature. For this reason, the temperature parameter needed to achieve the necessary exact junction depth has proven to be rather delicate. With a long drive-in time, the junction is deeper. PSG deposited during the pre-deposition step acts as an infinite phosphorus source. All these results confirm that the phosphorus profile is highly affected by the tube furnace conditions. Clearly, time and temperature must be considered carefully. In this work, firstly, the pre-oxidation is carried out with oxygen at a low temperature of 800 °C. Then, the first step of low-temperature POCl$_3$ deposition is carried out, and the impurities are redistributed at variable temperatures. Subsequently, the second step of high-temperature POCl$_3$ deposition is carried out, which is distributed with high-temperature P impurities. Finally, the third step is to deposit POCl$_3$ at a high temperature to cool down, and the impurities are redistributed. LHL diffusion process adopts low-high-low temperature and three-step diffusion with different POCl$_3$ flows. In the first step, the low temperature and high POCl$_3$ flow are the best to control the tail concentration of ECV curves. In the low-concentration tail, P diffuses into Si wafers primarily via interaction with Si self-interstitials [20,21]. In the second step, the high temperature and high POCl$_3$ flow control the kink of the slope. For high P concentration, a conversion from an interstitially to a slow vacancy-mediated process occurs, giving rise to anomalous P diffusion profiles [22]. In the third step, the low temperature and low POCl$_3$ flow can control the surface concentration of P doping. This method has the objective to decrease inactive phosphorus through an LHL step. Graphically this implies the reduction of the plateau width, which appears on the top of diffusion profiles near the high-phosphorus concentration zone. The low surface concentration of P doping could be beneficial to the V_{oc} and J_{sc} values of solar cells. However, it influences series resistance and FF values. Therefore, it is important to weigh the benefits against the risks.

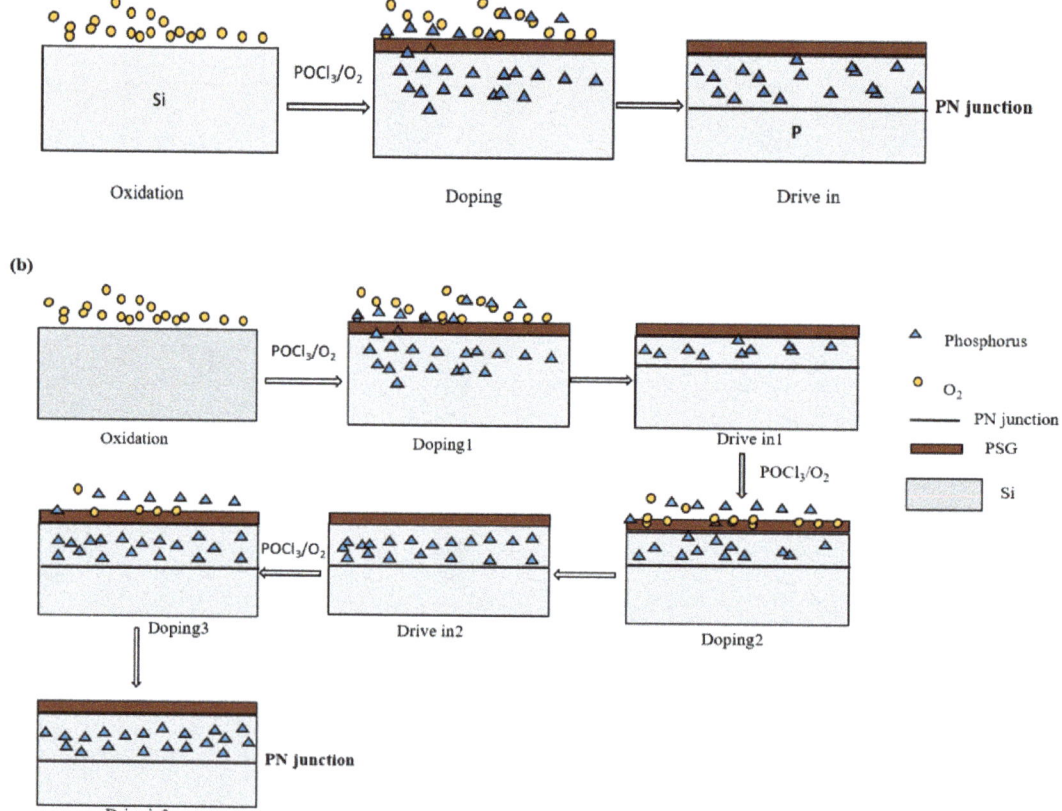

Figure 2. Schematic diagrams of the PN junction of solar cells obtained from (**a**) BKM and (**b**) LHL diffusion processes.

Figure 3 shows the sheet resistance box plots of solar cells and the ECV profiles of P doping for solar cells. The sheet resistance of solar cells was obtained (see Figure 3a). It can be observed that solar cells produced from LHL and BKM diffusion processes had the same sheet resistance of about 90 Ω/sq. However, the sheet resistance of solar cells from the LHL process was much more uniform than that of the cells from the BKM diffusion process. The results indicated that the LHL process could be beneficial for the FF and the series resistance of solar cells [17]. Figure 3b presents the P doping profile of solar cells produced by LHL and BKM diffusion processes. The solar cells obtained from the LHL diffusion process had a lower surface concentration of P doping, approximately 4.54×10^{20} fewer atoms/cm³ than those produced from the BKM diffusion process, which produced about 6.08×10^{20} atoms/cm³ at the junction depth of about 0.02 μm. For LHL and BKM diffusion processes, the solar cells had the same junction depth of around 0.3 μm at a dopant concentration of $N = 10^{17}$ atoms/cm³. During emitter formation and at high phosphorus concentrations, precipitates were formed on the silicon surface and promoted the existence of electrically inactive phosphorus which formed a dead layer at the silicon surface. This behavior is characterized by a kinked shape in the experimental profiles. This kink has a great impact on solar cell performance since it results in low collection efficiency near the front surface.

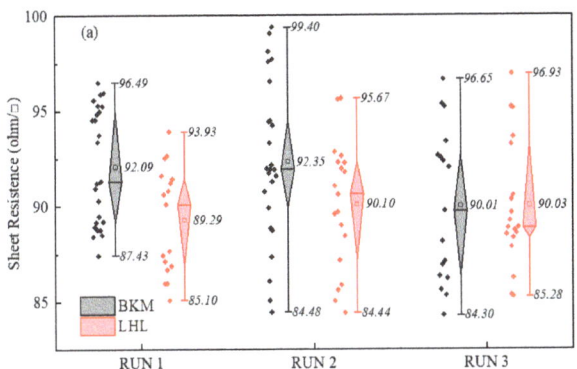 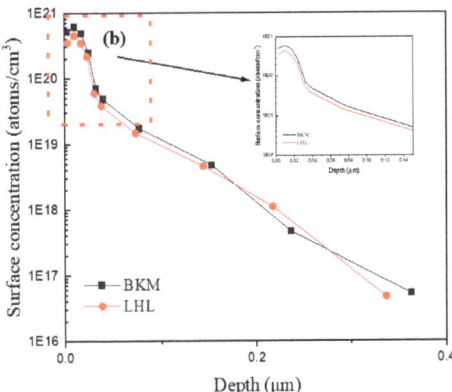

Figure 3. Sheet resistance box-plots of solar cells obtained from BKM and LHL diffusion processes (**a**), and the ECV profiles of P doping (**b**).

In the module of PV solar cell marketing, only the V_{oc} and FF values have more advantages on the power. In this study, the advantage of the LHL diffusion process could be beneficial to the increase in V_{oc} and FF values. Table 2 displays the gap of electrical characteristics of solar cells obtained by LHL and BKM diffusion processes. The solar cells obtained from the LHL diffusion process have an increase in median V_{oc} value of about 1 mV, compared with the BKM diffusion process. This increase might be due to the low surface concentration of P doping (see Figure 4b). At the same time, the median FF value is increased by 0.30%, which can be contributed to the strong impurity absorption effect of Si wafers and the decrease in inactive phosphorus in the LHL diffusion process.

Table 2. The gap in electrical characteristics for solar cells obtained by LHL and BKM diffusion processes.

Condition		V_{oc} (V)	J_{sc} (mA)	R_{ser} (ohm)	FF (%)	E_{ff} (%)
Gap (LHL-BKM)	Run1	0.0012	−0.03	0.0000	0.27%	0.08%
	Run2	0.0012	−0.05	0.0000	0.36%	0.08%
	Run3	0.0006	−0.04	−0.0001	0.27%	0.1%
Average		0.0010	−0.04	0.0000	0.30%	0.09%

Note: V_{oc}: open circuit voltage; J_{sc}: short circuit current; R_{ser}: series resistance; FF: fill factor; E_{ff}: efficiency.

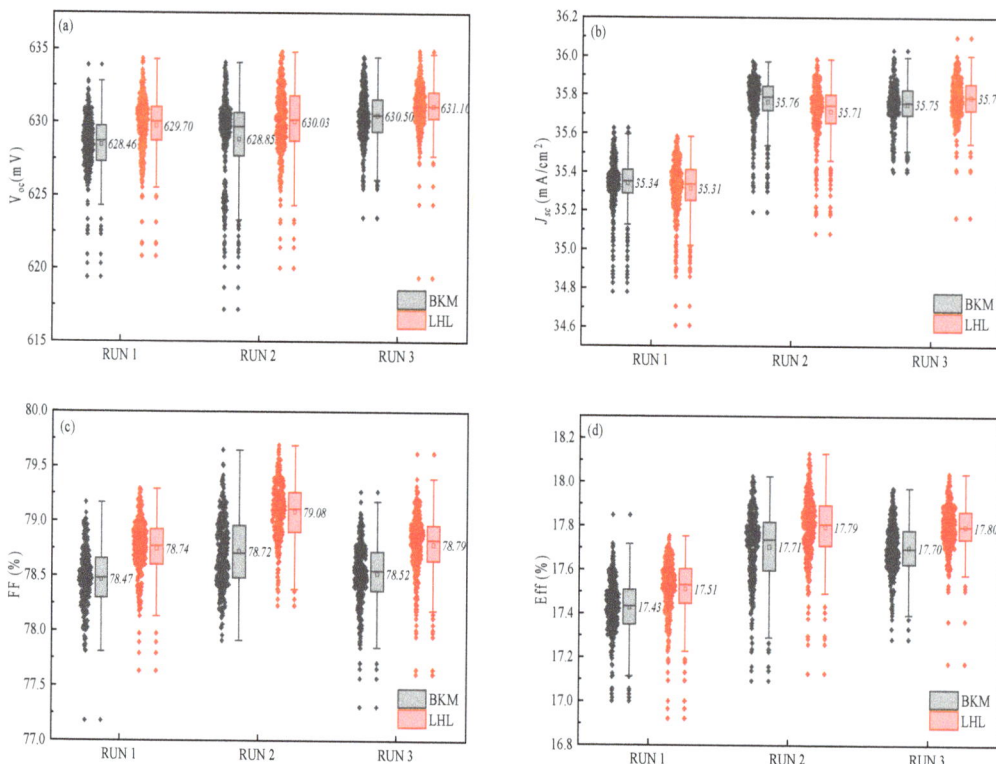

Figure 4. Box plots of the electrical characteristics of solar cells obtained from BKM and LHL diffusion processes (**a**) V_{oc}, (**b**) J_{sc}, (**c**) FF, and (**d**) E_{ff}.

Figure 4 shows the box plots of electrical characteristics of solar cells produced from BKM and LHL diffusion processes. The median V_{oc} values of the solar cells produced by LHL and BKM processes are 630 ± 1 mV and 629 ± 1 mV, respectively (see Figure 4a). For LHL and BKM diffusion processes, the median J_{sc} values of the solar cells are the same, about 35.55 ± 0.25 mA (see Figure 4b). The median FF values of the solar cells obtained by LHL and BKM diffusion processes are 78.9 ± 0.1% and 78.6 ± 0.1%, respectively (see Figure 4c). The median E_{ff} values of the cells produced by LHL and BKM diffusion processes are 17.65 ± 0.15% and 17.55 ± 0.15%, respectively (Figure 4d). The E_{ff} value of the solar cells from the LHL diffusion process is increased by 0.08–0.10%, which was mainly attributed to the increase in V_{oc} value of 1 mV and the FF value of 0.30%. These results show the convergence of the V_{oc}, FF and E_{ff} values of the solar cells obtained from the LHL diffusion process is better than the parameters of cells produced from the BKM diffusion process, which was attributed to the distribution and diffusion of impurities. At the same time, the effective doping concentration is effectively controlled, the generation of the dead layer is reduced, and recombination is reduced, resulting in the increase of the V_{oc} value. Furthermore, due to the LHL diffusion process with three-time temperature changes, it is more conducive to the precipitation of harmful impurities into the PSG layer, which improves the service life of the bulk, thus resulting in the improvement of FF values. The increase in Voc and FF can be explained by active phosphorus atoms and the optimized contact-formation process. Table S1 summarizes the performance of solar cells doped P by different diffusion processes (see Supplementary Material).

Figure 5 presents the typical top-view SEM micrographs of the front side of solar cells. It shows that the front busbar is of uniform height: about 12.4 μm (see Figure 5a), and the height of the front finger is also smooth: about 24.3 µm (see Figure 5b). However, the Ag paste does not uniformly corrode the P doping layer. In Figure 5c, the corroded depth is approximately 0.12 µm, so the effect of the surface concentration at the depth of 0.15 µm on the contact resistance (ρ_c) could be emphasized, which is important for the FF values. Figure 6 shows the contact resistivity of solar cells from two diffusion processes. The ρ_c values of the cells from LHK and BKM processes were 18.86 mΩ·cm^2 and 19.09 mΩ·cm^2. The LHL diffusion process has no additional costs. It would be a benefit for the large-scale industrial production of the P doping process of PV solar cells.

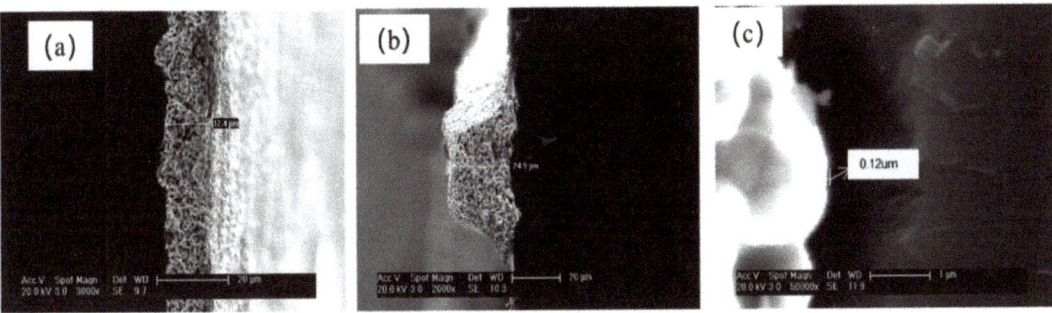

Figure 5. SEM images of (**a**) front busbar, (**b**) front finger, and (**c**) Ag-Si alloy.

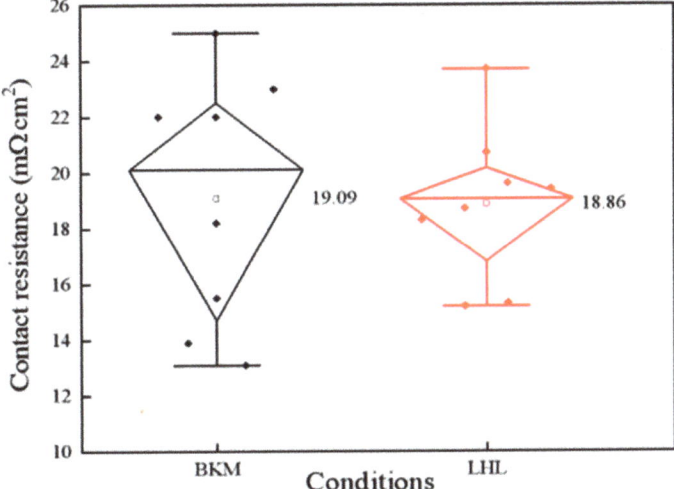

Figure 6. Contact resistivity of solar cells from LHL and BKM diffusion processes.

4. Conclusions

The non-uniformity of worm-shaped structures increases the difficulty of P diffusion. In addition, the phosphorus profile is highly affected by the tube furnace conditions. (Time and temperature must be considered carefully). A diffusion process featuring low-high-low temperature and three steps was used to diffuse P elements for solar cells with different POCl$_3$ flows in every step. This allows for systematic manipulation of doping profiles, especially for manipulation of the surface-active concentration of P doping, control of the doping depth, and reduction in the dead layer at the silicon surface, respectively. The solar cells with a low surface concentration of P doping of 4.54×10^{20} atom/cm^3 and junction depth of 0.31 µm at a dopant concentration of $N = 10^{17}$ atoms/cm^3 were obtained. The open-circuit voltage and FF values of solar cells increased up to 1 mV and 0.30%, compared with the online low-temperature diffusion process respectively, which can be contributed to the low surface concentration of P doping (decreasing inactive phosphorus) and the strong impurity absorption effect of Si wafers obtained from the low-high-low temperature diffusion process. The efficiency and the power of solar cells were increased by 0.1% and 1 W, respectively. The LHL diffusion process has no additional costs. It would be beneficial for the large-scale industrial production of the P doping process of PV solar cells.

Supplementary Materials: The following supporting information can be downloaded at: https://www.mdpi.com/article/10.3390/ma16051824/s1.

Author Contributions: X.X.: Data curation, Formal analysis, Investigation, Methodology; W.W.: Writing—original draft, Writing—review& editing, Supervision; Q.W.: Writing—review& editing. All authors have read and agreed to the published version of the manuscript.

Funding: This work has been partially supported by Jiangsu Province Cultivation base for State Key Laboratory of Photovoltaic Science and Technology (SKLPST 202201).

Informed Consent Statement: Not applicable.

Data Availability Statement: The data presented in this study are available upon request from the corresponding author.

Conflicts of Interest: We declare that we do not have any commercial or associative interests that represent a conflict of interest in connection with the work submitted.

References

1. Lee, S.H.; Min, K.H.; Choi, S.; Song, H.; Kang, M.G.; Kim, T.; Park, S. Advanced carrier lifetime analysis method of silicon solar cells for industrial applications. *Sol. Energy Mater. Sol. Cells* **2023**, *251*, 112144. [CrossRef]
2. Bennett, N.S.; Wight, N.M.; Popuri, S.R.; Bos, J.G. Efficient thermoelectric performance in silicon nano-films by vacancy-engineering. *Nano Energy* **2015**, *16*, 350–356. [CrossRef]
3. Kumar, P.; Pfeffer, M.; Willsch, B.; Eibl, O. Contact formation of front side metallization in p-type, single crystalline Si solar cells: Microstructure, temperature dependent series resistance and percolation model. *Sol. Energy Mater. Sol. Cells* **2016**, *145*, 358–367. [CrossRef]
4. Ghembaza, H.; Zerga, A.; Saïm, R. Efficiency improvement of crystalline silicon solar cells by optimizing the doping profile of POCl3 diffusion. *Inter. J. Sci. Technol. Res.* **2014**, *3*, 1–5. [CrossRef]
5. Wolf, A.; Kimmerle, A.; Werner, S.; Maier, S.; Belledin, U.; Meier, S.; Biro, D. Status and perspective of emitter formation by POCl3-diffusion. In Proceedings of the 31st European Photovoltaic Solar Energy Conference and Exhibition, Hamburg, Germany, 14–18 September 2015; pp. 414–419.
6. Khedher, N.; Hajji, M.; Hassen, M.; BenJaballah, A.; Ouertani, B.; Ezzaouia, H.; Bessais, B.; Selmi, A.; Bennaceur, R. Gettering impurities from crystalline silicon by phosphorus diffusion using a porous silicon layer. *Sol. Energy Mater. Sol. Cells* **2005**, *87*, 605–611. [CrossRef]
7. Shabani, M.; Yamashita, T.; Morita, E. Study of gettering mechanisms in silicon: Competitive gettering between phosphorus diffusion gettering and other gettering sites. *Solid State Phenom.* **2008**, *131–133*, 399–404.
8. Dastgheib-Shirazi, A.; Steyer, M.; Micard, G.; Wagner, H.; Altermatt, P.P.; Hahn, G. Relationships between Diffusion Parameters and Phosphorus Precipitation during the POCl3 Diffusion Process. *Energy Procedia* **2013**, *38*, 254–262. [CrossRef]
9. Kittidachachan, P.; Markvart, T.; Ensell, G.; Greef, R.; Bagnall, D. An analysis of a "dead layer" in the emitter of n/sup +/pp/sup +/ solar cells. In Proceedings of the Thirty-first IEEE Photovoltaic Specialists Conference, Lake Buena Vista, FL, USA, 3–7 January 2005; pp. 1103–1106. [CrossRef]
10. Book, F.; Dastgheib-Shirazi, A.; Raabe, B.; Haverkamp, H.; Hahn, G.; Grabitz, P. Detailed analysis of high sheet resistance emitters for selectively doped silicon solar cells. In Proceedings of the 24th European Photovoltaic Solar Energy Conference, Hamburg, Germany, 21–25 September 2009. [CrossRef]
11. Saule, W.; Delahaye, F.; Queisser, S.; Wefringhaus, E.; Schweckendiek, J.; Nussbaumer, H. High efficiency inline diffusion process with wet-chemical emitter etch-back. In Proceedings of the 24th European Photovoltaic Solar Energy Conference, Hamburg, Germany, 21–25 September 2009. [CrossRef]
12. Ostoja, P.; Guerri, S.; Negrini, P.; Solmi, S. The effects of phosphorus precipitation on the open-circuit voltage in N^+/P silicon solar cells. *Sol. Cells* **1984**, *11*, 1–12. [CrossRef]
13. Nobili, D. Precipitation as the phenomenon responsible for the electrically inactive phosphorus in silicon. *J. Appl. Phys.* **1982**, *53*, 1484–1491. [CrossRef]

14. Li, H.; Ma, F.-J.; Hameiri, Z.; Wenham, S.; Abbott, M. On elimination of inactive phosphorus in industrial POCl$_3$ diffused emitters for high efficiency silicon solar cells. *Sol. Energy Mater. Sol. Cells* **2017**, *171*, 213–221. [CrossRef]
15. Cui, M.; Jin, C.; Yang, Y.; Wu, X.; Zhuge, L. Emitter doping profiles optimization and correlation with metal contact of multi-crystalline silicon solar cells. *Optik* **2016**, *127*, 11230–11234. [CrossRef]
16. Cho, E.; Ok, Y.-W.; Dahal, L.; Das, A.; Upadhyaya, V.; Rohatgi, A. Comparison of POCl3 diffusion and phosphorusion-implantation induced gettering in crystalline Si solar cells. *Sol. Energy Mater. Sol. Cells* **2016**, *157*, 245–249. [CrossRef]
17. Ghembaza, H.; Zerga, A.; Saïm, R.; Pasquinelli, M. Optimization of Phosphorus Emitter Formation from POCl3 Diffusion for p-Type Silicon Solar Cells Processing. *Silicon* **2016**, *10*, 377–386. [CrossRef]
18. Li, H.; Kim, K.; Hallam, B.; Hoex, B.; Wenham, S.; Abbott, M. POCl3 diffusion for industrial Si solar cell emitter formation. *Front. Energy* **2017**, *11*, 42–51. [CrossRef]
19. Chen, J.; Xi, Z.; Wu, D.; Yang, D. Effect of variable temperature phosphorous gettering treatments on the performance of multicrystalline silicon. *Acta Energ. Sol. Sin.* **2017**, *28*, 160–164. [CrossRef]
20. Chen, Y.; Hao, Q.; Liu, C.; Zhao, J.; Wu, D.; Wang, Y. Effect of transition-metal contamination on minority lifetime in cast multi-crystalline silicon under rapid thermal processing. *Acta Energ. Sol. Sin.* **2019**, *30*, 611–614. [CrossRef]
21. Ural, A.; Griffin, P.B.; Plummer, J.D. Fractional contributions of microscopic diffusion mechanisms for common dopants and self-diffusion in silicon. *J. Appl. Phys.* **1999**, *85*, 6440–6446. [CrossRef]
22. Rousseau, P.M.; Griffin, P.B.; Fang, W.T.; Plummer, J.D. Arsenic deactivation enhanced diffusion: A time, temperature, and concentration study. *J. Appl. Phys.* **1998**, *84*, 3593–3601. [CrossRef]

Disclaimer/Publisher's Note: The statements, opinions and data contained in all publications are solely those of the individual author(s) and contributor(s) and not of MDPI and/or the editor(s). MDPI and/or the editor(s) disclaim responsibility for any injury to people or property resulting from any ideas, methods, instructions or products referred to in the content.

MDPI
St. Alban-Anlage 66
4052 Basel
Switzerland
www.mdpi.com

Materials Editorial Office
E-mail: materials@mdpi.com
www.mdpi.com/journal/materials

Disclaimer/Publisher's Note: The statements, opinions and data contained in all publications are solely those of the individual author(s) and contributor(s) and not of MDPI and/or the editor(s). MDPI and/or the editor(s) disclaim responsibility for any injury to people or property resulting from any ideas, methods, instructions or products referred to in the content.

www.ingramcontent.com/pod-product-compliance
Lightning Source LLC
LaVergne TN
LVHW070650100526
838202LV00013B/927